Susanne Duesterberg
Popular Receptions of Archaeology

Historische Lebenswelten in populären Wissenskulturen
History in Popular Cultures | Volume 14

Editorial

The series **Historische Lebenswelten in populären Wissenskulturen | History in Popular Cultures** provides analyses of popular representations of history from specific and interdisciplinary perspectives (history, literature and media studies, social anthropology, and sociology). The studies focus on the contents, media, genres, as well as functions of contemporary and past historical cultures.

Susanne Duesterberg (M.A.) is a staff member of the English Department and head of one of the examination offices at the University of Freiburg.

Susanne Duesterberg

Popular Receptions of Archaeology

Fictional and Factual Texts in 19th and Early 20th Century Britain

[transcript]

Bibliographic information published by the Deutsche Nationalbibliothek
The Deutsche Nationalbibliothek lists this publication in the Deutsche Nationalbibliografie; detailed bibliographic data are available in the Internet at http://dnb.d-nb.de

Cover layout: Kordula Röckenhaus, Bielefeld
Cover illustration: »Electric Lighting of the British Museum«, /Illustrated London News,/ February 8, 1890: 164.
Reproduced with kind permission of Mary Evans Picture Library.
Printed by Majuskel Medienproduktion GmbH, Wetzlar
Print-ISBN 978-3-8376-2810-4
PDF-ISBN 978-3-8394-2810-8

Table of Contents

Acknowledgements

This book is a revised version of my doctoral dissertation submitted to the University of Freiburg in 2012. The project was funded in part by the postgraduate programme »Geschichte und Erzählen«. I would like to thank my supervisors Professor Barbara Korte, Professor Ralf von den Hoff, and Professor Hans-Joachim Gehrke for their constructive and continuous support and advice. I would also like to thank the editors of the series »History in Popular Culture«. My special thanks go to Dr David Heyde, Doris Lechner, and Dr Isabel Seidel for their careful and valuable feedback on the manuscript. Furthermore, I am very grateful to Kathrin Göb, Jennifer Isele, and Georg Zipp for helping prepare this book for publication. My thanks also go to Hanna Kubowitz, Dr Ulrike Pirker, and Victoria Tafferner. My colleagues at the English Department of the University of Freiburg were a great help in completing this book. Last but not least, I would like to thank Heide Duesterberg, Michael Duesterberg, Artur Sälzle, and Lena Ketterer.

Fragments are surviving parts of lost wholes.

(GLENN W. MOST, »FRAGMENTS« 2010: 371)

It may, I think, with justice be said that archæology is at the present a popular and favourite study.

(P. H. DITCHFIELD, THE RELIQUARY, JUL 1891: 134)

Introduction

On January 5, 1867, the *Illustrated London News* informed its readers of the following devastating event:

»We regret to state that a fire broke out in the Crystal Palace at Sydenham on Sunday afternoon, which destroyed nearly all the north quarter of that magnificent structure, containing the Tropical Department; the whole of the Natural History Collection; the Assyrian, Alhambra, and Byzantine Courts; the Queen's Apartments; the Library and Printing Offices; the Indian, Architectural, Model, and Marine Galleries. But little was saved from those parts of the building which were separated by the screen of the Tropical Department from the main space of the interior. Our Illustration on the preceding page is a view taken from the garden terrace, showing the dismal aspect of the wreck of the north transept and the adjoining portions of the edifice, with the remains of the two gigantic Egyptian figures, and some ruins of the Art-Courts; […] At half-past three o'clock the end of the building gave way; the roof fell in as far as the north transept, and the flames shot up in a mass. The effect of this was so great that if the wind had not been blowing away from the central transept nothing could have saved the whole Palace from immediate destruction. The appearance of the burning end of the Palace at this moment was indescribably grand. The flames played along the red-hot girders in fantastic wreath, when suddenly the whole mass dissolved, and, sinking down, fell with a terrific crash. The main girders of the transept remained standing, and in the midst of their circling arches could be seen the colossal statues of the Egyptian Court sitting erect, with the flames leaping and playing around them.« (*Illustrated London News*, Jan 5, 1867: 22)

This report was accompanied by an illustration of the burning Crystal Palace, showing the ancient Egyptian statues surrounded by flames underneath the collapsing steel construction of the palace (cf. fig. 1) with a few lost-looking onlookers in the foreground, helplessly following the spectacle in front of them. A week later, a similar illustration was published by the paper, this time showing workmen sifting through the burnt-out ruins as the damaged Egyptian statues loomed above them (cf. fig. 2). The destruction of the Crystal Palace and the public mediation of the event through the *Illustrated London News* might be taken as symbolising the core issues of this book.

Fig. 1: »Fire at the Crystal Palace on Sunday Last: The Ruins as Seen from the Terrace«

Illustrated London News, January 5, 1867: 2. Reproduced with kind permission of Frankfurt University Library Johann Christian Senckenberg.

Fig. 2: »Ruins of the North Transept of the Crystal Palace Destroyed by the Late Fire«

Illustrated London News, January 12, 1867: 28. Reproduced with kind permission of Frankfurt University Library Johann Christian Senckenberg.

Constructed for the Great Exhibition in 1851, the Crystal Palace was the epitome of Britain's progress and prosperity. As a modern steel and glass building exhibiting the latest technology as well as exotic goods and antiquities, it represented British industrial, imperial, and political powers.[1] At the same time, the Crystal Palace Exhibition, or Great Exhibition, was also one of the first popular spectacles of its kind, attracting thousands of people from all social classes and thus mirroring the Victorian interest in public exhibitions and entertainments. Unlike anything before it, the Crystal Pal-

1 For a detailed discussion of the cultural meaning of the Great Exhibition, cf. Chapter 2.

ace Exhibition presented and represented modernity and antiquity, home and abroad, under one roof. The Crystal Palace Exhibition stood for Britain's belief in the present and the future but also for its concern with the past.[2] It is this seemingly paradoxical combination of a strong belief in progress on the one hand and a nostalgic occupation with the past on the other that is characteristic for the Victorian Age. In spite of Britain's economic, political, and imperial self-assurance, underlying fears of decline and degeneration had increased in the second half of the nineteenth century and climaxed towards its end.[3] Analogies between the home country and once powerful civilisations and empires of the past such as those of ancient Rome, Egypt, and Greece were frequent. So too was the allusion to their common destiny as great powers which had not been able to avoid their own demise.

Against this background, the burnt-out Crystal Palace becomes emblematic in that with the destruction of the palace, a sumptuous representation of Britain's power, the nation's innermost fears of doom and decline seem to have materialised. The choice of the *Illustrated London News* to accompany their reports about the destruction of the palace with illustrations of the Egyptian Court and its magnificent Egyptian statues is significant, since the two ancient Egyptian statues inevitably bring to mind the popular lines of Shelley's *Ozymandias* (1817):[4] »›My name is Ozymandias, king of kings: / Look on my works, ye Mighty, and despair!‹ / Nothing beside remains. Round the decay / Of that colossal wreck, boundless and bare / The lone and level sands stretch far away.« (Shelley 1855: 95) The Egyptian statues in the midst of the ruins of what used to be the representation of Britain's all-encompassing power become symbols of the nation's mutability and vulnerability. This impression is enforced by the vivid description of the fire in the *Illustrated London News*: »The main girders of the transept remained standing, and in the midst of their circling arches

---

2 The exhibition housed *inter alia* a Medieval Court, an Egyptian Court, and an Assyrian Court.

3 For a detailed discussion on fears of degeneration and decline in the Victorian Age, cf. Chapter 2.

4 Percy Bysshe Shelley is said to have been inspired to write his poem by the bust of Ramesses II brought to Britain and exhibited at the British Museum by Giovanni Battista Belzoni (cf. Chapter 3.3).

could be seen the colossal statues of the Egyptian Court sitting erect, with the flames leaping and playing around them.« (*Illustrated London News*, Jan 5, 1867: 22) The fire is presented as a blazing inferno[5] in which the only thing that endures is the »*colossal* statues of the Egyptian Court« (ibid., emphasis mine) as a reminder of yet another once-powerful empire now extinct. In the light of the power of such destruction, both the spectators of the fire in the first sketch and the workmen sifting through the burnt-down remains in the second sketch appear small, lost, and helpless.[6]

This book examines more closely the relationship between the »colossal [ancient Egyptian] statues« (ibid.) and the Victorians as well as the Edwardians. More precisely, I analyse the way in which archaeology was popularly received in the nineteenth and early twentieth centuries and show how this popular reception of archaeology can be read as mirroring contemporary discourses regarding the formation of identity at a time characterised by an underlying feeling of alienation.[7] For this purpose, I present a

5 The association with Dante's *Inferno* evoked here brings to mind underlying issues of guilt in connection with the British imperial behaviour, both in regard to the colonised people and their ancient relics, which is discussed in Chapter 7.

6 Interestingly, especially in the second sketch the image's foreground looks like a jungle and is reminiscent of the overgrown ancient remains of past civilisation (e.g., those of the vacant palaces of the Maya) and thus again can be seen as representing the powers of degeneration and decline, to which even the greatest empire is eventually not immune.

7 In this context, the *Illustrated London News* plays a major role inasmuch as it can be seen as indicative of the popularity of archaeology at the time. This is the case for two reasons: First of all, the newspaper was widely read and representative of contemporary concerns, as emphasised by its self-conception as the »biography of the world since 1842« (*Illustrated London News*, Special Number, May 13, 1967: n. pag.) (cf. also Christoph Hibbert (1975) The Illustrated London News' *Social History of Victorian Britain*; Virginia McKendry (1994) »*The Illustrated London News* and the Invention of Tradition«, and Peter W. Sinnema (1998) *Dynamics of the Pictured Page: Representing the Nation in* The Illustrated London News): »[I]n the midst of illiteracy and poverty – one in ten of the population were paupers – printer, bookseller, and newsagent Herbert Ingram had realised his life's ambition. At 31, he was creator of a new dimension in news gathering, architect of a revolution in journalism, and founder and

reading of popular archaeology based on interdisciplinary approaches to the topic, which situates this book within the topical study of memory research, the popularisation of knowledge, popular culture in general, and popular history with its related fields in particular.[8]

---

owner of the world's first illustrated weekly newspaper, *The Illustrated London News*. It was the tombstone of pioneer reporting and the cradle of modern journalism. In the history of his idea can be read a biography of the world since 1842.« (*Illustrated London News*, Special Number, May 13, 1967: n. pag.) Second, the *Illustrated London News* reported regularly on archaeological and antiquarian activities both at home and abroad. From January 28, 1865, to August 29, 1874, the column »Archæology of the Month« informed the readership about the latest archaeological news. In addition, all the major archaeological discoveries of the time were discussed and illustrated separately, as were the reports on the many minor excavations undertaken in Britain that mainly concentrated on Roman archaeology. In all these cases, the illustrations presented in the paper can be seen as particularly productive in regard to the popularisation of the subject (cf. Fischer 2007). Apart from the *Illustrated London News*, other contemporary newspapers and magazines also widely reported on and discussed archaeology (cf. PART II). Cf. also Leonard de Vries (1979) *History as Hot News 1865-1897: The Late Nineteenth Century World as Seen Through the Eyes of* The Illustrated London News *and* The Graphic.

8 For an overview of the different approaches, cf.: Astrid Erll (2006) *Kollektives Gedächtnis und Erinnerungskulturen*; Barbara Korte and Sylvia Paletschek (2009) »Geschichte in Populären Medien und Genres: Vom Historischen Roman Zum Computerspiel«. *History Goes Pop. Zur Repräsentation von Geschichte in populären Medien und Genres*. Eds. Barbara Korte and Sylvia Paletschek: 9-20; Hans-Joachim Gehrke and Miriam Sénécheau (2010) »Einleitung« *Geschichte, Archäologie, Öffentlichkeit. Für einen neuen Dialog zwischen Wissenschaft und Medien*. Eds. Hans-Joachim Gehrke and Miriam Sénécheau: 9-30; Andreas Daum (2002) *Wissenspopularisierung im 19. Jahrhundert. Bürgerliche Kultur, naturwissenschaftliche Bildung und die deutsche Öffentlichkeit 1848-1914*: 14-31; Angela Schwarz (1999) *Der Schlüssel zur Modernen Welt. Wissenschaftspopularisierung in Großbritannien und Deutschland im Übergang zur Moderne (ca. 1870-1914)*: 38-47 and 95-102; Carsten Kretschmann (2003) *Wissenspopularisierung. Konzepte der Wissensverbreitung im Wandel*: 7-21; DFG Research Group 875 »Historische Lebenswelten in populären Wissenskulturen

Although these approaches have been researched internationally, there have been different country-specific emphases (cf. Korte/Paletschek 2009: 11). Studies that have dealt with the reception of (popular) archaeology in Britain have focused on either a specific type of archaeology or aspect of archaeology, for example the reception of Egyptian archaeology, or a specific medium, such as the reception of archaeology in literature.[9] Other more general research on the topic concentrates on the public reception of archaeology and history in the twentieth and twenty-first centuries.[10] The majority of studies on popular archaeology have been undertaken by archaeologists,[11] while only a few studies have dealt with the relevance of the popular reception of archaeology in regard to Victorian and Edwardian British culture.[12] So far, there has been no extensive and comparative study of the discursive interrelation of archaeology and British literature and culture in the nineteenth and early twentieth century. The present book aims to close this research gap by analysing the reception of different archaeologies, namely Greek, Egyptian, and Zimbabwean, on the basis of a large text corpus comprising both fictional and factual contemporary texts.

Against and embedded within the historical background and context of Victorian and Edwardian times,[13] I will analyse various popular texts and

der Gegenwart« University of Freiburg; Gerda Henkel Stiftung Research Project »Geschichte für alle« University of Siegen.

9 Cf., for example, Shawn Malley (1996b), (2004), and (2012), who focuses on the reception of Assyrian archaeology in the Victorian press, while John Hines (2004) offers an exemplary analysis of the reception of archaeology in Medieval, Renaissance, and Victorian literature. For a discussion of the state of research regarding Greek, Egyptian, and Zimbabwean archaeology, cf. also Chapters 3.2, 3.3, and 3.4.

10 Cf. Gehrke/Sénécheau 2010: 10.

11 Cf. Browne (1991, 1993); Wallace (2004); Day (1997); Hudson (1981); Levine (1986); Gehrke/Sénécheau (2010); Samida (2010a, 2010b, 2010c, 2011a, 2011b, 2011c, and 2011d.

12 Cf. Malley (1997a, 1997b, 2004, and 2012); Daly (1999a); Korte (2000a) and (2000b); cf. also Chapter 3.

13 Although I will look at the popular representation of archaeology from the beginning of the nineteenth century to 1914, the focus will be on cultural texts and practices from 1850 to 1914.

practices[14] of the time dealing with archaeology. These texts and practices include both fictional and factual works such as novels and short stories, newspaper and magazine articles, and excavation reports as well as shows, spectacles, and exhibitions. Using a New Historical approach as its major framework, this book aims to show in what way archaeological texts and practices are both products and producers of their contemporary culture. Following Stephen Greenblatt, I have chosen a text corpus including various kinds of texts, regardless of their status concerning what used to be considered ›high‹ or ›low‹ literature:[15]

»In the analysis of the larger cultural field, canonical works of art are brought into relation not only with works judged as minor, but also with texts that are not by anyone's standard literary. The conjunction can produce almost surrealist wonder at the revelation of an unanticipated aesthetic dimension in objects without pretensions to the aesthetic. It can suggest hidden links between high cultural texts, apparently detached from any direct engagement with their immediate surroundings, and texts very much in and of their world, such as documents of social control or political subversion. It can weaken the primacy of classic works of art in relation to other competing or surrounding textual traces from the past. Or, alternatively, it can highlight the process by which such works achieve both prominence and a certain partial independence.« (Gallagher/Greenblatt 2000: 11)

In this context, Louis Althusser's concept of ideology as »represent[ing] the imaginary relationship of individuals to their real conditions of existence« (Althusser 2006: 338) is especially relevant:

»Ideology for Althusser is a particular organization of signifying practices which goes to constitute human beings as social subjects, and which produces the lived relations by which such subjects are connected to dominant relations of production in a society. As a term, it covers all the various political modalities of such relations,

---

14 For a definition of my understanding of popular text and practices, please see below. For a discussion of popular practices at the time, including exhibitions, shows, spectacles, panoramas, and dioramas, cf. Richard D. Altick's (1978) *The Shows of London*.

15 For a discussion of ›high‹ versus ›low‹ literature and culture, also cf. below.

from an identification with the dominant power to an oppositional stance towards it.« (Eagleton 1991: 18)

Consequently, in order to critically approach a text one has to deconstruct it and read it in regard to both what is present *and* absent. Interpreting Marx, Althusser introduces what he terms a »symptomatic reading« of a text: Althusser »called this method 'symptomatic reading', because it involved reading texts in a similar way to that in which psychoanalysts read the symptoms of their patients, namely for a meaning of which the patients – or the texts' authors, in Althusser's case – are unconscious« (Ferretter 2006: 51). Althusser's disciple Pierre Macherey elaborates this approach in his book *A Theory of Literary Production* (1978) by emphasising that

»[w]e must show a sort of splitting within the work: this division is its unconscious, in so far as it possesses one – the unconscious which is history, the play of history beyond its edges, encroaching on those edges: this is why it is possible to trace the path which leads from the haunted work to that which haunts it. Once again, it is not a question of redoubling the work with an unconscious, but a question of revealing in the very gesture of expression that which is not. Then, the reverse side of what is written will be history itself.« (Macherey 1978: 94)

In order to approach a text critically it is therefore necessary to reveal what Macherey refers to as the ›unconscious of the work‹ (cf. ibid.: 92): »The task of a fully competent critical practice is not to make a whisper audible, nor to complete what the text leaves unsaid, but to produce a new knowledge of the text: one that explains the ideological necessity of its silence, its absence, its structuring incompleteness – the *staging* of that which it cannot speak.« (Storey 2006: 60) In this process it is crucial to understand »what the work is compelled to say in order to say what it wants to say« (Macherey 1978: 94). I will use Macherey's approach to show how the popular reception of archaeology in the nineteenth and early twentieth centuries can be seen as disclosing underlying collective concerns.

Drawing on the cultural theorist Raymond Williams, who defines culture as comprising *cultural practices* and *cultural texts*,[16] I take a comprehensive and inclusive understanding of ›culture‹ as the basis of my work:

»Cultural history must be more than the sum of the particular histories, for it is with the relations between them, the particular forms of the whole organization, that it is especially concerned. I would then define the theory of culture as the study of relationships between elements in a whole way of life. The analysis of culture is the attempt to discover the nature of the organization which is the complex of these relationships.« (Williams 2006: 35)

In order to do justice to this understanding of culture, I use an approach of cultural studies that is based on the Italian Marxist Antonio Gramsci's concept of the »compromise equilibrium« (Gramsci 2006: 86). Unlike other cultural studies approaches that conceive popular culture as either imposed on the passively consuming masses by the ruling classes for means of manipulation[17] or as an authentic ›folk‹ culture originating with the ›people‹,[18] Gramscian cultural studies sees popular culture as a dynamic field characterised by the continuous underlying struggle between different forces:

»Undoubtedly the fact of hegemony presupposes that account be taken of the interests and the tendencies of the groups over which hegemony is to be exercised, and that a certain compromise equilibrium should be formed – in other words, that the leading group should make sacrifices of an economic corporate kind. But there is also no doubt that such sacrifices and such compromise cannot touch the essential; for though hegemony is ethical-political, it must also be economic, must necessarily be based on the decisive nucleus of economic activity.« (Ibid.: 86)

What Gramsci refers to as »hegemony« (ibid.) can thus be understood as a condition in which despite a particular dominant structure a society is still

---

16 As examples of popular cultural practices, Storey names the seaside holiday, Christmas, and youth subcultures, while he refers to soap opera, pop music, and comics as examples of popular cultural texts (cf. Storey 2001: 2).

17 This is best known from the Frankfurt School, prominently represented by Theodore Adorno and Max Horkheimer.

18 Cf., e.g., Raymond Williams (1958) »Culture is Ordinary«.

characterised by a certain degree of consent. Gramscian cultural studies conceives popular culture as made by the people »from the repertoire of commodities supplied by the culture industries« (Storey 2008: 52):

»The field of popular culture is structured by the attempt of the ruling classes to win hegemony and by forms of oppositions to this endeavour. As such, it consists not simply of an imposed mass culture that is coincident with dominant ideology, nor simply of spontaneously oppositional cultures, but is rather an area of negotiation between the two within which – in different particular types of popular culture – dominant, subordinate and oppositional cultural and ideological values and elements are ›mixed‹ in different permutations.« (Bennett 2006: 96)

This view thus sees the cultural field as characterised by a dynamic struggle for meaning between the different forces in which the result in the form of popular culture is continuously renegotiated. That this process is determined by the specific historical context is underlined by the diachronic change in the status of a certain culture, as pointed out by Storey:

»The process is historical (labelled popular culture one moment, and another kind of culture the next), but it is also synchronic (moving between resistance and incorporation at any given historical moment). For instance, the seaside holiday began as an aristocratic event and within 100 years it had become an example of popular culture.« (Storey 2006: 8)

Gramscian cultural studies thus underlines how what is conceived as popular culture at a certain point in time essentially varies. This is also true of the discourse of archaeology, which, as will be shown in Chapter 3, developed from a predominantly aristocratic occupation in the seventeenth and eighteenth centuries into a popular discourse in the nineteenth and twentieth centuries. In the following analysis I want to use a Gramscian approach to popular culture insofar as I understand popular archaeology as the product of a dynamic and conflicting process between different forces at a certain point in time. Closely connected to this and on the basis of theories of the popularisation of knowledge, I further see the emergence of a popular archaeological discourse in the nineteenth century as a complex and reciprocal interaction between various agents,[19] as programmatically formulated by Terry Shinn and Richard Whitley in *Expository Science* (1985):

»Alternative approaches to the study of the sciences [...] imply a richer and more sophisticated view of the popularisation process and its consequences for intellectual developments in different scientific fields. In this latter view, the dissemination of particular results and ideas to non-scientific publics is a more complex phenomenon, involving a variety of actors and audiences, that impinges upon the research process and cannot be totally isolated from it. Furthermore, the expansion and specialisation of scientific research in the past 200 or so years has resulted in many scientists popularising their work to other groups of scientists as well as to non-scientists – for a variety of purposes – so that the term has to be broadened beyond the simple traditional use.« (Whitley 1985: 4)[20]

---

19 These may include *inter alia* professionals, laymen, and the state.

20 This approach essentially differs from earlier models of the popularisation of knowledge in which the popularisers of knowledge, the professionals, were seen as responsible for the targeted production and distribution of knowledge, while the recipients, the lay masses, only passively consumed this processed knowledge. Due to its distinct hierarchical structure, Kretschmann refers to this earlier concept as a process of hierarchical knowledge transfer (cf. Kretschmann 2003: 9), conceiving producers and recipients in terms of active and passive entities. This is underscored by the assumed knowledge gap between the producers and recipients of knowledge, in that the producers of popular knowledge are seen as having an immense head start over the recipients with regard to knowledge. Since this approach to the popularisation of knowledge essentially denied any reciprocal exchange between producer and recipient, it is also referred to as a ›top-down notion‹ of popular science (cf. Topham 1998: 261) or a ›diffusionist model‹ (cf. Daum 2002: 27). Geoffrey N. Cantor and Sally Shuttleworth (2004) introduced the so-called ›conduit model‹ as a sophisticated version of the diffusionist model in regard to the popularisation of science in periodicals: »According to this model, each periodical fashions its response to science in the light of an intended readership.« (Cantor/Shuttlewort 2004: 4) For a discussion of the different approaches to the popularisation of knowledge, see Andreas Daum (2002) *Wissenspopularisierung im 19. Jahrhundert. Bürgerliche Kultur, naturwissenschaftliche Bildung und die deutsche Öffentlichkeit 1848-1914*: 14-31; Angela Schwarz (1999) *Der Schlüssel zur Modernen Welt. Wis-*

Following Carsten Kretschmann, who draws on Andreas Daum in understanding the popularisation of knowledge as a transformational process that eventually creates a new discourse (Kretschmann 2003: 15),[21] I consider the popularisation of archaeology as an emerging discourse that developed as a (collective) *reflex* to the drastic changes of the nineteenth and early twentieth centuries, affecting the different social classes in specific ways. Accordingly, I see the popular archaeological discourse of the time developing in a transformational and creative process as the result of a hegemonic struggle in the Gramscian sense between the different classes. It is crucial to note, however, that against this background popularisation must not be understood solely as a simplification process in regard to a certain knowledge, since popularisation constantly creates new discourses by including additional material provided and stimulated by the audience and the different media used for its transportation (cf. Ruchatz 2009: 102-103).

Combining this transformative and creative concept of the popularisation of knowledge with the Gramscian understanding of popular culture, I regard the popularisation of archaeology in the period under investigation as a heterogeneous and dynamic (cf. Kretschmann 2003: 21) process that concerned society as a whole and was determined by contemporary collective needs (cf. Korte/Paletschek 2009: 14). Consequently, popular archaeology in Victorian and Edwardian times can be conceptualised as a broad phenomenon. Within this broad reception of archaeology I understand the popular archaeological discourse as constituted by many different popular receptions of archaeology, that is, different archaeologies that can eventually be traced back to distinct underlying collective social concerns. Draw-

---

senschaftspopularisierung in Großbritannien und Deutschland im Übergang zur Moderne (ca. 1870-1914): 38-47 and 95-102; Carsten Kretschmann (2003) *Wissenspopularisierung. Konzepte der Wissensverbreitung im Wandel*: 7-21; Geoffrey N. Cantor and Sally Shuttleworth (2004) *Science Serialized: Representation of the Sciences in Nineteenth-century Periodicals*: 1-16.

21 Cf. also Korte and Paletschek: »In the various formats provided by the media, knowledge is not only conveyed and circulated but also always construed in a certain way. They are hence not only means of representation but also of knowledge production. Enquiry into the means and possibilities of representation of various media and genres thus proves to be essential for the study of historical culture(s).« (Korte/Paletschek 2009: 15, translation mine)

ing on a postmodern approach to cultural studies, which assumes that a differentiation between ›high‹ and ›low‹ culture is no longer possible, I read popular archaeology as an inclusive culture in which concepts of ›high‹ and ›low‹ culture are redundant due to their overlappings and their blurring of boundaries. Thus, my definition of popular archaeology is that of an inclusive and heterogeneous cultural field characterised by the Gramscian struggle for a »compromise equilibrium« (Gramsci 2006: 86).

On the basis of these theoretical assumptions, I argue that in its emergence as a discourse of the spatial and temporal ›other‹ popular archaeology functioned as an initially more or less undescribed parallel discourse to Victorian and Edwardian society and essentially lacked any structural predeterminations. This mirrors the character of archaeological fragments, which due to their decontextualisation are essentially ambivalent, indefinite, and arbitrarily arrangeable. From a structuralist point of view, this lack of structure necessarily entailed the absence of meaning. In order to create meaning, it was necessary to construct structure within this apparently blank archaeological space[22] by defining concepts in relation to each other through a series of articulatory practices. In this process of creating structure, however, meaning remained essentially ambivalent and indistinct. This is reflected in the unstable and floating relations and functions assumed and formed by concepts such as the familiar versus the unfamiliar, materiality versus spirituality, fact versus fiction, and madness versus sanity. By analysing the different representations of archaeology, I show how dominant Victorian and Edwardian concepts and assumptions were being deconstructed, challenged, and rearticulated in the formation of a popular archaeological discourse, which revealed in turn the social construction of reality in general (Berger/Luckmann 1966) and furthermore of concepts such as past/memory (Halbwachs 1950; Jan Assmann 1992; Aleida Assmann 1999), identity/alterity (Fludernik/Gehrke 1999; Storey 2003), gender/sex (Butler 1990), and race (Said 1978) in particular. At the same time, the unstable and floating character of identities within that archaeological space further emphasise the dependence of individual memory/identity on collective social frames of reference (Halbwachs 1925;

22 In the following I use the term ›archaeological space‹ synonymously with the archaeological discourse, which I define as including everything that has ever been thought, said and done in connection with archaeology.

Jan Assmann 1992; Aleida Assmann 1999). Without these social frames of reference, individual identity becomes essentially ambivalent and indefinite. Drawing on a model of the formation and performance of cultural identity through a complex interaction of memory, desire, and consumption presented by the cultural theorist John Storey (2003), I demonstrate how the contemporary popular discourse of archaeology in the Victorian and Edwardian Age reflected and (re)negotiated issues of individual and collective identity. I argue that the process of developing a popular archaeology as a space initially determined by ambivalence and indefiniteness was both fascinating and disturbing for contemporaries, since it essentially involved the continuous (re)creation and (re)definition of identit(y)(ies). At a time of profound sociocultural changes, which were essentially experienced as an alienation of the individual from a former more organic and holistic self, this parallel discourse of popular archaeology thus functioned as a space in which concepts of identity could be playfully (re)discovered, (re)negotiated, and (re)invented.

This further means that to a certain extent popular archaeology also always functioned as a projection surface for contemporary sociocultural concerns that were imported, transformed, and at times abandoned within the context of the parallel archaeological discourse. On this basis, and in analogy to major archetypical Victorian myths, I identify four dominant discursive threads in regard to the popular reception of archaeology, which I refer to in the following as (1) the *Frankenstein Discourse*, (2) the *Dracula Discourse*, (3) the *Jekyll and Hyde Discourse*, and (4) the *Sherlock Holmes Discourse*. Although manifested differently in cultural texts and practices dealing with the individual archaeologies, these four discourses can be seen as underlying the archaeological space in general with their prime occupation of negotiating different notions of identity and alterity. In particular, the major themes of the four discourses are the following: (1) *Frankenstein Discourse*: creation, origin, and belonging; (2) *Dracula Discourse*: invasion and inversion/subversion of identity, in particular in regard to gender roles; (3) *Jekyll and Hyde Discourse*: ambivalence and subversion of identity, double identities, and double standards; (4) *Sherlock Holmes Discourse*: (re)creation, invention, and narration of past identity. Characteristically, the ›protagonists‹ of these four discourses are all to a certain extent outcasts of society who commute between (the) inner and outer sphere(s) of the latter. Through their amorphous and floating identities, Victor Frankenstein and his creature, Count Dracula, Dr Jekyll/Mr Hyde, and Sherlock Holmes[23] reflect the ambivalence of the archaeological space, thus underlining the constructiveness and instability of concepts of identity/alterity and what Hans-Joachim Gehrke describes as follows:

> »As hard as societies, governments, associations, and other groups tried to achieve a ›clean‹ separation of identities and alterities and however many methods they applied to this endeavour, in pure and structural violence, in the ›production‹ of pasts and worlds of images – in the end they did not succeed. Oftentimes, it was art and literature that held things open at least in the *imaginaire* and made the other visible, also in the self.« (Gehrke 2004: 18, translation mine)

At the same time, the four discourses underscore the omnipresent feeling of alienation and insecurity characteristic of the time. It is for this reason that the four discourses prove so productive in regard to the study of popular archaeology in Victorian and Edwardian Britain. By analysing the popular reception of archaeology in nineteenth- and early twentieth-century Britain, I show how, at a time of profound and multiple changes, any sense of (collective) identity as such was deeply disturbed and needed to be renegotiated and re-established.

23 Although both Sherlock Holmes and Victor Frankenstein at first do not seem to fulfil the function of the outcast as much as Frankenstein's creature, Count Dracula, and Mr Hyde, their occupation turns them into eccentrics that commute between different spheres (within society and beyond), which is also reflected in their identities.

PART I: Preliminaries

1. Notions of Identity

> Traditionally, national identity has often been understood as something coherent and fixed, an essential quality of a group of people that is guaranteed by the ›nature‹ of a particular territorial space. However, although identities are clearly about ›who we think we are‹ and ›where we think we came from‹, they are also about ›where we are going‹. National identities are always a narrative of the nation becoming: as much about ›routes‹ as they are about ›roots‹. In other words, nations are never only invented once: invention is always followed by reinvention.
> (STOREY 2010: 13)

In his article »That Obscure Object of Desire: Victorian Commodity Culture and Fictions of the Mummy«[1] (1994), Nicholas Daly argues

»that the mummy is one of the figures through which changes in the material culture in Britain in the late nineteenth and early twentieth centuries were articulated. This period sees the shift in the British economy that caused people – intellectuals and

1 A revised version of this article was published as a chapter in Daly's book *Modernism, Romance and the* Fin de Siècle*: Popular Fiction and British Culture, 1880-1914* (1999).

popular readers alike – to conceive it in terms of *consumption rather than production*.« (Daly 1994: 26, emphasis mine)

Considering the treatment of mummies in both a literal and figurative sense in Britain over the centuries, he points out that »if the early modern period is marked by the literal consumption of mummies as medicine, and the early nineteenth century by the visual consumption of the mummy as a spectacle, the late nineteenth century is fascinated by the mummy as a sign that may be consumed in popular fiction« (ibid.: 25). I want to argue, however, that the popular consumption of the mummy and archaeology in the nineteenth and early twentieth centuries in general can be read as part of a larger process of production of identity. Drawing on the model used to explain the formation of cultural identity introduced by John Storey, I show in what way the popular reception of archaeology as a form of consumption became a means of producing collective identity.

Discussing cultural identity in connection with what he calls, following James Clifford (cf. Clifford 1997), ›roots‹ and ›routes‹ (cf. Storey 2008: 78-91), Storey conceives consumption as the underlying force of culture and identity formation, assuming that individuals express themselves through how they behave as consumers, that is, through how and what they consume:

»Consumption is a significant part of the circulation of shared and conflicting meanings we call culture. We communicate through what we consume. Consumption is perhaps the most visible way in which we stage and perform the drama of self-formation. In this sense, then, *consumption is also a form of production*.« (Ibid.: 78, emphasis mine)

For Storey, consumption thus becomes both the manifestation and realisation of identity. To explain this, Storey refers to Maurice Halbwachs and Jacques Lacan, introducing memory (which he calls the ›roots‹) and desire (which he calls the ›routes‹) as the two determining forces for the formation of identity: »In particular, I will explore the roots and routes of identities, that is how our identities are formed between memory and desire; between memory, with which we seek to ground ourselves in a known past, and desire, which props us through the present into an unknown future.« (Ibid.: 78) It is the combination of the past and the future (never to be attained) of

Storey's model that makes it applicable to the Victorian Age, since it very much reflects the latter's preoccupation with both past *and* future. Storey's model of the formation of identity touches and overlaps with other approaches concerned with the formation and performance of identity which form the basis of my analysis, such as Aleida and Jan Assmann's collective memory,[2] Pierre Nora's *sites of memory*,[3] Edward Said's *Orientalism* (cf. Said 1978), and Judith Butler's deconstruction of the categories sex and gender (cf. Butler 1990). Central to all these approaches is the assumption of identity as a cultural construction in the broadest sense, as pointed out by Aleida and Jan Assmann in regard to memory:

»The concept of the past that I adopt from Halbwachs is one that can be called socio-constructivist. What Peter L. Berger and Thomas Luckmann have shown to be true of reality as a whole was applied by Halbwachs, forty years earlier, to the past: it is a social construction whose nature arises out of the needs and frames of reference of each particular present. The past is not a natural growth but a cultural creation.« (Assmann 2011: 33)

2 Drawing on the Halbwachsian concept of the *mémoire collective*, Aleida and Jan Assmann introduced their influential model of a collective memory in the late 1980s, using the term collective memory as a generic term. For them, the collective memory includes the two different subcategories (1) the communicative memory, which comprises everyday communication, and (2) the cultural memory, which is based on symbolic cultural objectivations.

3 Cf. Pierre Nora *Les lieux de mémoire* (1984-1992). According to Nora, while in the nineteenth century there still existed a national memory that helped establish identity, this gradually disappeared in the twentieth century: »There are *lieux de mémoire* because there are no longer any *milieux de mémoire*.« (Nora 1990: 11, translation mine) Because of the lack of national memory, people are nowadays using sites of memory as a surrogate for the national memory. Consequently, sites of memory also always indicate the lack of a collective, national memory. According to Nora, there is no longer any such thing as a Halbwachsian collective memory in the twentieth century. (cf. Erll: 24-25). John Storey also sees what he calls the ›memory industry‹ as a site of memory (cf. below).

This social-constructivist approach to memory based on Berger and Luckmann (1966)[4] is also pivotal for the concept of *intentional history* introduced by Hans-Joachim Gehrke (cf. Gehrke 1994). Intentional history conceives a group's conception of the past as essentially dependent on what is needed by this group to sustain their collective identity in the present, as Gehrke points out:

»Against this backdrop, I would describe as *intentional history* those conceptions of pasts that are essential precisely for, even constitutive of, the identity of a group. It is precisely the traditions that are frequently adopted without reflection and the possibility, if need be, of committing oneself to them that guarantee the survival of the group or particular unit in question beyond the lifespan of the individuals that belong to it, and this is thus precisely what their identity is based upon. One could also [...] speak of ›self-history‹ but would then need to take into account the fact that the relation to the other has also entered into the *intentional history*. Hence, it includes both self-image and awareness of others; it is an elementary component of the ›imaginaire‹ of a collective.« (Gehrke 2004: 22-23, translation mine)

Gehrke emphasises that ›intentionality‹ for him in this context does not mean that an individual belonging to a certain group always deliberately and reflectively associates with that particular group. For him this intentionality works in regard to the collective consciousness insofar as the collective represents and generates particular discursive structures. This does not mean, however, that the group as such has to be aware of these

---

4 »The key assumptions of this book are expressed in the title and the subtitle, namely that reality is constructed by society – and – that it is the job of the sociology of knowledge to investigate the processes in which this occurs.« (Berger/Luckmann 1996: 1, translation mine) »Hence, we maintain that the sociology of knowledge should study everything that counts as ›knowledge‹ in a society, regardless of its absolute validity or invalidity. Inasmuch, namely, as all human ›knowledge‹ is ultimately developed, conveyed, and preserved in societal situations, the sociology of knowledge must endeavour to determine how knowledge developed, conveyed, and preserved in society congeals into ›reality‹ beyond all question for the man on the street. In other words, we maintain that the sociology of knowledge has the task of analysing the societal construction of reality.« (Berger/Luckmann 1996: 3, translation mine)

structures and consciously forms them. Usually, these structures are part of a tradition within which individuals of a certain group are socialised. In any case, there has to be a certain communal spirit for individuals to affiliate with a particular group, and this can also be used quite consciously, for example in certain political contexts (cf. ibid.: 21). For Gehrke, the concept of intentional history makes it possible to compare different historical phenomena, regardless of whether they are conceived as ›real‹ or ›fictional‹ events, whether they are traditionally considered as myths or history. Rather, intentional history focuses on the social function of history (cf. ibid.: 23). Furthermore, as Gehrke points out, intentional history assumes a similar role for the collective identity as a biography does for the individual (cf. ibid.: 25). As much as individual identity is dependent on an individual's biography, the collective identity of a group needs constant reassurance from a particular story of the past, that is, a history, of that group (cf. Gehrke 2001: 10).

Gehrke further connects this understanding of the past to the assumption that collective identity as defined by Berger and Luckmann is socially constructed and based on certain attributions of the group to the self and the ›other‹:[5]

»Let us now [...] assume that the collective identities are constructed; they are based on perception and condense it into clear images of the self and the other by means of appraisals and attributions. This is precisely where intentional history comes into play. It is thus, in the sense described by Berger and Luckmann, a societal construct that in turn creates societal coherence. [...] It is perceived – as also described by Berger and Luckmann – as concrete, reified, and no longer as a construct but rather as something objectively given, as physically present, and as such, as a fixed entity, it then, alongside the other facts of life, determines social existence. This shows itself precisely in intentional history. It is related especially to the present. The past and present tense are linked together and worked and woven into the tradition again and again. As a rule, however, the people concerned do not see things this way. *They absorb and understand the past, intentionally constructed or academically elucidated in this way, as a reality from which it is even possible to derive certain rights and binding obligations. And in doing so, they transform the constructs into real*

5 This was confirmed by the interdisciplinary research of Freiburg's Collaborative Research Centre *Identitäten und Alteritäten* (cf. below).

history, again and again. The clearest expression of this is the figure of physical descent, the talk of ancestors.« (Ibid.: 26, translation and emphasis mine)

In other words, although intentional history is first of all essentially a construct, constant repetition in the form of traditions and institutionalisations turns it into a seemingly objective truth that becomes a solid and determining parameter of any social existence.[6] This constructed past thus becomes a present reality that may even legitimise and demand certain behaviour in terms of specific rights and responsibilities of a group:

»It concerns a – at least in the case of myth creation – self-historisation of social relations, a form of detemporalisation created by self-observation that aims at increasing legitimacy. [...] By means of memory constructions, ›lasting‹ images are created against the background of the incessantly creeping changes and principal transformation – and these images are of course also commonly seen as virtually defining institutions.« (Rehberg 2004: 3-4, translation mine)

As Gehrke points out, this is particularly significant as far as a group's physical lineage is concerned and thus underlines once more how the process of remembering and creating history is very much determined by a group's present situation and its need to establish, sustain, and renegotiate collective identity (cf. Gehrke 2004: 26).

In the present book I will demonstrate in what way the concepts of collective memory, sites of memory, and intentional history introduced above functioned in regard to the popular reception of archaeology as a means of creating both individual and collective identity in the nineteenth and early twentieth centuries. I will delineate how the archaeological space became a site where, on the basis of the selective and constitutive processes of collective memory, histories were created in order to sustain a collective British identity. The constructed character of these histories and the immanent power structures underlying both individual and collective memory and oblivion[7] are pivotal to my analysis. Following Halbwachs, Assmann, and

6 This corresponds to what Aleida and Jan Assmann pointed out in regard to cultural memory.

7 Aleida Assmann emphasises the relevance of oblivion as an integral part of memory (cf. Assmann 2006: 104).

Gehrke, I will thus show in what way the popular reception of archaeology reflects the workings of remembering and forgetting in regard to the formation of identity, thereby emphasising the social conditionality of both individual and collective memory (cf. Gehrke 2001: 9).

As far as the reception of archaeology in Victorian and Edwardian times is concerned, this is particularly prominent in regard to the creation of a prestigious – in this case white, highly civilised, and long-standing – history on the basis of an alleged racial and cultural affiliation of the British with advanced civilisations of the past. Against this background, the concept of identity and alterity studies in general[8] are central to my analysis, since the creation and (re)negotiation of British identities within the archaeological space was strongly determined by and dependent on Orientalism, with its central assumption that the Western image of the Orient is essentially a construct used to define and secure a collective Western or European identity. In the following analysis of the different archaeologies, Said's *Orientalism* (1978) will play a major role in regard to the perception and reception of Egyptian, Zimbabwean, and Greek archaeology. I will show how texts dealing with these archaeologies strongly reflect what Said describes as the characteristic treatment of the Orient as an alterity functioning as a »surrogate and even underground self« (Said 1978: 3) in contemporary culture, politics, and academia in an attempt to sustain a white British identity. Said maintains:

8 Identity and alterity studies reached its first peak towards the end of the 1970s and was dealt with extensively by the aforementioned Freiburg Collaborative Research Centre *Identitäten und Alteritäten. Die Funktion von Alterität für die Konstitution und Konstruktion von Identität* (SFB 541, from 1997 to 2003), which concentrated on the study of collective identity. In an interdisciplinary study analysing different concepts and formations of identity and alterity in general, the focus of the special research project was on the processes of the formation of collective identity. The basic underlying assumption of the special research project, verified by the wide interdisciplinary research within the group over the years, was that identity as such is a construct which is the result of complex and intricate cultural perceptions, attributions, and assumptions that are essentially determined by the understanding of the ›other‹, i.e., in relation to an alterity.

»The relationship between Occident and Orient is a relationship of power, or domination, of varying degrees of a complex hegemony [...]. The Orient was Orientalized not only because it was discovered to be ›Oriental‹ in all those ways considered common-place by an average nineteenth-century European, but also because it *could be* – that is, submitted to being – *made* Oriental.« (Ibid.: 5)

In particular, the idea of the mysterious Orient and its various connotations of the exotic, voluptuous, and sexual becomes closely connected to archaeology as a spatial alterity, opposed to the western hemisphere, which on the one hand proves extremely rich in regard to the Westerner's experience, while on the other, through its associations with decadence and degeneration, it poses an acute threat to the Western self:

»For Orientalism was ultimately a political vision of reality whose structure promoted the difference between the familiar (Europe, the West, ›us‹) and the strange (the Orient, the East, ›them‹). This vision in a sense created and then served the two worlds thus conceived. Orientals lived in their world, ›we‹ lived in ours. The vision and material reality propped each other up, kept each other going. A certain freedom of intercourse was always the Westerner's privilege; because his was the stronger culture, he could penetrate, he could wrestle with, he could give shape and meaning to the great Asiatic mystery, as Disreali once called it.« (Ibid.: 44)

In the following analysis, Said's postcolonial approach will further help underline the different reception of the three archaeologies dealt with in this book. While Greek archaeology, representing what Uvo Hölscher has termed the *most familiar strangeness* (cf. Hölscher 1994),[9] was still considered part of the West's own collective identity, Egyptian archaeology was already perceived as part of the ›other‹, although it did offer certain

9 Uvo Hölscher coined the term the *most familiar strangeness* (›das nächste Fremde‹) in his eponymous book *Das nächste Fremde: von Texten der griechischen Frühzeit und ihrem Reflex in der Moderne* (1994), where he defines it as follows: »Rome and Greece are the most familiar strangeness for us, and what is so exceptionally educational about them is not their classical nature as well as their ›normality‹ but the fact that we meet in them that which is our own in another possibility, or even in the state of possibilities in general.« (Hölscher 1994: 278, translation mine)

aspects which contemporaries were happy to accept as part of their own identity.[10] Zimbabwean archaeology as encountered abroad, however, was perceived as the ›other‹ *per se* and thus had to be assimilated in the most radical sense, which is underlined by the forceful construction of a spurious white Western history by the British for Zimbabwe. With regard to the different functions of these three archaeologies, they can thus be categoryised in terms of (1) Greek archaeology – the familiar, (2) Egyptian archaeology – the familiar *and* the strange, and (3) Zimbabwean archaeology – the strange. Consequently, I want to adopt Höscher's concept of the *most familiar strangeness* for Greek archaeology and extend it by analogy to Egyptian archaeology as the *less familiar strangeness* and Zimbabwean archaeology as the *most unfamiliar strangeness*.

To round off the theoretical basis of my analysis, I want to return to Storey's assumption that identity is »formed between memory and desire; between memory, with which we seek to ground ourselves in a known past, and desire, which propels us through the present into an unknown future« (Storey 2008: 79) introduced at the beginning of this chapter.

In accordance with Halbwachs, Aleida and Jan Assmann, and Gehrke, Storey states that »memory seems to be at the very core of identity; it connects who we are to who we once were« (ibid.: 81). While memory, or rather what is experienced as memory,[11] thus constitutes the *root* of identities, relating peoples' identities to the past, desire is oriented towards the future in that it determines »who we want to be« or »who we think we should be in particular contexts« (ibid.: 86). To further explain his concept, Storey draws on Lacan's concept of the Real and the Symbolic. At the beginning, the child's world is dominated by what Lacan refers to as the Real, a condition in which the child's »union with the mother (or who is playing this symbolic role) is experienced as perfect and complete« (Storey 2006: 79). The condition of the Real gradually starts to lapse through the emergence of what Lacan refers to as the Symbolic, which essentially represents culture and is initiated by the mirror stage:

10 This is particularly prominent in the figure of the mummy representing the ambivalent materialisation of the self *in* the ›other‹ (cf. Chapter 3.3).

11 This includes any form of memory, regardless of whether it is ›real‹ or fictional.

»We have only to understand the mirror stage *as an identification*, in the full sense that analysis gives to the term: namely, the transformation that takes place in the subject when he assumes an image – whose predestination to this phase-effect is sufficiently indicated by the use, in analytic theory, of the ancient term *imago*.« (Lacan 2006: 288)

However, the self-image the child sees in the mirror is also essentially characterised by *mis*recognition, as it does not correspond to the way the child feels about herself. It is this moment, the child's recognition of the difference between subject and object, that is, the looking subject and the object looked at, that marks the alienation and the beginning of the realm of the Imaginary, which is characterised by the imaginary identification of the child in an attempt to return to the time before ›lack‹, that is, to the state experienced as a state of wholeness. The mirror stage is followed by the child's entry into language, which stands for the ultimate transition from the state of the Real to the state of the Symbolic:

»The division of the real into separate zones, distinct features, and contrasting structures is a result of the symbolic order, which, in a manner of speaking, *cuts into* the smooth facade of the real, creating divisions, gaps, and distinguishable entities and laying the real to rest, that is, drawing or sucking it into the symbols used to describe it, and thereby annihilating it.« (Fink 1997: 24)

Since it is essentially marked by difference and lack, the initiation into the Symbolic is disturbing:

»[A]lthough the child's subjectivity is made possible by language (without the symbolic realm of language the child would be unable to articulate its sense of the self), it is also undermined by being articulated in language; language remains a structure forever outside the child's being, belonging to others in the same way as it belongs to it.« (Storey 2008: 87)

It is thus through the transition from the Real to the Symbolic that desire develops out of the feeling of ›lack‹ experienced in the symbolic state:

»Desire is constituted and driven by ›lack‹ – the impossibility of closing the gap (and the continual need to try) between the child's ›self‹ and that which would make it whole again (the lost moment of plenitude when it existed in perfect union with the mother's body). The transition (although necessary) is so traumatic that we spend the rest of our lives trying to return; trying to get back to the fullness of the imaginary.« (Ibid.: 87-88)

For Storey, Lacan's desire – as opposed to Halbwachs's memory, which, forming our ›roots‹ relates back to the past – determines our ›routes‹ and is thus oriented towards the future, propelling an individual's performance in the present: »This endless quest organizes the very narrative of our lives, a narrative in which we pursue substitute objects and engage in *displacement strategies* in a hopeless attempt to find the completion we had once known in the imaginary.« (Ibid.: 88, emphasis mine) As an example of the attempt to overcome the chronic feeling of ›lack‹ Storey mentions the concept of romantic love: »The ideology of romantic love – in which ›love‹ is the ultimate solution to all our problems – could be cited as an example of this endless search. [...] Love in effect promises to return us to the Real: that blissful moment of plenitude, inseparable from the body of the mother.« (Storey 2006: 81)

The relevance of Storey's concept of identity formation for the popular archaeological discourse becomes explicit when we consider the various literal and figurative meanings associated with archaeology in general and the archaeological space in particular as reflected by the cultural texts and practices of the Victorian and Edwardian Age. Significantly, prominent connotations of archaeology and the archaeological space in the texts and practices analysed in this study are *inter alia* romantic love, sexuality, the mother/the womb, the subconscious, early history, pristine landscapes, indigenous people, and authentic culture.[12] In all these cases, the archaeological space thus connotes a state associated with a *time before* the present state, that is, before alienation, characterised by originality, authenticity, and organicity, in other words with a state of wholeness.[13] In this sense the archaeological search becomes a displacement strategy in that it appears to

12 Kuspit (1989), Pollock (2006, 2007a, and 2007b), and Orrells (2012) discuss archaeology in connection with the subconscious and psychoanalysis.

13 Although they do not represent a ›time before‹ as such, in Storey's terms romantic love and sexuality can be read as an attempt to achieve wholeness which stands for the time before alienation.

make the impossible possible, that is, the completion of the »endless quest in search of an imagined moment of plenitude« (ibid.: 79) to rehabilitate collective identity that has essentially been traumatised and destabilised in the wake of industrialisation and urbanisation.

That this state of wholeness is essentially unstable and can, if at all, be attained only momentarily, is underscored by the fact that although the encounters and experiences one has within the archaeological space at first seem to promise fulfilment, they eventually serve only to offer a glimpse at what could have been attained and leave the individual with a fundamental sense of lack. This is the reason why the archaeological experience is frequently perceived as insufficient and unsatisfactory in the long run, as it is underlined by the close interrelation between archaeology and both mental and physical diseases and degeneration. Against this background, the archaeological search thus becomes the endless search for origin, promising the formation and confirmation of identity in a state of completion.

In addition to romantic love, any kind of consumption can, according to Storey, be seen as a displacement strategy in the attempt to regain the Lacanian state of plenitude. At the same time, the way people behave as consumers essentially determines and forms their identity. In this context, consumption thus becomes a means of production, of producing identity. This is underscored by what Storey refers to as »memory industry« (Storey 2008: 85), which »produce[s] representations (›cultural memorials‹) with which we are invited to think, feel, and recognize the past« (ibid.). Analogous to Nora, Storey conceives the memory industry as a site of memory that creates ›fake‹ memories that function as a means of stabilising identity.[14] In this context, Storey refers to Alison Landsberg's concept of *prosthetic memory* (cf. Landsberg 1995: 175), which »describe[s] the ways in which mass media (especially cinema) may enable people to experience as memories what they did not themselves live« (Storey 2008: 85). Although Landsberg's concept of prosthetic memory was initially developed in regard to science fiction films, it has proved extremely helpful as far as the study of popular archaeology is concerned in that the latter offered people the chance to adopt and experience memories as their own which really

14 Apart from heritage sites and museums, Storey argues that mass media and popular culture in general also function as memory industries (cf. Storey 2008: 85).

were not. In her article »Prosthetic Memory: *Total Recall* and *Blade Runner*« (1995), Landsberg uses the film *The Thieving Hand* (1908), in which a beggar receives a prosthetic arm that once belonged to a thief, as an example to define her notion of prosthetic memory:

»By prosthetic memories I mean memories which do not come from a person's lived experience in any strict sense. These are implanted memories, and the unsettled boundaries between the real and simulated ones are frequently accompanied by another disruption: of the human body, its flesh, its subjective autonomy, its difference from both the animal and the technological. Furthermore, through the prosthetic arm the beggar's body manifests memories of actions that it, or he, never actually committed. In fact, his memories are radically divorced from lived experience and yet they motivate his actions. Because the hand's memories – which the beggar himself wears – prescribe actions in the present, they make a beggar into a thief. In other words, it is precisely the memories of thieving which construct an identity for him. We might say then that the film underscores the way in which memory is constitutive of identity. This in itself is not surprising. What is surprising is the position the film takes on the relationship between memory, experience and identity.« (Landsberg 1995: 175)

Landsberg's prosthetic memory can in a broader sense also be applied to the popular reception of archaeology in that people used ›borrowed‹ memories evoked through archaeological excavations, provided by newspapers, magazines, books, exhibitions, and shows of the time, to create and sustain individual and collective identity.

Finally, central to Storey's concept of cultural identity is its performative character, in that »our identities are made from a contradictory series of identifications, subject positions and forms of representation which we have made, occupied and been located in as we constitute and are constituted by *performances that produce the narrative of our lives*« (Storey 2006: 91, emphasis mine). Following Judith Butler, who in *Gender Trouble: Feminism and the Subversion of Identity* (1990) reveals the cultural constructiveness of the categories sex and gender, Storey argues that »[o]ur identities are not the expression of our ›nature‹, they are a performance in culture« (ibid.). For Butler, (gender and sexual) identity as such are not the origin – or what she refers to as the ›cause‹ – of dominant discursive structures based on the binary oppositions of ›women‹ and ›men‹ but rather a reaction to and result of them:

»The tactical production of the discrete and binary categorization of sex conceals the strategic aims of that very apparatus of production by postulating ›sex‹ as ›a cause‹ of sexual experience, behavior, and desire. Foucault's genealogical inquiry exposes this ostensible ›cause‹ as ›an effect,‹ the production of a given regime of sexuality that seeks to regulate sexual experience by instating the discrete categories of sex as foundational and causal functions within any discursive account of sexuality.« (Ibid.: 32)

By revealing the constructed character of gender and sex, Butler thus emphasises the relevance of performance for gender identity. It is this central role of performativity that Storey assumes for identities in general (cf. Storey 2006: 91).

The relevance of performativity for the constitution of identity becomes particularly clear in regard to the archaeological space in the texts and practices analysed in the present study. As pointed out above, popular archaeology can be characterised as a liminal space in which identities and alterities were renegotiated and reinvented. Due to the indefiniteness and ambivalence of this archaeological space, archaeology was experienced as threatening and promising at the same time, functioning as a counter-discourse to the patriarchal Victorian and Edwardian discourse essentially determined by the default setting of white healthy middle/upper-class heteronormative masculinity. This is underlined by the many instances where the archaeological space becomes the space within which gendered and racialised identities are questioned, dissolved, inverted, and reinvented, resulting in a deconstruction of the conventional concepts of identity. On the basis of Storey's concept of the formation of cultural identity with its emphasis on memory, desire, and performativity, I will thus show how the archaeological space reflects contemporary concerns in regard to individual and collective British identity.

2. Victorian and Edwardian Britain

The emergence of a popular archaeological discourse in the nineteenth century was the result of various and complex contemporary developments and insights that had a lasting influence on British society and culture. In order to illustrate the interrelations of these different factors, this chapter delineates the major advances of the Victorian and Edwardian era and their effect on people's conception of time, space, and identity.

2.1 AGE OF AMBIVALENCE: THE RISE OF MASS CULTURE

> For people who cannot read a great many books, and whose education has not been such as to give them a comprehensive view of the relations of the different sciences and of the chief branches of history and literature to each other, inspecting the contents of a museum is a salutary lesson. [...] The attitude of intelligent curiosity, delighting in the vast abundance of intellectual wealth for mankind, while gratefully receiving the minutest fragment added to one's own share, and desiring ever to increase it, is the happiest disposition for each of us, when we enter the British Museum. Among the many thousands of ordinary Londoners, who throng the halls and galleries on a Bank holiday, a large proportion being of the working classes, this disposition is quite as likely to prevail as in the generality of visitors on other days, belonging to the more leisured classes of society.
> (*ILLUSTRATED LONDON NEWS*, APR 9, 1887: 411)

The nineteenth century stands out as a period of radical transitions in Britain. At no other time in English history had the country and its people experienced such drastic reforms and changes as during the reign of Queen Victoria. Industrialisation, urbanisation, population growth, progress, and the Empire were the determining parameters of this era, entailing far-reaching consequences not only for British economics and politics but also for the human individual, as formulated by Leavis:

»It seems, then, not unnecessary to restate the obvious. In support of the belief that the modern phase of human history is unprecedented it is enough to point to the machine. The machine, in the first place, has brought about change in habit and the circumstances of life at a rate for which we have no parallel.« (Leavis 2006: 13)

As the Romantic concern for nature and preoccupation with the condition of men in the cities exemplifies, industrialisation had already started in Britain well before the end of the nineteenth century. Morgan points out that the first offshoots of the Industrial Revolution appeared as early as the 1760s, when »Britain gradually achieved rates of economic and industrial growth that reached a 2-3 per cent annual increase by the early nineteenth century« (Morgan 2004: 3). Even though these initial changes were minor compared to those of the following century, they offered a taste of what people were to expect in the years to come. An increasing productivity in agriculture, which first began in the sixteenth and seventeenth centuries, had caused drastic population growth and became a driving force of the Industrial Revolution.[1] Since Britain's economy remained dependent on agriculture during the first two decades of the nineteenth century, its productivity continued to be a crucial factor. This gradually changed through the establishment of the first modern factories producing textiles and crucible steel.[2] Although agriculture continued to be essential well into the 1850s in Britain, the new industries drew people away from the country to the cities, providing work in the factories and smaller workshops. Mainly responsible for this development was the shortage of work in the country, caused by continuing population growth and technological improvements, which gradually substituted manpower.

However, the growing industrialisation and urbanisation meant not only a drastic change in people's working conditions but also the emergence of a new social class in the cities, the urban industrial working classes. This new social class soon came to be perceived as an increasing problem by the middle and upper classes, when it became obvious that their lives, though

1 Morgan notes that »[t]he population of England and Wales rose from 5.7 million in 1751, to 8.7 million in 1801, to 16.8 million by 1851« (Morgan 2004: 23).

2 Although crucible steel had already been invented by Benjamin Huntsman during the 1740s, it was not in demand in Britain until the turn of the century, when it soon became an important economic factor.

primarily centred in the factories, were not restricted to them. Consequently, it was the time beyond the hours spent at the factory, the ›leisure‹ time, when the working classes were left to their own devices, indulging in their »rough amusements« (ibid.: 18), that was viewed with concern by the middle and upper classes:

»The large working-class districts had formed in a rather haphazard and unregulated manner, a phenomenon which initiated attempts by the middle class to catalogue and observe the social character of the larger urban developments. Moreover, the working class was forging a culture which, despite the overcrowded and poor sanitary conditions, utilised the benefits of living in close proximity to one another. [...] Significantly, along with the local public house and corner shop, the street became an important crucible for sociability and recreation, a factor which ensured that working-class culture, particularly among the young, became increasingly visible to the social observer.« (Beaven 2005: 89)

The middle and upper classes regarded the leisure activities of the working classes as a threat to society and civilisation as such: »The fear that working men were unable to act with restraint away from the workplace, particularly with the emergence of mass commercial leisure, was a constant concern across the political spectrum.« (Ibid.: 48) These contemporary anxieties of cultural degeneration and political instability[3] caused by the urban working classes are prominently dealt with by Matthew Arnold in *Culture and Anarchy* (1869). Introducing his concept of culture, Arnold propagates the cultivation of the working classes:

»[Culture] does not try to teach down to the level of inferior classes; it does not try to win them for this or that sect of its own, with ready-made judgements and watchwords. It seeks to do away with classes; to make the best that has been thought and known in the world current everywhere; to make all men live in an atmosphere of sweetness and light, where they may use ideas, as it uses them itself, freely – nourished and not bound by them.« (Arnold 2006: 7)

3 These anxieties were caused *inter alia* by the radical upheavals abroad, such as the French and American revolutions and the colonial conflicts Britain had to deal with.

According to Arnold, the working classes, which he refers to as Populace, had been alienated from their folk culture in the course of nineteenth-century industrialisation and urbanisation and created a culture of ›anarchy‹ characterised by »raw and uncultivated« (ibid.: 176) practices and performances. For Arnold, in order to ensure social stability and harmony it should be incumbent upon the state to educate and cultivate the masses for the sake of society as a whole: »Through culture seems to lie our way, not only to perfection but even to safety.« (Ibid.: 10) As the different educational reforms of nineteenth-century Britain show, Arnold's endeavour is representative of the time. Various societies concerned with the education of the working classes, such as the Society of the Diffusion of Useful Knowledge,[4] emerged.[5] In addition, the general educational system was improved, eventually culminating in Forster's Elementary Education Act in 1870.[6] Furthermore, a number of cheap newspapers and magazines aimed at the working classes were published at the time.[7]

The comprehensive endeavour to educate and cultivate the working classes was also prominently reflected by the various institutions that started opening their gates to a broader public. This becomes particularly clear in regard to the museum movement,[8] as pointed out by Daly: »For much of its history the British Museum authorities were more concerned with keeping people out than luring them in, but the museum movement of the second half of the nineteenth century reimagined the storehouse-museum as an instrument of popular education.« (Daly 1994: 31) Thus,

4 The Society of the Diffusion of Useful Knowledge (SDUK) was a Whiggish organisation founded in 1826 with the aim to provide affordable scientific, historical, and other texts for the working classes.

5 Cf. Jonathan Rose (2001) *The Intellectual Life of the British Working Classes*. At the same time, the working classes organised themselves in labour unions and socialist parties.

6 Cf. Philip Gardner (2006) »Literacy, Learning and Education«. Frequently these societies and schools concerned with the education of the working classes had a strong Christian alignment.

7 A prominent example is the illustrated *Penny Magazine*, founded in 1832 and published weekly by the Society of the Diffusion of Useful Knowledge.

8 Cf. Christopher Whitehead (2005) *The Public Art Museum in Nineteenth Century Britain: The Development of the National Gallery*.

rather than excluding the masses from certain spheres, the middle and upper classes had come to understand that it would eventually be safer to include the working classes to a certain degree: »Educational recreation was to draw the public from less acceptable pursuits; indeed not to do so was to risk the museum's transformation into a forbiddingly exotic, even Gothic space.« (Ibid.: 32)

One of the earliest and most outstanding examples in this respect was the Great Exhibition of 1851. As pointed out by Gurney, »[t]he relationship between the working class and the Great Exhibition was widely discussed at this time; [...] many liberal intellectuals hoped that the Great Exhibition, as an ambitious model of ›rational recreation‹, would fulfil a wider educative function and exert a civilising influence on the majority« (Gurney 2001: 115). In spite of the prominent fear of the working-class mob, which was particularly strong in the 1840s due to commercial and agricultural instabilities, people from all classes were finally allowed access to the exhibition, as Auerbach outlines in his analysis of the Great Exhibition:

»[T]he exhibition was an event that cut across traditional class boundaries and distinction, both in its formulation and in its popularity. [...] Most prognosticators had anticipated that there would be social segregation at the Crystal Palace [...]. The reality, however, was of social integration, as the Crystal Palace brought together all classes under one roof.« (Auerbach 1999: 2)

Yet, even though the working classes were admitted to the exhibition, entry fee regulations kept the less privileged classes away from the exhibition for the first few weeks.[9]

The concern for the cultivation and stabilisation of society as a whole was also strongly represented in General August Pitt Rivers's attempt to diffuse culture amongst the working classes by making his anthropological collection publicly accessible in London and Oxford later in the century (cf. Daly 1994: 32):

»Public education was the ultimate purpose of all the General's efforts in archaeology, just as it was the ultimate purpose of all his anthropological work. The motiv-

9 From May 26, 1851, the exhibition opened to the one-shilling visitors, thus becoming affordable for the lower classes.

ation was his belief in political and social evolutionary progress. [...] The study of anthropology and archaeology, like the study of English literature [...], was to be the force of the maintenance of Victorian social order.« (Bowden 1991: 141)

Stimulated by these institutionalised recreational offers, a new leisure culture developed, forming a combination of common working-class activities and the more traditional recreational concerns of the middle classes. A case in point is the Egyptian Hall at Piccadilly Circus, which opened in 1812. Commissioned and owned by William Bullock, it first displayed Bullock's private collection, until it became an exhibition hall housing various public exhibitions and shows in 1819. As illustrated by a contemporary description of the Egyptian Hall, its shows were aimed at entertaining its audience and tended to have a spectacular rather than an educational character:

»There are of course different kinds of mob. There is the musical mob, the political mob, and the ecclesiastical mob. They are going to Monday Evening Concerts, or Reform Meetings, or great Protestant demonstrations. Then there is the mob for the Egyptian Hall, who are going to see a giant or a dwarf as the case may be, or some entertainment of some kind.« (*London Society* 1867: 272)

The Egyptian Hall, however, was also a place where spectacular entertainment and education fused, as prominently represented by the popular Egyptian exhibition of the former showman and explorer Giovanni Batista Belzoni in 1821:[10] »The Egyptian Hall's prestige remained high the next year when an odd stroke of luck determined that it be occupied by, of all things, a sensational exhibition of Egyptian art and artefacts. The story behind this show was as romantic, in its way, as the exhibition itself.« (Altick 1978: 243) Belzoni's exhibition thus combined both the spectacular entertainment of fairs favoured by the working classes (cf. Donajgrodzki 1977: 18) and the more ›cultivated‹ recreational activities associated with the middle and upper classes. In this context, Belzoni's exhibition might be read as pivotal in regard to the development of a popular culture at the time, in that both Belzoni's professional career from showman to ›archaeologist‹ and his exhibition represent the merging of what was traditionally conceived as ›high‹ and ›low‹ culture into a new popular discourse.

10 Belzoni's Egyptian exhibition will be discussed in more detail in Chapter 3.3.

In addition to the development of a popular working class culture, the popularity of other nineteenth-century exhibitions and shows underlines the general interest in these kinds of public amusements, which was also prominent within the middle and upper classes. While in the eighteenth century the middle classes tended to spend their recreational time reading periodicals and books in domestic circles, the nineteenth century was dominated by public amusements: »The combination of time, leisure and money led to the increase of leisured activity, and the invention of popular pastimes to fill the newfound vacant hours.« (Wilson 2002: 408) Coffee houses, clubs, theatres, pleasure gardens, public shows, museums, panoramas, and theatres flourished at the time,[11] and the lending libraries became one of the major suppliers of books for a growing reading public.[12] Newspapers and periodicals offered information and entertainment for the masses, and by the mid-century the first public libraries had opened in Britain:

»The chair of the ›Arts and Manufactures‹ Select Committee was the radical Member of Parliament William Erwart. In 1845 he successfully guided through parliament a Museums Act, the wording of which, imaginatively interpreted, enabled three towns – Canterbury, Warrington and Salford – to provide ›free‹ libraries attached to their local museums. Along with public museums and art galleries, Erwart viewed public libraries as a means of elevating public taste and of inculcating artisans with an appreciation of good design and the skills necessary to carry it out. This would contribute in a small but significant way to increased prosperity and a civilisation of the masses. A similar mix of worldly knowledge and civilising culture was evident in the work, in 1849, of a Select Committee on Public Libraries, established to find the best means of extending the establishment of libraries freely open to the public especially in large towns in Great Britain and Ireland.« (Black/Hoare 2006: 22)

11 Cf. Richard D. Altick (1978) *The Shows of London* and Ralph Hyde (1988) *Panoramania! The Art and Entertainment of the »All-embracing« View.*

12 This reflects the general mass media production and consumption determined by innovative production technologies of the time. Cf. John O. Jordan (1995) *Literature in the Marketplace: Nineteenth-Century British Publishing and Reading Practices* and Jerry Don Vann and Rosemary T. van Arsdel (1994) *Victorian Periodicals and Victorian Society.*

As these changes in regard to the leisure activities of all social classes underline, in the nineteenth century recreation was characterised by a move away from the private space to the public space. What had previously been confined to individual circles amongst family and friends was now experienced in a collective as part of an emergent mass culture.

However, all these developments in the leisure activities of the different social classes in combination with the drastic changes in peoples' working and everyday lives were not without psychological consequences for Victorians of all classes. This was particularly true in regard to the contemporary perceptions of time. The increasingly capitalist system meant that people of all classes were bound to rigid and mandatory time frames and no longer lived in accordance with nature and its pristine rhythm.[13] This becomes particularly clear in regard to work at the factories and the latest technological innovations:

»In the nineteenth century, working time became a central theme in the ›social question‹ that divided political movements and parties. The struggle over the normal working day, the spreading notion of working time as a factor of production that was scarce and therefore had to be put to effective use, turned many old problems of justice, distribution, and money into problems of time. Railway traffic and the telegraph had put the fact that urban, regional, and national times of day were not synchronous on the political agenda. The movements of the sky became problematic as an absolute frame of reference for scientific and technical time-measurement when astronomers, using progressively better clocks, discovered a growing number of irregularities. Old interests in chronology, the science of time-reckoning and time-measurement, were reawakened in the wake of the debates over national and international time conventions.« (Dohrn-Van Rossum 1996: 1-2)

Fuelled by capitalist production, time as such became the governing principle of the Victorian Age, determining the habits of a whole society and leading to a collective synchronisation (cf. Lübbe 1992: 379): »The Victorians, at least as their verse and prose reveal them, were preoccupied almost obsessively with time and all the devices that measure time's flight.« (Buckley 1967: 2) In addition to a collective synchronisation resulting from

13 This was especially the case for the urban working classes that had formed out of people moving from the countryside to the cities.

fixed time schedules governing the public sphere, recent technological innovations had their lasting effect on the Victorians' conception of time.

The most revolutionary achievement in this context was undoubtedly the construction of the railway. Even though there had been wagon ways for moving coal in mining areas, there was not any public railway until the beginning of the nineteenth century, when horse-drawn railways became popular. The first public steam railway, though still interrupted by a horse-worked section, opened in northern England in 1825, connecting Darlington and Stockton-on-Tees. In September 1830, the first ›inter-city‹ railway line fully operated on steam was established by the Liverpool and Manchester Railway Company. Although the opening ceremony of the new line was overshadowed by the lethal railway accident of William Huskisson, a Member of Parliament, the railway was extremely popular and enjoyed an immediate boom in Britain. The rapidly growing railway system and its popularity in the 1840s gave rise to the term ›railway mania‹.[14] In addition to the birth of a modern railway system, transport in general was revolutionised in Britain at this time by the extension of the road and canal network throughout the country. Omnibuses, though still horse-drawn, improved transport within the cities, and the first steamboats carried passengers from Britain to Europe and the Middle East:[15]

14 Cf. Michael Freeman and Derek H. Aldcroft (1985) *The Atlas of British Railway History*.

15 One of the first to offer travels to the Middle East was Thomas Waghorn (1800-1859), whose life-long project was to establish a quick steamship route between England and India via Egypt. Since the 1830s Waghorn's company had built guesthouses and set up transportation in Egypt to improve travelling from Cairo to Suez. Even though Waghorn's company was very successful, it was taken over by the Peninsular and Oriental Steam Navigation Company in 1847, which continued to transport thousands of people to the Middle East. Another pioneer of the tourism industry was Thomas Cook (1808-1892), who started his business in the 1840s, organising inner-British trips, such as those for visitors to the Great Exhibition. In 1869, Cook took the first travellers through Egypt to Jerusalem and from the 1880s regularly transported tourists to the Middle East and Egypt. From 1890 Cook also offered luxurious steamboat cruises on the Nile.

»Where the Grand Tour and seasons in fashionable resorts had been an elite and exclusive privilege, tourism became another commodity for middle-class consumers, and desirable places to visit – pleasurable beaches and resorts, and romantic countrysides – developed facilities such as places of amusement and hotels to respond to that.« (Hines 2004: 202)

All these changes and improvements in public transport of course meant an increase in speed and thus brought with them both an objective and subjective shortening of travel time for the passengers. Distances, which had previously taken days or even weeks, could now be covered in only a few hours. Consequently, the subjective conceptions of both time and space underwent a drastic change, as prominently pointed out by Schivelbusch:

»Traffic engineering is the material substrate of making something available or accessible, i.e., it is just as much the material substrate of space-time perception in travelling. Changing an essential element of a particular sociocultural space-time structure will have consequences for the entire structure. Space-time consciousness loses its usual orientation.« (Schivelbusch 1993: 38, translation mine)

The present was no longer conceived as an extended experience but rather as a brief and transient moment. Other inventions in the communication sector, such as the electric cable, which connected England, Ireland, and France from 1851 and Liverpool and New York from 1866, as well as the telegraph and telephone, further influenced people's perception of time:

»In practical affairs we are constantly part of an objective time order measured quantitatively and uniformly according to the behaviour of objects in nature; at the same ›time‹, we are also conscious that these events have an entirely different quality in that they are part of the subjective time order of personal experience. Time as experienced exhibits the quality of subjective relativity, or is characterized by some sort of unequal distribution, irregularity, and nonuniformity in the personal metric of time. This quality differs radically from the regular, uniform, quantitative units characteristic of an objective metric.« (Meyerhoff 1955: 13)

The consequences of these overall spatial and temporal transformations affecting society as a whole were complex; they might, however, be characterised as a general feeling of collective alienation. Dohrn-Van Rossum describes this feeling as an alienation of the individual in the present from a romantically transfigured past in which people were imagined to still have lived in harmony with themselves and their surroundings: »As contemporaries looked at their own national history as well as foreign cultures, the more recent developments struck them as the progressive loss of the possibilities of individual control of time, an alienation from its natural and humane rhythms.« (Dohrn-Van Rossum 1996: 2)[16] The past became a place associated with stability and continuity which could no longer be found in the fast-moving life of the present, as emphasised by Buckley: »If the present seemed wavering and amorphous, the past at least was fixed and definite.« (Buckley 1967: 97)[17] Seemingly paradoxically, the changed perception of time characterised by the elusive qualities of the moment caused this past to appear to be within reach of the present, while at the same time the drastic and overall changes and innovations made it seem very distant:[18]

16 This is further emphasised by the prominent feeling of nostalgia characterising the Victorian Age (cf. Colley 1998).

17 »History proved a consistently popular pursuit precisely because of its seeming ability to provide a framework for justified beliefs; the action of selective memory governed the historical topics most appropriate to the institutions and ideals dominant in Victorian England.« (Levine 1986: 4-5) This is further underlined by the popularity of the work of Victorian historians, such as Thomas Babington Macaulay's *The History of England from the Accession of James the Second* (1849-1861) (cf. Sullivan 2009: 315). In addition, the popular interest in the national history of England was spurred by the works of prominent historians, such as William Stubbs's *The Constitutional History of England in Its Origin and Development, Edward Augustus Freeman History of the Norman Conquest* (1867-1876) and James Anthony Froude's *History of England from the Fall of Wolsey to the Defeat of the Spanish Armada* (1856-1870). Cf. also John W. Burrow's (1981) *A Liberal Descent: Victorian Historians and the English Past.*

18 For a discussion of aspects responsible for the popular interest in history arising in nineteenth-century Germany and Britain, cf. also Korte and Paletschek (2009) *History goes Pop: Zur Repräsentation von Geschichte in populären Medien und Genres*: 18-20.

»It was an age shaped, perhaps more than any other previous period, by the forces of transition. The predominantly feudal and agrarian past had disintegrated under the action of democracy and industrialism; and yet the final consequences of these truly revolutionary processes remained unclear. All that people knew was that a gulf was opening up with the past.« (Cox and Gilbert 2003: ix)

New concepts of time and a nostalgic view of the past caused by drastic social and economic changes in the present were thus major psycho-sociological factors responsible for an emergent collective Victorian fascination with the past.

An event which like no other represented the contemporary concern with the past, present, and future as well as the popularity of public exhibitions across classes, was the Great Exhibition of 1851.[19] Although there had already been exhibitions in the 1840s organised by the Society for the Encouragement of Arts, Manufactures, and Commerce, displaying »paintings and ›useful inventions‹ in the Society's rooms at Adelphi« (Auerbach 1999: 9), the Great Exhibition was the first of its kind in Britain. Probably inspired by the success of the French Industrial Exposition of 1844 in Paris, the Great Exhibition became a symbol of Victorian progress and prosperity, displaying Britain's international economic and political power: »The fruits of Free Trade, the thought of constitutional reform achieved without violent revolution, nearly forty years of peace, and a growing Empire overseas, all combined to give the Victorian a sense of optimism and a certain self-satisfaction.« (Wood 1960: 176) The British architect Joseph Paxton had won the international call for designs issued by the Royal Commission after having already successfully popularised his design of the Crystal Palace by publishing it in the *Illustrated London News*.[20] In spite of many dissenting voices, conceiving the building as the blight of Hyde Park, construction work started immediately after Paxton's design had been chosen by the commission. When the building was finally finished, shortly before the opening day of the Great Exhibition on May 1, 1851, the Crystal Palace

19 There were two catalogues that guided visitors through the Great Exhibition: *Official Descriptive and Illustrated Catalogue: Great Exhibition of the Works of Industry of all Nations* (1851) and *The Crystal Palace Exhibition, Illustrated Catalogue* (1851).

20 *Illustrated London News*, Jul 20, 1850: 53-54.

with its steel and glass structure came to stand as the epitome of Victorian modernity and progress. Even though at first the public was not too keen on the exhibition (cf. Auerbach 1999: 54), it eventually turned out to be a magnificent success. From its opening day on May 1 to its end on October 15, 1851, more than six million people, one-fifth of the British population at the time, came to visit the Great Exhibition. Its first day alone attracted more than 20,000 people (cf. ibid.: 9, 137).

While the Crystal Palace itself had already surpassed anything that had been seen before, the Great Exhibition topped this with the unprecedented abundance of exhibits from various countries and sectors: »More than 100,000 exhibits [were displayed], sent in by almost 14,000 individuals and corporate exhibitors, selected by hundreds of committees from Britain, its colonies and dependencies, and numerous other countries.« (Ibid.: 92) The British part was divided into the four sections raw materials, machinery, manufactures, and fine arts, which again were classified in thirty subcategories.

Apart from objects representing contemporary industry and technology, the Crystal Palace also accommodated a Medieval Court. This court, furnished in the Gothic style, presented people with an idealised past, commemorating craftsmanship and at the same time functioning as a foil to the progress and efficiency of the modern Victorian present. This demonstrates how the two opposing concepts of the Victorian Age, the celebration of the future through progress on the one hand and the fascination with an idealised past on the other, coexisted and became palpable at the Great Exhibition.

Much as Britain's past was juxtaposed with its present through the different sections of the Great Exhibition, foreign exhibits, arranged separately from the British sections according to their geographic provenance, were functionalised as counterfoils to Britain's economic and industrial progress and prosperity, emphasising British supremacy: »The Crystal Palace, like the British Museum and India House, presented and represented British-controlled territories throughout the world in such a way as to project an image of wealth and control.« (Ibid.: 102) Asia and Africa in particular were conceived of as backward and inferior to Britain, only there to serve as sources of raw materials for British industry (cf. ibid.: 101). The segregation and the public reception of foreign exhibits displayed at the Crystal Palace both mirrored the British colonial attitude, characterised by a genuine interest in the exotic on the one hand and a prominent impulse to control it and present it as allegedly inferior to British culture on the other.

Thus, aside from the glorification of progress and the future, the Empire discourse played a significant role at the Great Exhibition. Moreover, it anticipated the exhibition at Sydenham Crystal Palace, which opened in June 1854 and exceeded the Great Exhibition in extent and magnitude. While the Great Exhibition had concentrated on the presentation and glorification of industry, the later exhibition at Sydenham focused on amusement and entertainment of the masses:

»Whereas the Hyde Park exhibition had symbolized the mixing of the classes and the masses, the Sydenham site was clearly for the masses; whereas the former, at least on the surface, promised peace and international understanding, the latter instead carried forward the Great Exhibition's nationalistic undercurrent; and whereas the Great Exhibition was designed to educate British men and women about industrialization and tasteful consumption, the Sydenham Crystal Palace was, quite frankly, designed to amuse. In the final analysis, the arts manufacturers that were so prominently on display in 1851 were the concern of the few, not the many, and so the Crystal Palace was transformed from a school to a playground.« (Ibid.: 200)

Consequently, the focus at Sydenham was no longer on the presentation and celebration of British industry. Even though machinery and carriages were still displayed and science also played a prominent role, natural history and theme courts, »each illustrating the art and architecture of a great period in history« (ibid.: 201), were among the popular departments of this new exhibition. An average of two million visitors came to Sydenham every year to marvel at the works of bygone civilisations such as ancient Egypt and Assyria (cf. ibid.: 202). What had been displayed at places like the British Museum and the Egyptian Hall, was now accessible to an even broader public. People of all classes came into direct contact with archaeological artefacts and architecture which they had heard and read about and seen pictures of in contemporary magazines and newspapers. The character of the Great Exhibition and its sequel at Sydenham, in combination with their public success, thus represents an emergent popular culture celebrating the past, present, and future.

In addition to the ambivalent Victorian preoccupation with the past and the future and the development of a popular leisure culture in the wake of the Industrial Revolution, recent discoveries, insights, and changes regarding history, science, the Empire, and gender roles further influenced the development of a popular archaeological discourse in nineteenth-century Britain.

2.2 Discovering New Territories: History, Science, Empire, and Gender

Origin of Species. – In the new numbers of the *North American Review* and *American Journal of Science* we have the opinions of two American savants on this question. In the former, Mr. Darwin's theory is examined in a very thoughtful manner. Its excellence as a work of imagination is admitted, but its truth denied. In the latter journal the eminent naturalist Dr Asa Gray speaks much more favourably of the work. He says ›it is a scientific one, rigidly restricted to its object, and by its science it must stand or fall‹. He reserves his judgement as to the theory.
(*Illustrated London News*, May 5, 1860: 434)

While the first half of the nineteenth century in Britain was still characterised by the struggle for economic stability, the position of mid-nineteenth-century Britain was distinguished by prosperity and global power. The economic and political strength of Britain during that period also enhanced several developments and changes as far as science, society, and the British Empire were concerned.[1] Since many of these deployments had an enduring influence on the genesis and popular reception of archaeology in Britain, the relevant factors in regard to them will be discussed in the following.

---

1 For a transnational approach to the nineteenth century, cf. Jürgen Osterhammel (2009) *Die Verwandlung der Welt: Eine Geschichte des 19. Jahrhunderts.* Cf. also Charlotte Trümpler (2008) *Das Große Spiel: Archäologie und Politik zur Zeit des Kolonialismus (1860-1940).*

While the First British Empire of the seventeenth and eighteenth centuries had focused on the Americas, the West Indies, and India, the Second British Empire concentrated on Asia and Africa. Although India continued to be the core of the British Empire – Victoria became Empress of India in 1875 – Africa became increasingly important for the British. Since the late eighteenth century, Britain's interest in Africa had grown, with the Cape Colony and Egypt being particularly significant for the British trade routes to India. The Cape Colony was first seized by the British from the Dutch in 1795 and finally taken by British troops in 1806. As far as Egypt was concerned, the British had been present in the country for years and were in close trade relations with Egypt when it was occupied by Napoleonic troops in 1798. For the British this invasion meant not only a loss of control over Egypt but also a threat to their Indian colonies, since Egypt was part of the overland route to India. Consequently, the defeat of the French by the British army in 1801 and the restoration of British control in Egypt were celebrated as a huge success of British imperial politics. However, the British were forced to submit to Muhammad Ali and his army in 1807 and did not regain power in Egypt before the second half of the century, when Disraeli bought the Khedive's holding of the Suez Canal Company in 1875 and thus became the largest single shareholder of the company. In 1882 British troops were sent to Egypt to suppress protest initiated by the Arabs against the French and the British, and quickly defeated the Arabs. Thus, although Egypt was not officially a British colony, the British controlled and dominated the country politically and militarily.

The British rule in Egypt reflects Britain's aggressive New Imperialism, the radical expansion of the British Empire in the last third of the nineteenth century culminating in the so-called ›Scramble for Africa‹, in which Britain, France, and Germany were the main competitors in the struggle for African colonies. The three colonial powers met during the Berlin West Africa Conference (1884-1885) to partition the African colonies, and Britain was awarded Kenya, Uganda, Ghana, and Nigeria. Apart from the colonies in Eastern and Western Africa, Britain established further colonies in southern Africa during the 1880s, which were of particular interest due to the discovery of gold and diamond deposits in the 1860s.[2]

2 As Philippa Levine points out, the colonisation of southern Africa was mainly commercial, emphasised by the many British companies founded in Africa at

A person embodying the aggressive British imperial attitude in connection with colonial enterprise was Cecil Rhodes (1853-1902). Rhodes had been sent to Africa in his early teens to live with his brother and became a prominent figure in southern Africa in the 1880s and 1890s, first in the diamond industry and later in politics as a member of the Cape Colony Parliament. Rhodes launched the De Beers Mining Company in 1880 and founded the British South Africa Company in 1889. Both companies came to dominate the South African diamond and gold market. In 1890 Rhodes became Prime Minister of the Cape Colony, an office he was forced to relinquish in 1896 after having been involved in a revolt against the Transvaal, the Jameson Raid. Nevertheless, Rhodes continued to have a strong influence on South African commerce and politics. By means of the British South Africa Company, Rhodes controlled the region[3] and drove British expansion northwards with the ultimate aim of creating a British Empire including eastern Africa from the Southern Cape to Cairo.

»Mr. Cecil Rhodes, sprung from the ranks of private citizenship to the leadership of the powerful southern colony, is making history faster than any man of his generation, and is great with the greatness of those who can breathe a living purpose into the uniformed clay of the human masses. From his masterful nature came the impulse, which has welded the heterogeneous population of the South African dependencies into a living organism, and carried the British flag at one stride across a thousand miles of desert from the Limpopo to the Zambesi.« (*Dublin Review*, Jan 1894: 145-146)

Rhodes had already been securing land towards the north as early as 1885 with the support of the British government and eventually succeeded in expanding British territory northwards through the annexation of Bechuanaland, contemporary Botswana. British colonists followed Rhodes and

that time: »From the late 1880s, three new chartered companies – the Royal Niger Company in West Africa (1886), the British East Africa Company (1888) and the British South African Company (1889) – paved the way for imperial expansion in Africa, a method much cheaper for government than formal direct rule. The British North Borneo Company, a far smaller operation, had been established a little earlier, in 1881.« (Levine 2007: 93)

3 With the exception of the Boer republic of the Transvaal.

settled in the area of present-day Harare, which became a British protectorate in 1891. After various fights between British settlers and native tribes during the 1890s, Rhodes's company finally brought the region under British control. In 1895 the territory then known as South Zambezia was renamed after Rhodes as Rhodesia, present-day Zimbabwe. In 1901 and 1911 the southern and northern territories became known under the names Southern and Northern Rhodesia (present-day Zambia), respectively: »You can measure Rhode's achievement by surveying the map of Southern Africa in 1870 and comparing it with the same in 1895, when he had annexed, with the blessing of the British government and Crown, Mashonaland, Matabeleland, and all the territory which is now Zambia and Zimbabwe.« (Wilson 2002: 603) Although Rhodes died in 1902, his imperial ideology and his politics, climaxing in his vision of an Anglo-Saxon master race, paradigmatic for the nineteenth century and early twentieth century, continued to influence and represent the general British imperial attitude:

»Why should we not form a secret society with but one object, the furtherance of the British Empire and the bringing of the whole uncivilised world under British rule for the recovery of the United States for making the Anglo-Saxon race but one Empire. What a dream, but yet it is probable, it is possible. […] Africa is still lying ready for us [sic] it is our duty to take it. It is our duty to seize every opportunity of acquiring more territory and we should keep this one idea steadily before our eyes that more territory simply means more Anglo-Saxon race, more of the best of the most human, most honourable race the world possesses.« (Rhodes 2010: 222)

As reflected by Rhodes's fantasy, the imperial venture was driven by the idea of the unquestioned British superiority and predestination in regard to the allegedly ›primitive‹ people to be colonised. The colonies, and the African colonies in particular, were perceived as a blank and uncivilised space waiting to be appropriated by the British imperialists. In the process of colonisation, African cultures and their past were completely blanked out in that they were either declared to be uncivilised and lacking a history, as was particularly prominent in regard to southern Africa,[4] or they were por-

4 This is reflected by the reception of Zimbabwean archaeology (cf. Chapter 3.4 and PART II).

trayed as the degenerated[5] descendants of once highly advanced societies, as was the case in regard to modern Egyptian society (cf. Davis 1993: 196).[6] In both cases, the African space was conceived by the British colonisers as a sphere that had to be disclosed and explored in order to (re)discover and (re)establish culture and civilisation, as emphasised by the large number of British African explorers (cf. Levine 2007: 93).

In this context, Rhodes again serves as a prime example, since in addition to his aggressive imperial behaviour focusing on the present and the future, he was also very engaged in creating a white history for southern Africa to legitimise British supremacy. The conception of Africa as the ›dark continent‹[7] that needed to be enlightened by Western civilisation also dominated British culture at home (cf. Porter 2001: 1):

»The great explorers' writings are nonfictional quest romances in which the hero-authors struggle through enchanted or bedevilled lands toward a goal, ostensibly the discovery of the Nile's sources or the conversion of the cannibals. But that goal also turns out to include sheer survival and the return home, to the regions of light. These humble but heroic authors move from adventure to adventure against a dark, infernal backdrop where there are no other characters of equal stature – only bewitched or demonic savages. Although they sometimes individualize their portraits of Africans, explorers usually portray them as amusing or dangerous obstacles or as objects of curiosity, while missionaries usually portray Africans as weak, pitiable, inferior mortals who need to be shown the light.« (Brantlinger 1985: 176-178)

5 For a discussion of contemporary theories of degeneration, cf. below.

6 »When a learned Orientalist traveled in the country of his specialization, it was always with unshakable abstract maxims about the ›civilization‹ he had studied; rarely were Orientalists interested in anything except proving the validity of these musty ›truths‹ by applying them, without great success, to uncomprehending, hence degenerate, natives.« (Said 1978: 52)

7 Colonial histories have a long-standing tradition of associating darkness with African spaces, a metaphor which is steeped in racist ideology. Cf. Patrick Brantlinger (1985) »Victorians and Africans: The Genealogy of the Myth of the Dark Continent« and Patrick Brantlinger (1990) *Rule of Darkness: British Literature and Imperialism, 1830-1914.*

Besides the imagined necessity to civilise and missionise the African continent, adventure and romance played a prominent role in the Empire discourse, as reflected by the popularity of contemporary explorers such as David Livingstone and Henry Morton Stanley:[8]

»Curiosity about strange lands and peoples was nothing new to the British: it was at least as old as the empire itself. What was new, however, was the ease with which that curiosity could be satisfied from the safety of British shores. People with imperial tales to tell were more plentiful than ever before. Missionaries on home leave detailed their struggle to convert the heathens of distant lands from the pulpits and in the periodicals of their churches and chapels. Explorers like Henry Morton Stanley became public celebrities, their lecture tours attracting huge audiences and their books becoming best-sellers – *In Darkest Africa* (1890) sold 150,000 copies almost immediately. A new breed of war correspondents provided dramatic first-hand accounts of colonial campaigns, dispatching them by telegraph for immediate publication in their newspaper. The military expedition sent to save Gordon had twenty reporters in its entourage. And the mass reproduction and dissemination of images in lantern slide presentations and illustrated newspapers brought the colonies and their inhabitants to life in ways that words never could.« (Kennedy 2002: 19)

This immediate and extensive exposure to the Empire discourse at home also meant that British contemporaries were, more than ever before, confronted with the vastness of the newly disclosed colonial space. In the context of the drastic changes regarding the perception of time and space in the wake of industrialisation and urbanisation discussed above, the expansion of the Empire had a lasting influence on the collective British identity. Together with earlier insights in biology and geology,[9] the Empire discourse thus dramatically influenced contemporary temporal and spatial conceptions that were strongly intertwined with the popularisation of archaeology in the nineteenth century.

In the field of science, biology and geology came up with the most provocative results as far as Victorian society in general and religion in particular were concerned. Although Charles Darwin had published pioneering

8 The popularity of Livingstone and Morton Stanley is reflected by the various reports on their lives and expeditions in the *Illustrated London News*.

9 Cf. below.

insights in geology, biology, and anthropology gained during his expeditions on the *Beagle* on different continents as early as 1839 in his *Journal and Remarks*, his most ›disturbing‹ works did not appear until the late 1850s. However, once *The Origin of Species* (1859) and *The Descent of Man* (1871) were published, his impact on contemporary science on the one hand and society on the other was immense. The two basic strands of Darwin's theory, natural selection and evolution, had a lasting effect on the Victorians. For one thing, the theory of natural selection implied that certain understandable processes were responsible for the regulation and generation of the animal population, rather than an almighty creator:

»As many more individuals of each species are born than can possibly survive; and as, consequently, there is a frequently recurring struggle for existence, it follows that any being, if it vary however slightly in any manner profitable to itself, under the complex and sometimes varying conditions of life, will have a better chance of surviving, and thus be *naturally selected.* From the strong principle of inheritance, any selected variety will tend to propagate its new and modified form.« (Darwin 1959: 74)

Even though in *The Origin of Species* Darwin does not yet explicitly relate his interferences regarding the natural world to men, the book still meant a radical departure from what had been assumed thus far. Whereas nature and the animal world had previously been seen as controlled by man and in a more abstract sense by God, through Darwin's theory they were now ascribed an independent self-dynamics. Consequently, both human superiority and sovereignty were put into new perspective, which was experienced as a threat by many contemporaries, as Janet Radcliffe Richards notes: »It is common knowledge [...] that Darwin's theory was horrifying to most of his contemporaries because, like Copernicus's, it presented a direct threat to received religious belief.« (Richards 2000: 20) However, what Sigmund Freud later was to consider the ›three blows‹ to human narcissism (cf. Homas 1989: 75) was asserted by Darwin in *The Descent of Man*, where he argues: »Nevertheless the difference in mind between man and the higher animals, great as it is, certainly is one of *degree* and not of kind.« (Darwin 1989: 130, emphasis mine) In other words, Darwin postulated that not only were men subjected to the same natural processes as animals, but they were themselves animals, since men and animals were of the same »kind« (ibid.). This claim, of course, was outrageous and deeply disturbing at the same time. Notwithstanding the recognition and support Darwin's theories received from men like Joseph Hooker, Thomas Henry Huxley, and the American Asa Gray, as well as from the so-called X Club, a dining club which advocated scientific liberalism, the Church and the conservative public felt affronted by Darwin's claims. As a consequence, a furious debate started over Darwin's theory on natural selection and evolution, culminating in a meeting of the British Association for the Advancement of Science in Oxford where Darwinists and their opponents brought forward their arguments in favour of and against Darwin's theory, respectively. While this debate was restricted to academic circles, the popularity of Darwin's books and the many reactions of the press, reviewing, commenting, satirising, and caricaturing Darwin's theories, speak for a wide public interest in this controversy.[10]

Behind this public dispute over Darwin's theories, they had a strong impact on traditional belief systems, in particular in connection with the crisis of the Anglican Church:[11]

»In the context of the third quarter of the century, the proponents of scientific naturalism, as well as the spokesmen for other new theologies or new reformations, could in many respects seem to have been claiming the traditional mantle of the Anglican Church. Their cultural claims constituted a virtual mirror image of those of the established church. They were restoring real religion in place of the nominal religion of a divided, strife-torn Anglican Church. The new theology could now successfully oppose materialism, religious enthusiasm, and the resurgence of Roman Catholicism. Furthermore they could embrace new learning and knowledge. They alone were the protectors of stability, moderate religion, and Protestantism. And in the context of the debates of the 1870s those claims were intended to establish the scientists and secular laymen as the natural successors to the Church of England as the proper educators of the nation.« (Turner 1990: 19)

---

10 Cf. Alvar Elegård (1990) *Darwin and the General Reader: The Reception of Darwin's Theory of Evolution in the British Periodical Press 1859-1872* and Eve-Marie Engels and Thomas F. Glick (2008) *The Reception of Charles Darwin in Europe*.

11 Cf. also Richard J. Helmstadter and Bernad Lightman (1990) *Victorian Faith in Crisis: Essays on Continuity and Change in Nineteenth Religious Belief.*

Similar to the trauma people had suffered through the radical changes and transformations caused by industrialisation, religious beliefs were now challenged and effectively shaken by Darwin's ideas:

»In early Victorian Britain most people – at least according to censuses and on Sundays – were Christians; and since their Christianity tended to fundamentalism, most would probably have claimed to believe in the literal truth of the Genesis creation story. According to this account, God had made the world in six days, and created animals as distinct species in (it was taken for granted) their modern forms. And although the Bible did not say anything directly about the age of the earth, Archbishop James Ussher in 1654 had done calculations on the basis of Old Testament genealogies, and his conclusion that the earth was created in 4004 BC – on 23 October – was generally accepted. By the nineteenth century these beliefs were already under threat from the work of geologists who were beginning to have ideas of evolution and a much older earth. But nobody had any real idea how evolution could have occurred, and fundamentalists resisted the new ideas with explanations of their own – such as that fossils could be the remains of creatures who had died in Noah's Flood, or other that had been a series of creations and annihilations. The reason why Darwin's theory presented a serious threat to fundamentalism was that it made the heretical idea of evolution seem plausible and comprehensible for the first time.« (Richards 2000: 20-21)

As pointed out by Richards, Darwin's evolutionary theory strongly supported uniformitarianism, a scientific assumption introduced by James Hutton and John Playfair at the end of the eighteenth century. In contrast to catastrophism, favoured by the geologists of the Wernerian school,[12] which assumed that the earth had initially been created by divine forces and then was formed through a series of natural disasters, uniformitarianism argued that the development of the earth involved slow-moving and continuous processes over a prolonged span of time. Since uniformitarianism assumed that »present-day causes are the same as those that have always been operative[, this] [...] position was later called Uniformitarian« (Dean 1985: 112). While catastrophists almost unanimously propagated the biblical conception of Genesis, uniformitarianists, like Hutton, rejected religious

12 The Wernerian school was named after the German mineralogist Abraham Gottlob Werner.

beliefs in divine Creation and the Flood. Consequently, the catastrophist view was widely accepted by Victorians, whereas the uniformitarian position was seen as blasphemous and thus untrue by most.[13] For many, uniformitarianism, as presented by Hutton in his »Theory of the Earth; or an Investigation of the Laws observable in the Composition, Dissolution and Restoration of Land upon the Globe« (1788),[14] meant a direct attack on the prevailing traditional conceptions of history, as formulated by the Irish Archbishop Ussher in the seventeenth century:

»Until the late eighteenth century, most scientists believed in the literal historical truth of Genesis, Chapter 1. God had created the world and its inhabitants in six days. The story of Adam and Eve provided an entirely consistent explanation for the creation of humankind and the peopling of the world. Seventeenth-century cleric Archbishop James Ussher of Armagh, Ireland, used the Old Testament genealogies to calculate the world was created on the night preceding October 23, 4004 B.C. Ussher's chronology became theological dogma, allowing only 6,000 years for all of human history.« (Fagan 1996: 35)

Prehistory, the history before the written history, was thus nonexistent at the time, since it was incompatible with contemporary religious beliefs. This gradually changed in the course of the nineteenth century through insights gained by scientists such as William Smith (1769-1839) and Georges Cuvier (1769-1832). Smith, who studied geological strata and animal fossils around the turn of the century, inferred from his findings that different strata are characterised by distinct fossils, a fact that could help identify subsequent periods of history. Smith's discoveries and insights, which earned him the name ›Strata Smith‹ (cf. *Gentleman's Magazine*, Jan

---

13 »The first British geologist of central importance to the Victorians was William Buckland (1784-1856), whose lectures to the undergraduates at Oxford (from 1820) were formative. His Inaugural Address as reader of geology, *Vindiciae Geologicae: or, The Connexion of Geology with Religion Explained* (1820), affirms Creation and the Flood while rejecting the efficacy of present-day causes and Huttonian attempts to divorce geology from the Bible.« (Dean 1985: 112)

14 »Theory of the Earth; or an Investigation of the Laws observable in the Composition, Dissolution, and Restoration of Land upon the Globe« was first held by Hutton as a lecture at the Royal Society of Edinburgh in 1785.

1840: 98), influenced the French naturalist and zoologist Georges Cuvier. Although himself a strong advocate of catastrophism, a fact which made him popular in Britain at the beginning of the nineteenth century, Cuvier was one of the first to reconstruct different prehistoric periods.

While Smith and Hutton were avant-gardists in regard to their theories, uniformitarianism was eventually popularised by the geologist Charles Lyell (1797-1875) in his groundbreaking book *Principles of Geology*, which appeared in three volumes between 1830 and 1833 and was later republished in eleven editions. Central to Lyell's argument favouring uniformitarianism was the assumption that geological processes in the present should be analysed in analogy to and in connection with the past and that they could thus help provide information about the formation of the earth:

»By these researches into the state of the earth and its inhabitants at former periods, we acquire a more perfect knowledge of its *present* condition, and more comprehensive views concerning the laws *now* governing its animate and inanimate production. When we study history, we obtain a more profound insight into human nature, by instituting a comparison between the present and former states of society.« (Lyell 1997: 5)

Lyell's theory of uniformitarianism strongly influenced Darwin's studies and exploration on the *Beagle* and eventually stimulated the formation of Darwin's evolutionary theory:

»[*Principles of Geology*] was not only an original scientific treatise directed at other geologists; it was also at the same time a work in what the French recognized as a distinctive genre, that of *haute vulgarisation* or authoritative high-level popularization. The construction of the first edition, and its modification in subsequent editions, should therefore be seen as Lyell's lifelong contribution to continuing debates, not only about the changing profile of geological knowledge and its basic ›principles‹, but also about the place of such knowledge in a scientific view of the world.« (Rudwick 2005: 2)

Central to both Lyell and Darwin's theories is the approach of explaining the present through the past and the rejection of the omnipresence of an almighty creator in favour of natural processes: »The systematic use of physical and biological processes observable in the present world came to be taken for granted by geologists as the necessary first step towards interpreting the always fragmentary traces of past geohistorical events.« (Ibid.: 19) In addition to Lyell and Darwin's theories, the ›discovery‹ of prehistory,[15] which went hand in hand with the revolutionary geological and biological insights, strongly influenced both historical studies and the public perception of the past from the mid-century on. Until then, the human past had been conceived as covering no more than the time span documented by the Bible: »Antiquities were perceived as being ancient, but they belonged to a human past constrained by the 6,000 years of biblical time said by theological dogma to have elapsed since the Creation and the Garden of Eden.« (Fagan 1996: 286) This general perception of history gradually changed during the mid-century, after influential work from Denmark and France had been discovered and taken up by British scientists.

Crucial for the intellectual and archaeological coverage of prehistory was the model of the Three Age System introduced by the Danish museum curator Christian Jürgensen Thomsen (1788-1865). The Three Age System was first presented in a guidebook to the National Museum in Denmark issued in 1836, in which Thomas classified antiquities of the museum according to the three categories Stone Age, Bronze Age, and Iron Age. It was not until 1848, however, that Thomsen's book was translated into English and published under the title *A Guide to Northern Antiquities*. His pupil Jens Jacob Asmussen Worsaae, who excavated Danish burial mounds and proved Thomsen's Three Age System on the basis of stratigraphy, verified Thomsen's theory.

Thomsen's Three Age System shortly preceded Jacques Boucher de Perthes's archaeological findings in the Somme Valley in France in 1837, which eventually made material access to prehistory possible. De Perthes discovered human bones and artefacts and the remains of extinct animals in the same stratum, which proved that the two had populated the earth contemporaneously in a time *before* the past ascribed to human antiquity. In 1856 quarrymen in the Neander Valley in Germany excavated human bones clearly differing in shape from what had previously been known and identified as human remains, a finding that further emphasised the old age of

15 The term ›prehistory‹ originated from the French word ›pré-historique‹ and was first introduced into English in 1851 by the Scottish archaeologist Daniel Wilson in his book *Archaeology and the Prehistory Annals of Scotland.*

man. Excavations in France, Britain, and Switzerland offered further proof of a prehistoric human past, which was finally also recognised in Britain. In 1859 the British archaeologists John Prestwich and John Evans visited de Perthes in France. When they returned to Britain, Evans presented his conviction to the Society of Antiquaries in London that »in a period of antiquity remote beyond any of which we have hitherto found traces, this portion of the globe was peopled by man« (Evans 1860: 14-15). The various prehistoric archaeological findings throughout Europe and their introduction to Britain by men like Evans, Darwin, and T. H. Huxley eventually established prehistory as a period in historical and archaeological studies. The idea of prehistory was further popularised in Britain by John Lubbock's (1834-1913) best-selling book *Prehistoric Times* (1865) (cf. Patton 2007: 95), in which the author introduced the terms ›Palaeolithic‹ and ›Neolithic‹ to distinguish between the Old and New Stone Age, respectively. The Three Age System and Lubbock's subcategorisation of the Stone Age into Palaeolithic and Neolithic were confirmed by archaeological excavations throughout Europe and broadly accepted by the end of the century.

The groundbreaking insights gained by biologists and geologists at the time influenced the development of science in that they enhanced, for example, the genesis of archaeology from an amateur avocation to a scientific discipline.[16] Furthermore, they also affected Victorian society in general, which was both disturbed and fascinated by these recent discoveries and insights. Although people felt their traditional Christian belief system questioned and threatened, the idea of prehistory captured the interest of many Victorians (cf. Freeman 2004: 148). Beyond the past, which had traditionally been studied and passed on over generations, an unknown deep time suddenly appeared out of darkness:

»During the early decades of the nineteenth century traditional understandings of the Mosaic cosmogony had come under increasing strain form the encounter with new geological data. An ever-expanding time frame put more and more pressure on standard chronologies of the earth, and a growing fascination with deep time by visual representations of primitive life forms raised the public's consciousness of a

16 Cf. also Phillipa Levine (2002) *The Amateur and the Professional: Antiquarians, Historians and Archaeologists in Victorian England, 1838-1886.*

distant primeval world. Engravings and illustrations, together with the famed models of dinosaurs erected in Crystal Palace, which were opened to the public in 1854, raised in some minds the question of why so much history had elapsed before the appearance of the crowning of creation – the human race.« (Livingstone 2008: 82)

Prehistory appeared as a new discourse which hitherto had not been part of Victorian society. As a consequence of the new time dimension disclosed by the ›discovery‹ of prehistory, people felt closer to their comparatively more recent and familiar past, as Arthur Evans's famous letter (1859) commenting on the excavations in the Somme Valley illustrates:

»Think of their finding flint axes and arrowheads at Abbeville in conjunction with bones of elephants and rhinoceroses forty feet below the surface in a bed of drift! In this bone cave in Devonshire, now being excavated by the Geological Society, they say they have found flint arrowheads among the bones, and the same is reported of a cave in Sicily. I can hardly believe it. *It will make my ancient Britons quite modern if man is carried back in England to the days when elephants, rhinoceroses, hippopotamuses, and tigers were also inhabitants of the country.*« (Evans, quoted in Evans 1943: 100, emphasis mine)

This change in perception in regard to the past also entailed a readjustment in people's conceptual borderline between past and present. The more recent and national past now became part of what people considered as their own collective identity, while prehistory assumed the role of the unfamiliar and unknown ›other‹:

»The significance of these changes was widely understood by those living through them, and stimulated one of the great passions of Victorian society – a contemplation of the antinomies between progress and degeneration, of perfectibility and original sin, of the particulars of history and the great generalizations about the course of human history. Yet these antinomies reflected a more encompassing uncertainty which stemmed from the sheer scale of terrestrial history, and the yawning gulf which seemed to separate a richly-textured *knowable* present and a shady, insubstantial and potentially *unintelligible* prehistoric past.« (Murray 1993: 176)

It was in this context that discourses of degeneration and diffusionism became prominent in nineteenth-century Britain. Anxious to establish a borderline between modern man and the temporal and spatial ›other‹,[17] people commonly defined and strengthened their own identity against whatever alterity presented itself by using theories of degeneration and diffusionism.[18] Although theories of degeneration, emerging in the 1850s, were influenced by various contemporary trends, they were dominated by the recent insights drawn from evolutionary biology. However, Pick remarks the following in regard to these theories:

»There is no real sense of a ›founding text‹ of degeneration or atavism in England, like Morel's *Treatise* (1857) or Lombroso's *Criminal Man* (1876); it is more difficult to coordinate the precise moments of a theorisation of degeneration which runs alongside, but also within, the terms of Darwinian evolution and its more obvious publication ›landmarks‹ – 1859 and 1871.« (Pick 1993: 176)

In spite of the lack of a »›founding text‹« (ibid.) of degeneration in Britain, however, the prevailing idea of degeneration corresponded to those on the continent. In analogy to evolutionary biology, theories of degeneration essentially assumed that if humankind could develop and progress over time, the reverse in the form of devolution and degeneration must be possible as well. While according to these theories the white Western race was conceived as having achieved the highest level of evolution so far, other non-European cultures were seen as inferior and primitive, representing earlier stages of human development corresponding to that of prehistoric ancestors. These assumptions were programmatically connected to theories of diffusionism by Edward Tylor in his book *Primitive Culture: Researches into the Development of Mythology* (1871):

»The two theories which thus account for the relation of savage to cultured life may be constructed according to their main character, as the progression-theory and the degradation-theory. Yet of course the progression-theory recognizes degradation, and the degradation-theory recognizes progression, as powerful influences in the course of culture. Under proper limitations the principles of both theories are con-

---

17 This also involves the ›other‹ in one's own society (cf. below).

18 Malley names the anthropologist Sir Edward Tylor and the writer and anthropologist Andrew Lang as prominent representatives of diffusionist theories in nineteenth-century Britain (cf. Malley 1997).

formable to historical knowledge, which shows us, on the one hand, that the state of the higher nation was reached by progression from a lower state, and, on the other hand, that culture gained by progression may be lost by degeneration. If in the enquiry we should be obliged to end in the dark, at any rate we need not begin there. History, taken as our guide in explaining the different stages of civilization, offers a theory based on actual experience. This is a development-theory, in which both advances and relapse have their acknowledged places. But so far as history is to be our criterion, progression is primary and degradation is secondary; culture must be gained before it can be lost. Moreover, in striking a balance between the effect of forward and backward movement in civilization, it must be borne in mind how powerfully the diffusion of culture acts in preserving the results of progress from the attack of degeneration. A progressive movement in culture spreads, and becomes independent of the fate of its originators.« (Tylor 1871: 34)

Basing his theory on the study of what he refers to as ›primitive culture‹ and history, Tylor defines progression and degradation as coexisting, and he sees progression as the inevitable path to the ultimate achievements of civilisation. According to this approach, cultures considered less advanced are thus either the mark of a lower state of development or the degenerated descendants of once higher civilisations. Although both assumptions were popular at the time, the first clearly prevailed. In this context, diffusionism became the safeguard against degeneration and at the same time a welcome way of locating the origin of Western civilisation in the advanced cultures of ancient Egypt, Greece, and Rome:

»Over the centuries Egyptology has attracted a large number of eccentrics and disciples of bizarre theories. Among them have been the pyramid cultists […] and the cultural diffusionists. […] The cultural diffusionists believed either that every civilization originated with an Egyptian colony or that wandering Egyptians greatly influenced all the other ancient nations.« (Wortham 1971: 46)

These popular assumptions of diffusionism were a mixture of myths[19] and contemporary archaeological and ethnographical theories:

---

19 The most prominent example as far as Britain is concerned is the legend of Brutus of Troy, which functioned as a founding myth for Britain (cf. Chapter 3).

»There were, in the second half of the nineteenth century [...] two explanations current as the origin of the cultural changes which were manifest in the archaeological and ethnographical record, of apparently the same cultural features. One explanation was that these features had evolved separately and developed independently. The other was that a feature had spread from one area to another, that it had in fact been diffused – by trade, by the movement of people, or by cultural contact. It is these two explanations of culture change that we refer to when we talk of evolution versus diffusion in the study of human cultural development.« (Daniel/Renfrew 1988: 81)[20]

No matter how far-fetched these theories of popular diffusionism were, they served as a grounding and stabilisation of Western collective identity against a temporal and spatial alterity. However, in addition to degeneration associated with the temporal or spatial ›other‹, contemporaries also saw the danger of degeneration as coming from *within* their own culture. This was particularly the case for individuals and groups of people that were considered as deviating from the ›norm‹.[21] In all these cases, contemporary anthropology, medicine, and criminology[22] were ready to provide examples to support these beliefs and thus legitimise the alleged superiority of the white sane/rational heterosexual[23] race.[24] At the same time, under the sur-

20 Diffusionism culminated in the hyperdiffusionist school represented by Grafton Elliot Smith at the end of the nineteenth century, which assumed that Ancient Egypt was the birthplace of civilisation. According to Smith, »[t]here was no civilization before Egypt, at least no civilization that was not derived from Egypt. [...] Everything started in Egypt – everything. When [Smith] was once asked what was taking place in the cultural development of the world when Egypt was allegedly laying the foundations of civilisation, he answered at once, ›Nothing‹« (Daniel/Renfrew 1988: 85-86).

21 This included, for example, physically and mentally affected people, homosexuals, and criminals, but also women and the working classes in general. Cf. Kelly Hurley (1996) *The Gothic Body: Sexuality, Materialism, and Degeneration at the* Fin de Siècle.

22 The Italian physician Cesare Lombrose (1835-1909) was the first to connect criminal potential to a degeneration of the brain.

23 In addition, social class was also a determining factor in this context.

face of all these theories, people's immanent fear of their own degeneration was lurking. This became particularly strong towards the end of the nineteenth century when

»it also began to seem to some Victorians that they might have been shooting themselves in the foot ideologically with degeneration theory. For any who took it to mean that civilized humanity, including themselves, could still actually go backwards and deteriorate to the level of naked savages, whom many Victorians saw as fundamentally (biologically, intellectually, and morally) inferior to themselves (not just Europeans groping their way through the dark ages), this might not have sat well with an even more pervasive inveterate Victorian article of faith, that of the inevitability of ›progress‹, or improvement inherent in the human condition.« (Patterson 2011: 35)

One of the most influential figures in Europe in this context was Max Nordau, who in his book *Entartung* (1892), translated into English as *Degeneration* in 1895, postulated the degeneration of modern society:

»Having borrowed various contemporary terms and ideas from the works of Morel, Lombroso, Mausley, Taine, Charcot and others, Nordau argued that modern society was witnessing a terrible crisis born out of the growing division between the human body and social conditions. […] Nordau was obsessed by the relation between fin-de-siècle culture and hysteria. He found massive obfuscation and disorders of speech in famous writers and painters.« (Pick 1993: 24)

As opposed to approaches that locate the sources of degeneration in the realm of the temporal and spatial ›other‹, Nordau's theory supports the assumption that degeneration had become an inherent part of modern society and was thus threatening it from within. This essentially reflected prominent contemporary collective anxieties in British society. While far into the second half of the nineteenth century progress and prosperity had

24 In almost all cases, degeneration was somehow connected to the phenotype of a group or an individual. In this context the Victorian obsession with phrenology and craniology played a crucial role. Cf. John van Wyhe (2004) *Phrenology and the Origins of Victorian Scientific Naturalism* and Patrick Joyce (1994) *Democratic Subjects: The Self and the Social in Nineteenth-Century England.*

dominated Victorian society, towards the end of the century theories of degenerations were mirrored by fears of decline and doom prevailing in Britain, as summed up by Arata:

»The decay of British global influence, the loss of overseas markets for British goods, the economic and political rise of Germany and the United States, the increasing unrest in British colonies and possessions, the growing domestic uneasiness over the morality of imperialism – all combined to erode Victorian confidence in the inevitability of British progress and hegemony. Late-Victorian fiction in particular is saturated with the sense that the entire nation – as a race of people, as a political and imperial force, as a social and cultural power – was in irretrievable decline.« (Arata 1990: 622)

Accompanying the decrease of political power and the stagnation of economic growth was a collective perception of a moral decline associated with the destabilisation of traditional belief systems and social structures. The emergence of the New Woman and the decadent Dandy was perceived as the ultimate confirmation of this:

»The rise of the New Woman at the *fin de siècle* was symptomatic of an ongoing challenge to the monolithic ideological certainties of mid-Victorian Britain. [...] The collision between the old and the new that characterized the *fin de siècle* marks it as an excitingly volatile transitional period; a time when British cultural politics were caught between two ages, the Victorian and the modern; a time fraught both with anxiety and with an exhilarating sense of possibility. The recurrent theme of the cultural politics of the *fin de siècle* was instability, and gender was arguably the most destabilizing category. It is no coincidence that the New Woman materialized alongside the decadent and the dandy. Whilst the New Woman was perceived as a direct threat to classic Victorian definition of femininity, the decadent and the dandy undermined the Victorians' valorisation of a robust, muscular brand of British masculinity deemed to be crucial to the maintenance of the British Empire.« (Ledger 1995: 22)

Both the New Woman and the Dandy embodied the contrary of established gender identities, such as the Victorian ›angel of the house‹ female and the patriarch, and thereby questioned and subverted them.[25] Both were also commonly associated with a deviant sexuality that posed a threat to hetero-normative, and especially male, society:

»If the New Woman was considered to be and often caricatured as a sort of masculine female, the Dandy was described as an effeminate male and was thus almost synonymous with the homosexual. Dandy and homosexual were at least as dangerous for the patriarchal order of Victorian culture as the New Woman and the prostitute. Like the prostitute, the homosexual was often connected with degeneration, debauchery, and venereal diseases; like the New Woman, the Dandy was often seen as sterile and asexual. Contradictory as these stereotypes may seem, they all converge in a deviant, unproductive attitude towards sexuality that subverts gender boundaries and was thus considered a threat to the male-dominated symbolic order of Victorian culture.« (Meier 2002: 119)

In the case of the New Woman, the threat associated with this independent femininity was transformed into the luring *femme fatale* of *fin-de-siècle* literature, who threatened male identity. The general feeling of moral decline and degeneration from within Victorian society also materialised in

25 Traditional gender roles were also challenged by various Women's Acts passed by Parliament during the reign of Victoria and particularly by the suffragette movement with its early roots in the 1860s, as pointed out by Mayhall: »Suffragettes were therefore believed to have been motivated to move from constitutionalism to various forms of anarchy because seemingly having exhausted the legislative possibility for women's enfranchisement, they fully rejected government's authority. An alternative explanation accepts the pervasiveness of accumulating frustration – forty years of political agitation with limited success could, and did, operate as a rallying cry – but qualifies it. Those women who became suffragettes had, in fact, come to believe that the legislative avenues available to them had been exhausted. Far from rejecting constitutionalism, however, suffragettes – at least until 1910, and many until 1914 – pursued strategies of resistance that drew upon the constitutional idiom. As had liberals, radicals, and conservatives throughout the nineteenth century, suffragettes staked claim to inclusion in the political nation by insisting that the British constitution live up to its own ideals.« (Mayhall 2003: 41)

fantasies of invasion and in what Arata terms *reverse colonisation*[26] towards the end of the century in his study of Bram Stoker's *Dracula*:

»*Dracula* enacts the [Victorian] period's most important and pervasive narrative of decline, a narrative of reverse colonisation. Versions of this story recur with remarkable frequency in both fiction and nonfiction texts throughout the last decades of the century. In whatever guise, this narrative expresses both fear and guilt. The fear is that what has been represented as the ›civilized‹ world is on the point of being colonized by ›primitive‹ forces. [...] In each case, a terrifying reversal has occurred: the colonizer finds himself in the position of the colonized, the exploiter becomes exploited, the victimizer victimized. Such fears are linked to a perceived decline – racial, moral, spiritual – which makes the nation vulnerable to attack from more vigorous, ›primitive‹ peoples.« (Arata 1990: 632)

Collective anxieties of invasion were further spurred by recent technological inventions, such as the microscope and improved photographic techniques, which made visible what had previously been either not visible at all or only transiently.[27] This was particularly the case for the recent discovery and identification of germs and bacteria, such as anthrax and the tuberculosis bacillus by the German physician Robert Koch.[28] Scientific investigation in the pollution of water and air further stirred the public fear of the potential contamination of these elements by invisible and hostile forces (cf. Worboys 2000: 158).[29] The fact that the physiologist John Burdon-Sanderson stated in 1878 that »germs did not just fall into wounds from the air, being far too few and fragile; rather, they mainly came from other diseased sources and had to be ›introduced‹ into the body« (ibid.) did

26 In the following, the term *reverse colonisation* will be used synonymously with the term *reverse invasion*.

27 Cf. Philipp Sarasin (2007) *Bakteriologie und Moderne. Studien zur Biopolitik des Unsichtbaren 1870-1920* and Nancy Tomes (1990) »The Private Side of Public Health: Sanitary Science, Domestic Hygiene, and the Germ Theory, 1870-1900«.

28 Koch identified the bacillus anthracis in 1876, the tuberculosis bacillus in 1882, and the vibrio cholerae in 1884.

29 The first contemporary scientific attempts to determine air and water pollution were carried out, but failed (cf. Worboys 2000: 158).

not ease these anxieties but rather gave rise to popular attempts to determine the »diseased source« (ibid.).[30] In her article »Photography as Witness, Detective, and Impostor: Visual Representation in Victorian Science« (1997), Jennifer Tucker shows how Victorians used photography in an attempt to capture and document not only microbes, but also meteorological phenomena and even sprits as the transient and invisible. In spite of the objectivity generally attributed to photography, Tucker also emphasises that »[t]he role and integrity of photography was interrogated not only in realms such as spiritualism, but in contexts where photographs are not usually thought to have been controversial, such as meteorology and bacteriology« (Tucker 1997: 403). This might be read as pointing towards the general collective Victorian uncertainty in regard to the borderline and the validity of the material versus the invisible, abstract, and spiritual, which is prominently reflected in the popularity of ghost stories[31] and detective fiction[32] towards the end of the century.

---

30 The fear of the body being penetrated by an invisible diseased source is popularly dealt with in *Dracula* and the mummy curse (cf. also Chapter 7.1). In addition, the vampire bites in *Dracula* can also be read as symbolic contamination through a syringe, reflecting contemporary distrust in vaccination, culminating in the anti-vaccination movement of the 1870s and 1880s (cf. Brunton 2008: 1).

31 In *Night Visitors* (1977), Julia Briggs argues that the ghost story, which was often set in a remote past, offered people stability beyond the modern material reality characterised by instability.

32 »Nineteenth-century psychology would define individual character as something constructed by the accumulation of sensations and impressions, effectively rendering all human perception ›subjective‹ and therefore suspect from an evidentiary point of view. In courts of law, accordingly, the application of rational principles of evidence to verbal testimony increasingly required substantiation by material and circumstantial evidence. […] At least in the popular imagination, the nineteenth-century figure who most elaborately and successfully stages his transformation is not the lawyer, but the literary detective. He may be an attorney, a physician, a drawing master, or a professional detective; but his central action is the transfer of power over establishing a person's true identity to an authority outside the individual self by revealing how suspect every character in the story is.« (Thomas 2001: 182)

At the same time, the fact that microbes and spirits were investigated in a similar manner emphasises the as yet indeterminate character of supernatural phenomena in the context of the emerging scientific discourse. At a time of groundbreaking scientific insight that revolutionised people's perception of the world around them, everything seemed to be possible. It is in this context that the Victorian fascination with spiritualism,[33] materialising in the popularity of Victorian séances and the foundation of various occult societies, has to be read:[34]

»Modern spiritualism arrived in England from its birthplace in the United States in the early 1850s and during the next twenty years took the country by storm. Like mesmerism, spiritualism held huge appeal for women and men of all classes and shades of belief, and similarly eschewed supernatural explanation. Spiritualists explained spirit communication and phenomena by proposing a hitherto undiscovered form of rarefied matter that allowed spirits to manifest on the worldly plane. But for many of the thousands of Victorians who got caught up in the spiritual craze, explanation was of secondary importance. They went wild for tables that jumped and cavorted as spirits made their presence known and obligingly ran through the repertoire associated with a first rate séance. By the time spiritualism reached its Victorian heyday in the 1870s there can hardly have been a household in the land that had not been touched in some way by the spiritualist fever.« (Owen 2007: 18)

After the death of Queen Victoria in 1901 and the succession of her son Edward, the main tendencies of the Victorian era continued to determine the Edwardian period, which has been described as both »some kind of golden age« (Read 1982: 14) and »as an age of accumulating crisis« (ibid.). In any case, it is true that the Victorian concerns and developments discussed so far, in particular those regarding economic, social, and political factors, were aggravated in the Edwardian period.[35] While towards the end of Victoria's reign the optimism felt in regard to progress and the economic

33 Cf. Robert A. Gilbert (2001) *Rise of Victorian Spiritualism.*

34 Note that spiritualism is very much directed towards the past, which is one of the reasons why ancient Egypt with its death cult was so popular at the time (cf. Glover 1995 and 1996).

35 Cf. Paul R. Thompson (1975) *The Edwardians. The Remaking of British Society.*

boom was overshadowed by fears of decline, depression, and decadence, the accession of Edward VII in 1901 represented for many contemporaries the beginning of a new era which promised a return to Britain's former political and economic strength (cf. ibid.). At the same time, social and gender inequalities, Britain's declining imperial power, and the political and military threat perceived as coming from abroad, particularly from Germany, became central themes of focus (cf. Porter 1982: 129, 133). In addition, modernisation continued to shape people's lives and their perception of time and space, characterised by the transitory on the one hand and vastness on the other.

As the present outline with an emphasis on the Victorian Age has shown, both the Victorian and Edwardian period were essentially characterised by *ambivalence*. Although the degree to which the various determining social, economic, and political factors manifested themselves at certain times and changed in the course of the nineteenth and early twentieth centuries, the dominating dichotomies might be subsumed under the categories present versus past, progress versus decline, materiality versus spirituality, rationality versus irrationality, reality versus fantasy, and visibility versus invisibility.

3. The Genesis of a Popular Archaeological Discourse in Britain

In the previous chapter major changes and advances of Victorian and Edwardian society and culture were presented. In order to show how these developments fostered the emergence of a popular archaeological discourse, this chapter first delineates the genesis from early antiquarian studies to archaeology as a scientific discipline in Britain. The second part of the chapter focuses on the formation and reception of Greek, Egyptian, and Zimbabwean archaeology in nineteenth-century Britain, forming the basis of my analysis in PART II.

3.1 FROM ANTIQUARIANISM TO ARCHAEOLOGY

> Archaeology in the true sense of the word [...] did not exist between the fifth and the fifteenth centuries; in the centuries which followed, however, scholars began to turn more and more to the earliest history of Central Europe.
> (SKLENÁR 1983: 6)

The early roots of archaeological interest in Britain reach back into the sixteenth century when, »[a]s a result of Henry VIII's religious revolution and the dissolution of the monasteries, the countryside [was] [...] littered with ruins« (Wallace 2004: 132). As palpable marks in the British landscape, the ruins of the once powerful monasteries naturally attracted the attention of passers-by. Soon the inspection and unearthing of the ruins and burial sites of the old monasteries became popular, even though it meant breaking taboos, since »[t]he antiquaries [had to] breach the rules of everything that was previously thought holy« (ibid.) by invading and intruding into sacred territory and ground. In particular, the excavation of graveyards meant a break with contemporary religious and moral rules. Nevertheless, antiquarianism started to evolve from these initial activities concentrating on these more recent ruins of the monasteries. However, as the extensive, posthumously published work of John Leland, »the first official antiquary in England« (ibid.),[1] shows, not only ruins and burial sites but also abbeys, castles, and churches were among the objects of interest for the emerging amateur antiquary. Other contemporaries well known for their local field studies in ›archaeology‹ were William Lambarde, Humfrey Lhwyd, and William Camden. Apart from being attracted to buildings and ruins from a more recent past, these early British antiquaries also displayed an interest in Roman antiquities and prehistoric artefacts in their homeland. Responsible for this unspecified interest in ruins, regardless of their cultural

1 John Leland was appointed King's Antiquary in 1533 (cf. Trigger 1989: 47).

and chronological affiliation, was certainly the lack of experience, qualification, and specialisation of men like Camden.

In spite of the humble professional knowledge of these early antiquaries, quite a few books of the time discussed British antiquities. A prominent example is William Camden's popular *Britannia* (1586), which through the study of antiquities attempted to draw a picture of the British past. Besides publications on antiquities, the first antiquarian society for the »preservation of national antiquities« (Daniel 1950: 18) was formed in 1572. Even though this first antiquarian society in Britain was abolished by James I in 1604, antiquarianism continued to flourish during the seventeenth century. This was particularly true of the time after the Restoration of the Stuart line in 1660, the year which also saw the foundation of the Royal Society of London by Charles II.

The Royal Society was one of the first societies to support and publish the work of antiquarians such as John Aubrey (1626-1697) and William Stukeley (1687-1765), who both studied the prehistoric sites at Avebury and Stonehenge in southern Britain. Aubrey, one of the most famous antiquaries of that time, was the first to ›discover‹ and depaint the prehistoric stone monuments of Avebury and Stonehenge, which he attributed to druidical tribes by »using comparisons with living non-Western peoples such as Native Americans to describe both sites as cult monuments« (Fagan 1996: 286). Often the work of these early antiquarians was a combination of historical and topographical research rather than of a purely antiquarian character (cf. Trigger 1989: 48). This was also true of a group of antiquaries who, working along the lines of the Royal Society, first met in London in 1707 and would finally found the Society of Antiquaries of London, the successor of the first antiquarian society mentioned above, in 1718.

While these antiquarian activities were restricted to the British countryside, early antiquaries were also interested in relics of the past found in foreign countries. As early as in the sixteenth century, the *Grand Tour*[2] had taken members of the British aristocracy to classical sites in the Mediterranean region, confronting them with antiquities of this area. As a result, from the seventeenth century onwards, collecting archaeological artefacts

---

2 Cf. Andrew Wilton and Ilaria Bignamini (eds.) (1996) *Grand Tour: The Lure of Italy in the Eighteenth Century* (1996).

from the Mediterranean became fashionable among the British upper classes, who had the leisure time and the finances to travel abroad. Other prominent men like Charles I, the Earl of Arundel, and the Duke of Buckingham had their agents sent to Greece to provide their palaces and manors in Britain with classical artefacts.

The first travel guides, such as Jonathan Richardson's *An Account of the Statues, Bas-reliefs, Drawings, and Pictures in Italy, France, etc.* (1722) and Joseph Addison's *Remarks on Several Parts of Italy* (1705),[3] appeared on the market, offering information and instructions for those travelling in the Mediterranean region. Apart from the material souvenirs, these early tourists and antiquaries also brought records and depictions of their encounters abroad home with them, as illustrated by the many popular travelogues appearing at the time by writers such as James Boswell, Edward Gibbon, William Beckford, Tobias George Smollett, and Laurence Sterne.[4] In 1732, the Society of Dilettanti[5] was founded in London under the leadership of Francis Dashwood to support the study of classical Greek and Roman art through travel, collection, and excavation of these antiquities.

When the spectacular excavations of Herculaneum and Pompeii began in 1738 and 1748, respectively, constituting »landmarks in the history of archaeology« (Fagan 1996: 274), the *Grand Tour* had been established as part of the education of young men of the British aristocracy. In the meantime, various British noblemen had moved to Italy and settled down in Rome and Florence to study and collect antiquities on site. Additionally, news from the sites at Pompeii and Herculaneum also reached a wider public in Britain and Europe. This is illustrated by a variety of contemporary newspapers and magazines that reported and discussed the excavations, such as the following example from the *Penny London Post*:

»As we had lately an Account from Naples, of some new Discoveries in relation to the ancient City of Heraclea (or Herculaneum) which was destroyed above 1600

3 Although not primarily a travel guide, Addison's *Remarks* were used as such by many British gentlemen travelling abroad (cf. Geyken 2002: 69).

4 Cf. Barabara Korte and Catherine Matthias (2000) *English Travel Writing: From Pilgrimages to Postcolonial Explorations*.

5 The Society of Dilettanti can be seen as the root of the British amateur antiquarian tradition (cf. below).

Years ago by an Earthquake, when there happened a great Eruption from Mount Vesuvious; the two following Letters, in Confirmation of that Account, cannot but be acceptable to our Readers. [...] I have seen what may be esteemed a Singularity, in History; the City of Heraclea, of which Pliny speaks in his Letters, that by an Eruption of Mount Vesuvius, was covered many Feet deep under Cinder, and has been by Degrees discovered at a Place called Portici, a Country Palace of the King of The two Sicilies. The City is entire the Houses have been found perfectly furnished and the Furniture well preserved. I have seen every Thing prepared for Dinner at the Time the Eruption happened.« (*Penny London Post or The Morning Advertiser*, Jul 18, 1748)[6]

In 1772, the first collection of Pompeian artefacts was purchased by the British Museum from Sir William Hamilton, English Consul in Italy at the time and member of the Society of Dilettanti. Hamilton's private collection was also documented in the catalogue *The Collection of Etruscan, Greek, and Roman Antiquities from the Cabinet of the Honourable William Hamilton* (1766-1767) by Pierre Francois Hughes d'Hancarville. Other books on the discoveries in Pompeii and Herculaneum had been translated into English and published earlier the same century, such as Marcello Venuti's *A Description of the First Discoveries of the Ancient City of Heraclea* (1748) and the eight-volume folio *Le Antichità di Eroclono esposti* (1757-1792), which appeared in English under the title *The*

6 Another characteristic report on these excavations was published by the *Newcastle General Magazine*: »An Account of the late Discoveries of Antiquities at Herculaneum, Pompeii, &c. in a Letter from Camillo Paderni, Keeper of the Museum Herculanei, dated Naples, June 28, 1755. His majesty the king, my master, is always increasing his taste for matters of antiquity, which he loves with zeal of the most passionate antiquary; for he not only makes all the necessary trials and enquiries in these cities, which have been covered by mount *Vesuvius*, but extends his researches into other parts of his kingdom. [...] In April, his majesty was acquainted, that a little beyond La Terre della Nunziata, where stood the ancient Pompeii in digging near the amphitheatre, there was discovered a marble capital of the Cornithan order, and that it was necessary to examine farther into what might be there. His majesty had formerly caused some workmen to dig in his place.« (*Newcastle General Magazine*, Aug 1775)

Antiquities of Herculaneum in 1773. Between 1791 and 1795 Hamilton published his own book, *Collection of Engravings from Ancient Vases*.

Apart from the positive reception of the excavations at Pompeii and Herculaneum, these early archaeological activities also gave rise to the first critical voice to comment on the excavation methods at the sites, namely that of the German art historian Johann Joachim Winckelmann (1717-1768).[7] Winckelmann visited the excavations at Pompeii in 1758 and 1762 and »was the first to study changes in classical art styles through time« (Fagan 1996: 286). In 1762, Winckelmann criticised the excavations undertaken by the royal house of Naples and Palermo, the Bourbons, at Pompeii and Herculaneum, in his »Open Letter on the Discoveries made at Herculaneum«:

»On his first visit [to Pompeii], Winckelmann was discouraged from participating in the excavations as the Bourbon Court feared that he might object to the state of archaeological activities. As a result, Winckelmann spent most of his time in the Naples museum, deciphering papyri with Camillo Paderni, the museum's director. His second visit, four years later, confirmed his suspicion that the excavations were being mishandled. Work gangs were largely made up of captured pirates and convicts chained together while the excavations themselves were generally random and efforts to loot the site of potentially valuable antiquities. Winckelmann published his ›Open Letter on the Discoveries made at Herculaneum‹ later that year. While the letter addressed various matters surrounding Pompeii and Herculaneum, including an overview of Vesuvian topography and the artefacts in the museum, the letter is essentially a public castigation of the bureaucrats in charge.« (Colby 2009: 21)

Instead of unsystematic, clumsy, and opportunistic approaches to archaeological sites, Winckelmann postulated certain mandatory criteria for archaeological excavations, which, according to Fagan, led to the adoption of the following principles in later years:

»(1) [E]xcavation must concentrate on topographic unities that would eventually be joined, (2) wherever possible, buildings would not be covered over after the excavation but left for visitors to see, (3) the destruction of wall painting considered to be

7 For a further discussion of Winckelmann, cf. Chapter 3.2.

inferior to that desired by the Royal Museum would be halted, and greater attention would be given to architectural context.« (Fagan 1996: 275)

Not least for the postulation of these early ›archaeological‹ maxims, Winckelmann has often been called ›the father of archaeology‹ (cf. Daniel 1955: 17).[8]

The foundation of a second antiquarian society, the Society of Antiquaries of Scotland, in 1780, further reflects the persistent and increasing enthusiasm for local antiquarianism, which continued to exist in Britain parallel to antiquarian activities abroad. What is noteworthy, since it is essential for the further development of the field from antiquarianism to archaeology, is that both ›branches‹ of antiquarianism, local and Mediterranean, stand for a move away from theoretical research towards fieldwork, the study of the material object *in situ*.[9]

Before a second wave of interest in archaeology abroad swept over Britain in the aftermath of the Napoleonic expeditions to Egypt at the turn of the century, the Romantic movement was responsible for a renewed awareness of national antiquities at home. With their emphasis on nature and the human individual as such, the Romantics elevated the past, associated with the authenticity of both man and nature, to one of their major ideals. As mirrored by the many publications on local history and topography by men like Richard Fenton, Sir Henry Englefield, and Richard Colt Hoare, the concern for a British past was by no means restricted to the Romantic poets but marked a general tendency prominent in contemporary society. Apart from these theoretical approaches to local history, archaeological excavation became very popular in Britain towards the end of the eighteenth century. British people were now deeply interested in the material remains of their non-Roman ancestors and took pride in their own Anglo-Saxon past rather than in that of classical Greece and Rome: »Scholars turned away from the classical light to barbarian gloom, and

8 Alternatively, Winckelmann has been referred to as the ›father of *classical* archaeology‹ (cf. Colby 2009: 20).

9 Even though people had studied archaeological remains in Britain and abroad before (cf. above), early antiquarianism had still been very much characterised by the collection and description of antiquities rather than their excavation.

romanticised the Ancient Britons and the Druids, and the local British antiquities attributed to them.« (Daniel 1955: 22)

While throughout the Renaissance and the Augustan Age ancient Rome had been idealised, the Romantic movement rejected these neoclassical traditions and ideas. Instead of classical decorum, artificiality, smoothness, and clarity, the Romantic poets celebrated spontaneity of emotions, imagination, the human individual, and nature in their pristine state. Wordsworth formulates this ideal programmatically in his preface to *Lyrical Ballads* (1802):

»Low and rustic life was generally chosen, because in that condition, the essential passions of the heart find a better soil in which they can attain their maturity, are less under restraint, and speak a plainer and more emphatic language; because in that condition of life our elementary feelings co exist in a state of greater simplicity, and consequently, may be more accurately contemplated, and more forcibly communicated; because the manners of rural life germinate from those elementary feelings; and, from the necessary character of rural occupations, are more easily comprehended; and are more durable; and lastly, because in that condition the passions of men are incorporated with the beautiful and permanent forms of nature.« (Wordsworth 2000: 241)

Very much in the way in that Wordsworth emphasises »low and rustic life« (ibid.) as a source of inspiration which eventually leads to a fusion of men and nature, the past came to be seen as an original and pure place where men were still unspoilt by modernity. The interest in an allegedly unspoilt human nature associated with the Anglo-Saxon past, together with concepts of the picturesque and the sublime, kindled once more in British history an enthusiasm for ruins and graveyards in the countryside. Instead of the preoccupation with classical relics imported from abroad, a rising national awareness and sense of belonging led to the glorification of the remains of ancient Britain:[10]

10 It is important to note, however, that particularly Byron and Shelley also celebrated classical Greece, which they conceived as pure and clear in contrast to the alleged decadence of ancient Rome (cf. Chapter 3.2).

»The popularity of antiquarian nationalism in Romantic-period Ireland is a case in point. More than in England or Scotland, these nationalisms delighted in the discovery and cataloguing of artefacts, material or linguistic, which adherents presented as surviving evidence of an ancient nation, and which spurred practitioners to reconstruct the national traditions they declared themselves to have lost. In engaging in these acts of reconstruction, antiquarian nationalists collapsed space and time into the small room they called nation, scanning landscapes as historical evidence and reading history as a set of enforced disjunctions made visible, even allegorical, by impact on the land.« (Klancher 2009: 92)

Although particularly strong in Ireland, this return to non-Roman roots helped construct a strong national identity in Britain in general, justified through continuity and tradition, which people now tried to find in a long-forgotten past. As pointed out by Levine, this continued to be the case in Victorian Britain:

»History offered Victorians not merely the chance to revel in images of their own past but to interpret their own age in terms of that past, to regard themselves as the natural and rightful heirs to long and fine traditions which they sought to uphold and sustain. History was not the servant but the master of a powerful ideology, based on exploiting the reconstruction of tradition and change.« (Levine 1986: 176)

Especially at a time of political terror and turmoil on the continent, these ruins, graveyards, and stone monuments not only served as a poetic inspiration for a small group of men, but through their sheer age also offered a collective sense of stability and continuity. At the same time, historical studies were now seen as a means of identifying the different historical periods in their singularity (cf. Korte and Paletschek 2009: 19).

What becomes clear is that in spite of a decline in intellectual interest in antiquarianism during the last few decades of the eighteenth century (cf. Trigger 1989: 66), the number of excavations undertaken in the British countryside was high, and barrow digging became a popular pastime, stimulated by the Romantic movement. One of the first barrow diggers was Reverend Bryan Faussett, who excavated more than 800 barrows in south-eastern Britain between 1757 and 1773. Other successful and laborious barrow diggers were William Cunnington and Colt Hoare, who excavated innumerable burial places in Wiltshire and Dorset, and who, according to

Daniel, »may very properly be called the fathers of archaeological excavation in England, just as John Aubrey was the father of field archaeology« (Daniel 1950: 31). Apart from bones, the barrow diggers also hit upon stone and metal artefacts during their excavations, which, however, due to their lack of knowledge, they were unable to classify. Even though antiquaries of this time made an attempt to distinguish between different types of barrows, they did not come up with any system capable of explaining the different materials and layers of their findings: »In spite of their hard work and devotion antiquaries like Colt Hoare and Cunnington failed to find any way of breaking down the apparent contemporaneity of pre-Roman remains.« (Ibid.)

As this outline of the development of antiquarianism from the sixteenth to the eighteenth century shows, the character of antiquarianism underwent a significant change over the centuries. While the early antiquarian activities were characterised by collecting and describing ancient artefacts, antiquarianism at the end of the eighteenth century was dominated by fieldwork and excavation. This growing significance of excavation during the last decades of the century marks a noteworthy change in the genesis of antiquarianism, since it anticipates its differentiation from a primarily antiquarian to an archaeological occupation in the Victorian period. In addition, this differentiation was paralled by an increasing professionalisation and popularisation of archaeology in the course of the nineteenth century. As Philippa Levine shows in her study *The Amateur and the Professional: Antiquarians, Historians and Archaeologists in Victorian England, 1838-1886* (1986), although the three strands of antiquarian, historical, and archaeological studies continued to engage a »body of enthusiasts and committed devotees« (Levine 1986: 7), interest in antiquarianism decreased and was eventually seen as synonymous with amateurism, whereas archaeology and history became more and more professionalised (cf. ibid.: 173-174).

Fuelled by major archaeological discoveries and excavations abroad, such as in Assyria, Egypt, and the Mediterranean area, in the course of the nineteenth century archaeology emerged as a scientific discipline, with John Disney establishing the first chair of archaeology in Britain in 1851 (cf. Renfrew 1994: 58-59). In spite of this growing professionalisation of archaeology, it is crucial to note that most archaeologists of the time were occupied in other fields before they devoted themselves to the study of archaeology:[11]

»Unlike the historians, early archaeologists were rarely university educated; John Evans had been entered at Oxford but did not go, Flinders Petrie was educated at home, Rawlingson was schooled by the East India Company, George Smith had begun life apprenticed to be a banknote engraving firm, and Pitt Rivers was sent to Sandhurst to follow in the footsteps of his military father. A remarkable number did succeed though, in making archaeology or more precisely excavation, a full-time occupation, if only for a portion of their adult lives.« (Levine 1986: 32)

This shows how the amateur tradition of seventeenth- and eighteenth-century antiquarianism, prominently represented by the Society of Dilettanti, was perpetuated far into the nineteenth century, having a lasting effect on the genesis of archaeology. This amateur tradition is reflected in archaeologists, such as Giovanni Batista Belzoni and Heinrich Schliemann, who were particularly popular in nineteenth-century Britain as embodiments of the self-made man. It is in the context of this inherent amateur tradition of archaeology in Britain, creating a reciprocal permeability from inside and outside the archaeological discourse, that in combination with the original professional backgrounds of the amateur archaeologists[12] a popular archaeological discourse developed in nineteenth-century Britain. In order to show how a popular archaeology emerged as a product of a reciprocal exchange between specialists and the Victorian public, I will focus in the following on the genesis of Greek, Egyptian, and Zimbabwean archaeology, with an emphasis on the reception of legendary Homeric Troy, Egyptian mummies, and the ruin complex of Great Zimbabwe.

11 As Levine points out, »[i]n England, archaeology was more likely to remain an amateur preoccupation. [The antiquary] Thomas Bateman, having inherited a vast family fortune, was able to dig barrows whenever he desired, but for most of those domiciled in England archaeology remained the more usual leisure-time pursuit« (Levine 1986: 32).

12 Belzoni was a showman and Schliemann a merchant before they started their archaeological careers (cf. below).

3.2 Greek Archaeology: ›Ubi Troia Fuit‹ or ›Ubi Britannia Est‹

> Strange relics these of a race long since dust: / Quaint cups once brimming with the grape's sweet / juice Drained amidst laughter and with jests profuse / And passionate vows of lasting love and trust; / Gauds that once glittered, spite their present rust, / In the ears of maidens fair as could be seen, / Sprightly voiced maidens of soul-gladdening mien, / Now, a mere part of earth's strange-kneaded crust; / Swords too and spear-heads that once dealt out death; / Ay, and the charms they wore for fear of ill; / Even the idols unto whom they knelt / And cried, when anguished, with wild frantic breath, / If haply their gods' hearts for ruth might melt. / O race long dust, we see, we hear you still!
> (Hales, »Troja Fuit, Suggested by the Hissarlik Remains«, 1878: 588)

In his article »Schliemanns Troia« (2006), Justus Cobet argues that the public perception of archaeology has been and continues to be strongly determined by associations with ancient Troy, which is popularly connected to the German archaeologist Heinrich Schliemann (cf. Cobet 2006: 146). I want to extend this assumption by claiming that the popular reception of Schliemann in Victorian and Edwardian Britain was strongly influenced by the chain of associations elicited by Troy, Homer, and ancient Greece, with the latter functioning as a ›defining culture‹.[1] In this context, Schliemann

1 Although the main emphasis will be on the analysis of the reception of Schliemann's excavations at Hisarlık, the following analysis also discusses the popular

does not only stand between Homer and us, as proposed by Cobet for the twenty-first century (cf. ibid.), but by extension also functioned as a pivotal figure between his Victorian and Edwardian contemporaries and ancient Greece. Closely connected to Schliemann's popular reception in nineteenth- and early twentieth-century Britain are thus several factors that converge in the nineteenth century.

Generally speaking, the reception of classical antiquity can be seen as characterised by two central lines of reception, namely a theoretical consideration versus a material preoccupation with antiquities. While the theoretical concern with antiquity had dominated the material reception in earlier times, this changed increasingly towards the eighteenth century. The question as to the location of Homer's legendary Troy was crucial in this respect. Accompanying the growing concern for material remains in Britain in the eighteenth century, the occupation with classical antiquity experienced a shift of interest from the preoccupation with ancient Rome to a dominant interest in ancient Greece. It is against the background of this twofold change, characterised by the growing concern for material remains on the one hand and an increasing importance attributed to Greek over Roman antiquity on the other, that Schliemann's enormous popular reception in nineteenth-century Britain has to be seen. In the following, I will briefly outline the individual developments that created the grounds for Schliemann's popularity in Victorian Britain.

The Homeric legends of the Trojan War[2] have captivated the imaginations of people throughout the centuries, both in the eastern and western hemisphere.[3] This is also particularly true of Britain with its long-standing classical tradition, as emphasised by Gehrke:

»While the Troy myth had in this way become an integral part of Roman identity, this image of the past also shaped identity in the Romanized parts of the empire. It was widespread and virulent, and it was precisely in Gaul and Britannia that people

reception of the Mycenaean excavations Schliemann carried out from 1874 to 1876.

2 The *Iliad* and the *Odyssey*.

3 Cf. Joachim Wohlleben (1990) »Homer vor Schliemann«, Martin Zimmermann (2006) *Der Traum von Troia: Geschichte und Mythos einer ewigen Stadt*, and Michael Siebler (1990) *Troia – Homer – Schliemann* (Siebler 1990: 36-53).

in many places, even at the local level, saw themselves explicitly as being in the Trojan tradition and of Trojan descent. In this regard, they had thus been related to the Romans since ancient times. This is no doubt the reason why the notion of a Trojan line of descent also gained currency among the Franks and Britons from late antiquity on.« (Gehrke 2006: 218, translation mine)

The reception of Homer in Britain goes back to the Middle Ages (cf. Allen 1999: 39; Cobet 2006: 153; Lascarides 1977: n. pag.) and reached its first climax in the seventeenth and eighteenth centuries with poets such as Alexander Pope, Samuel Johnson, John Dryden, and John Milton. In addition to a theoretical concern with Homer, however, pragmatic questions as to where the ancient site of Troy was to be located also occupied the minds of innumerable scholars and amateurs throughout history:

»The magic touched off by the legend of Troy held an unbroken fascination on the mind of the man in Antiquity and the Middle Ages. Homer's epic had been the source and centre of a whole literature on the destruction of Troy and the later fate of the Trojans. The tradition was greatly amplified by Virgil, kept alive in Apollodorus' mythology, in Proclus and other later obscurer writers. The texts were transmitted through the Middle Ages in the *Ilias Latina* of Baebius Italicus, through the Latin text of the so-called Dictys of Crete and by the medieval poetry of Benoît de Saitne Maure and Guido delle Colonne. The early Renaissance produced its own garland of Trojan epics such as Jacques Millet's *La Destruction de Troye la Grant* (1452). The net effect of this continued interest in the great war of Troy on the reading public was eventually to kindle the question: what was fact and what was fiction? Where was Troy?« (Lascarides 1977: n. pag.)[4]

4 The question of Troy's location was closely intertwined with the question of the historicity of the Trojan War, as summed up by Cobet: »The question of the right place was connected since the end of the 18th century increasingly with the question of whether the events described in the *Iliad* corresponded to historical reality at all. Responsible for this was the criticism of the early transmission of the poem inflamed by Friedrich August Wolf's *Prolegomena* to a Homer edition published in 1795, and the search for the place was charged with the desire to secure the historicity of the myth and thus distinguish an early Greek history with the weight of the Homeric poetry. At this point we can already make out

For a considerable time the ruins of Troy were thought to lie under the Greek and Roman city of Ilion.[5] The geographer Strabo (c. 63 BC-AD 20) was one of the first to contest this assumption. In the following centuries the alleged location of ancient Troy varied,[6] without any definite proof for the site of the city being produced.[7] During the 1780s and 1790s the French diplomat Marie-Gabriel-Florent-Auguste de Choiseul-Gouffier undertook the first systematic exploration of the Troad with a team of professionals and finally narrowed down the possible sites of ancient Troy to Pınarbaşı and Hisarlık. Jean-Baptiste Le Chevalier, who was part of Choiseul-Gouffier's exploration, was convinced that Troy was to be found at Pınarbaşı.[8] Others, such as Gilbert Wakefield, William Vincent, J. B. S. Morritt, James Dallaway and William Francklin, and William Martin Leake, supported Le Chevalier's Pınarbaşı thesis[9] (cf. ibid). Eventually,

the intellectual constellation to which Schliemann thought he had contributed an irrefutable argument with his spade.« (Cobet 2006: 154, translation mine)

5 Herodotus, Xenophon, Arrian, and Plutrach shared this opinion (cf. Allen 1999: 40).

6 One of the first Englishmen who, travelling the Hellespont in 1103, felt certain of having discovered the site of ancient Troy, was the merchant and pilgrim Saewulf (cf. Gehrke 2006: 153).

7 Amongst the people speculating on the site of ancient Troy were also quite a few British, including Lady Mary Wortley Montagu: »Travellers took early on the opportunity of visiting what looked so easily like the ruins of Homeric Troy. First Pierre Belon in 1546, followed by William Lithgow (1610) and Petro della Valle (1614) recorded their enthusiasm or nostalgia on treading this hallowed ground and dreamed of the heroes who had fought and died on these walls. Some objecting voices were raised early enough: George Sandys in 1610 already sees a hill near Chiblak as the likelier site of Troy, recognizing days of Alexandreia Troas. With Spon and Wheler (1675) and Lady Mary Wortley Montagu (1718) the days of Alexandreia Troas as a substitute for Troy are finished. Richard Pococke, visiting the Troad in 1740, records the long-lived tradition that Troy is under the ruins of Ilium and definitely points at Hissarlik as the location of the elusive city.« (Lascarides 1977: n. pag.)

8 As pointed out by Traill, by 1868 the majority of scholars supported the assumption that Pınarbaşı was the site of ancient Troy (cf. Trail 1995: 53).

9 Chevalier (1791).

»[Sir William] Gell published his superbly illustrated folio Topography of Troy (1804) as a crowning banner of the Bunarbashi contingent« (cf. ibid.). The German Franz Kauffer was eventually the first to discover ancient remains at Hisarlık in 1793 (cf. Cook 1973, Easton 1991).

The increasing interest in locating the site of ancient Troy in the eighteenth century marks a general trend away from the preoccupation with written texts to the material study of antiquity, which in Britain up to this point was dominated by a concern with Roman antiquities. Although the Neoclassical Period, and the Augustan Age in particular, were still very much concerned with the textual study of ancient Roman authors, there had also always been a strong material interest in ancient Rome in Britain, as represented by the various collections and excavations of Roman antiquities both at home and abroad depicted above: »Roman remains in England had been described with varying degrees of understanding and accuracy in early topographical works, such as John Leland's *Itinerary* (undertaken 1535-43) and William Camdem's *Britannia* (1607), originally published in Latin.« (Vance 1997: 22) Abroad, the interest in the Roman past was further spurred on by the aforementioned discoveries at Heraculaneum in 1732 and the subsequent excavations at Pompeii in 1748,[10] which attracted many British travellers on the *Grand Tour*.[11]

Central in regard to the changes in the study of classical antiquity is Winckelmann, who stands for a merging of theoretical and material concern with ancient Rome and Greece. On the basis of the study of classical artefacts, Winckelmann, who as a papal antiquary had access to a wide range of antiquities, installed Grecian art as the aesthetic ideal, celebrating

10 In spite of the unprofessional and crude approaches characterising the archaeological work at Pompeii and Herculaneum at the time, which were later criticised by Winckelmann, who visited the sites in 1758 and 1762, accurate architectural plans were created of the cities and, *inter alia*, the Temple of Isis (1764), the Odeon (1764), and the Villa of Diomedes (1771) successfully unearthed. Archaeological work further continued at the sites during the short French occupation under Napoleon (1798-1799), before it was finally taken over by the Italian archaeologist Giuseppe Fiorelli (1823-1896) in the 1860s.

11 Since the beginning of the nineteenth century this also included members of the middle classes, who now had the means to undertake travels abroad (cf. Chapter 3.1).

its ›noble simplicity and calm grandeur‹[12] in his publication *Geschichte der Kunst des Altertums* (1764):

»For Winckelmann, stylistic analysis was the key to aesthetic understanding. So rather than just explaining works of art as artefacts, Winckelmann sought to explain a culture by its works of art. In Greek art he found an idealised pagan soul, a noble simplicity and calm grandeur (the famous ›*eine edle Einfalt und eine stille Grösse*‹). Greece was to be seen as the childhood of Europe, the foundation from which all European culture sprang.« (Shanks 1996: 55)

Although Winckelmann has been referred to as the ›father of archaeology‹, as pointed out by Whitley,

»his achievement was more theoretical than empirical. He imposed order and system to the chaos of antiquarian scholarship, which had been more concerned with the details of iconography and authorship than with the overall development of Greek art. This is not to say that he was indifferent to aesthetic questions – far from it, since much of the History[13] is given over to extended, lyrical descriptions of such famous works as the Laocoön, the Apollo Belvedere and the Niobe.« (Whitley 2001: 21)

In addition to the fact that Winckelmann's occupation with ancient Greece was still very much focused on the study of texts, he himself, like many of his German contemporaries, never went to Greece and studied Roman casts instead of the Greek originals (cf. Shanks 1996: 57).

In contrast to this, the British Society of Dilettanti had sent the painter James Stuart and the architect Nicholas Revett to Athens to paint the local ruins as early as 1751. Their drawings were first published in 1762, in the first volume of *The Antiquities of Athens*, which was an immediate success. In 1772, the potter Josiah Wedgwood first decorated his vases, which later were to become highly popular, in Greek style, using William Hamilton's illustrations as a model. To the mind of Hamilton and other intellectuals of

---

12 »The general distinctive feature of the Greek masterpieces is ultimately a noble simplicity and quiet grandeur, both in posture and expression.« (Winckelmann 1808: 31, translation mine)

13 *Geschichte der Kunst des Altertums* (1764).

the time, Greek art and sculpture were suitable »as paradigms for modern art« (Fagan 1996: 289). Accordingly, Lord Elgin (1766-1841), British ambassador to Turkey since 1798, travelled to Athens in order to make casts of the statues of the Acropolis and bring them home to Britain. However, at the time of Elgin's arrival at Athens in 1798, the French were about to persuade the Turks to allow them to transport the Parthenon Marbles to France.[14] Eventually the Turkish government denied the French their request and instead granted it to the British after British troops defeated Napoleon in Egypt in 1799. The British Museum, however, was at first reluctant to purchase the Marbles from Elgin, since it was unclear whether they would comply with what Joshua Reynolds, member of the Royal Academy and the most influential painter of the eighteenth century in Britain, propagated as the ideal of ›grand style‹ (cf. Reynolds 1842: 36-51). It was not before 1816, after public debate in Parliament, that the Parthenon Marbles were finally bought and displayed by the British Museum.

The hesitation of the British Museum to purchase the Elgin Marbles is significant, since it emphasises how even as late as at the turn of the century Greek antiquities had still not achieved a prestigious status in Britain. However, the shift of interest from Roman to Greek antiquity had clearly begun in Britain, represented in the works of the Romantic poets with their affiliation to ancient Greece:

> »By the turn of the [eighteenth] century it was possible to feel with some confidence that much more was known about Greece than in the time of Milton or of Pope. The cultural supremacy of Rome had been challenged by Greece; although the first generation of Romantic poets (with the striking but ambivalent exception of Blake) would not transfer their allegiance, Shelley and Keats had no hesitation in preferring Greece, while Byron was politically and emotionally committed to the ideal of Greek independence from Turkish hegemony but less convinced of the virtues of ancient Greek literature and mythology.« (Webb 1993: 162)

Consequently, as Turner points out, while »[i]n the eighteenth century Virgil and Horace served as prescriptive literary models, and Cicero was regarded by many (though certainly not by all) as a sound philosopher of public life and a sober religious thinker, [...] already in the eighteenth

14 Greece was still part of the Turkish-ruled Ottoman Empire at this time.

century Homeric studies had begun to generate broad interest, and the next hundred years might be considered Homer's century« (Turner 1989: 62-63). As further underlined by Fagan, the increasing interest in ancient Greece over Rome was strongly characterised by a growing identification with the Greek past:

»Most important was the shift among intellectuals from seeing themselves as Christians to seeing themselves as Europeans. European-ness began not with Rome but Greece, which was reconstructed within the growing Romantic movement as a pure, young, vivacious, and spontaneous civilization, unlike the powerful but decadent civilizations of Rome and Egypt.« (Fagan 1996: 289)

Soon, ancient Greece became the ›defining culture‹ for Victorian Britain: »Greek culture became *à la mode*; art, archaeology, history and philology would have a nucleus around which they could gather, each stimulating interest in the others.« (Jenkyns 1980: 5)[15]

Although the study of classical antiquity had traditionally been pursued by the upper classes in Britain, this also changed in the course of the nineteenth century, as Jenkyns emphasises when he ascertains that »ancient Greece preoccupied many of the finest minds of the last century, and thus, directly and indirectly, it became a pervasive influence, reaching even to the edges of popular culture. Even those who hated the Greeks or detested the system of classical education were affected willy-nilly« (ibid.: x).[16] In

15 This is further emphasised by the various buildings in the Greek style designed and constructed by architect and archaeologist William Wilkins and architect Robert Smirke in London at the time: The Theatre Royal, Covent Garden (1808-1809), the General Post Office (1824-1829), the British Museum (1823-1848), Wilkins University College London (1826-1830), and the National Gallery (1832-1838).

16 This is underlined by the preoccupation with ancient Greek in various areas, as Dowling sums up: »Serious archaeological study was sufficiently new in the nineteenth century that Pater himself could be called ›the father of archaeological teaching at Oxford‹ on the basis of a series of demonstrably unscientific lecturers on archaic Greek art he offered in 1878. Heinrich Schliemann's excavations at Troy and Mycenae took place in the same decade. The Elgin marbles had been opened to the British public's view as recently as 1816. The impres-

other words, by the time of the nineteenth century ancient Greece had eventually displaced ancient Rome as the aesthetic ideal predominating education, art, and literature both in Britain and abroad and turned into what Hölscher refers to as the *most familiar strangeness*. In Britain the preoccupation with ancient Greece was prominently reflected by the Hellenism[17] of Matthew Arnold, Walter Pater, and Algernon Charles Swinburne,[18] as emphasised in the chapter of Pater's *Renaissance: Studies in Art and Poetry* dedicated to Winckelmann:

»Soon afterwards we find Winckelmann in the library at Nöthenitz. Thence he made visits to the collection of antiquities at Dresden. [...] And now a new channel of communion with Greek life was opened for him. Hitherto he had handled the words only of Greek poetry, stirred indeed and roused by them, yet divining beyond the words some unexpressed pulsation of sensuous life. Suddenly he is in contact with that life, still fervent in the relics of plastic art. Filled as our culture is with the classical spirit, we can hardly imagine how deeply the human mind was moved, when, at the Renaissance, in the midst of a frozen world, the buried fire of ancient art rose up from under the soil. Winckelmann here reproduces for us the earlier sentiment of the Renaissance. [...] Here, then, in vivid realisation, we see the native tendency of

sion of insuperable classical grandeur that these sculptures afforded was soon after leavened with a piquant impression of classical quotidian life with the publication of William Gell's *Pompeiana* (1817-32).« (Dowling 1988: 222)

17 The Greek revival also found many supporters of Philhellenism in Britain, a movement favouring Greek independence and backing the Greek Revolution (1821-1829) both personally and financially. The most prominent British supporter of Philhellenism was Lord Byron, who fought in and died during the Greek War of Independence.

18 »A third indication of the predominant stature of Greek studies is the public and scholarly distinction of the persons conducting them. One need only mention the names of Connop Thirwall, George Grote, William Gladstone, Matthew Arnold, Benjamin Jowett, John Ruskin, Alexander Grant, Walter Pater, Samuel Butler, Edward Caird, John Burnet, A. E. Taylor, and Alfred Zimmern to establish the eminence of the coterie of Victorian Hellenism. All of them knew that the pursuit of Hellenism held the possibility of influencing their contemporaries in a manner in which Roman studies did not.« (Turner 1989: 63) Cf. also Richard Jenkyns (1989) »Hellenism in Victorian Painting«.

Winckelmann to escape from abstract theory to intuition, to the exercise of sight and touch.« (Pater 1980: 146-147)

The merging of a theoretical and material affiliation with ancient Greece here alluded to by Pater in regard to Winckelmann emphasises the amalgamation of a theoretical and concrete concern with the Greek past which gained increasing importance in nineteenth-century Britain. This was particularly the case in regard to alleged cultural and racial connections established by contemporaries between ancient Greece and Britain:

»Many Victorians liked also to explain history in racial terms. All over Europe nationalism was waxing strong; while in imperial England it was hard to resist the sense of a special Anglo-Saxon destiny. Thomas Arnold used his inaugural lecture from the Regius chair at Oxford to argue that civilization progressed by being transmitted from one race to another: Greece fed the intellect, Rome established the rule of law, Christianity gave the perfection of spiritual truth. […] Arnold was typical of his age, both in his sense of a Germanic distinctness and in his desire to claim a kinship with the old Mediterranean world.« (Jenkyns 1980: 166)[19]

Moreover, in the process of establishing a lineage from the past to the present, people did not desist from drawing on distant myths and legends. In Britain the legendary Brutus of Troy assumed a central role in this context:

»In the West, with the incipient rise of nationalism, nations again strove to construct genealogies that connected them with the illustrious heroes of Troy, inventing heroes such as Brutus, descended from Aeneas, who had fled Rome for Britain, leading the last of the Trojans to that western island and calling it ›Britain‹ from his own name. Another even went so far as to claim that this Brutus, descended from Ilius, founder of Troy, had founded London as ›Troynovant‹, or New Troy. Throughout England and Wales, early mazes and labyrinths called ›the walls of Troy‹ further celebrate this kinship. So the British had established a national affiliation with the Trojans long before the Tudors embraced the Trojans as their ancestors and Shakespeare wrote them into his plays.« (Allen 1999: 39)

19 This is also reflected by theories of diffusionism introduced in Chapter 2.2.

As delineated by Allen, the creation of the mythical figure of Brutus assumed the function of a founding myth for Britain and was adopted for British culture to establish a prestigious and long-standing tradition by connecting the ancient Greek to a more recent Anglo-Saxon past. In addition to the racial, cultural, and mythical affiliation with ancient Greece, Turner points out that Victorians might have also identified with ancient Greece on the grounds of its discontinuous reception in Britain, which reflected peoples' disruptive relations to their own past:

»Around the close of the eighteenth century British scholars not only came to reinterpret that classical universe, but they sharply shifted their field of interest from Rome to Greece. This development was not inevitable; nor was it, in 1750 predictable: it is an event within intellectual history that begs for an explanation. The nineteenth-century interest in things Greek did not, of course, preclude concern for Roman culture. [...] [T]o appeal to Rome was to draw upon a line of continuous cultural influence within Europe; to appeal to Greece was to appropriate and domesticate a culture of the past with which there had been, particularly in Britain, a discontinuous relationship. And that very discontinuity may have been part of the attraction for nineteenth-century writers who regarded much of their own experience as discontinuous with the recent past.« (Turner 1989: 61)

It is within the context of this emergent Victorian Graecomania and the accompanying attempts to form lineages going back to the Greek past that the interest in locating the site of ancient Troy continued to be spurred on in the nineteenth century. In the aftermath of Le Chevalier and Kauffer's eighteenth-century excavations, the debate as to the location of ancient Troy had focused on the sites of Pınarbaşı and Hisarlık.[20] This assumption was also adopted by the Scottish editor and amateur archaeologist Charles Maclaren, who started excavations at Hisarlık in 1821 and published his results in *The Plain of Troy Described and the Identity of Ilium of Homer with the New Ilium of Strabo Proved by Comparing the Poet's Narrative with the Present Topography* (1863).[21] Maclaren strongly influenced the

20 Cf. above.

21 Other amateur archaeologists who considered Hisarlik the site of ancient Troy were Henry Owgam, Gustav von Eckenbrecher, and George Grote (cf. Traill 1995: 53).

young British expatriate Frank Calvert (1828-1908), who lived with his family in the Dardanelles and had been interested in the archaeology of the Troad from early on. Calvert, convinced that Hisarlık was the site of ancient Troy, carried out archaeological explorations there for several years. However, Calvert failed to assert and promote his Hisarlık thesis to experts and the public on a large scale and had to witness how Schliemann, the wealthy German amateur, won the glory and reputation that should have been his.[22] Accompanied by his young Greek wife Sophia, Schliemann started excavations at Hisarlık in 1870 without any official excavation permit. Anxious to find supporters all over Europe, Schliemann made sure to inform the public of his excavations at Hisarlık even before he had made any significant discoveries at the site:

»[Schliemann] had a real desire to enquire, to investigate, to communicate his results. All this he did with great energy, at no expense spared, in the glare of publicity. In time he became an expert. And he had an instinct for the big question [...]. If by ›hero‹ you mean a faultless idol, a clean-cut guy, a paragon of all necessary virtues, then Schliemann fails the test by a considerable margin. But if you mean a flawed human being, sometimes confused, sometimes mistaken, dishonest, inadequately equipped, who sets all his energies to one great end and who, despite his faults, changes the picture in a whole subject and leaves behind a lasting legacy of information and enthusiasm – then I think he might pass.« (Easton 1998: 343)

Unlike in other European countries, Schliemann's public reception in Britain had been largely positive right from the start of his archaeological career in the early 1870s.[23] As soon as Schliemann publicly announced his intention to excavate ancient Troy, the British public was able to eagerly follow the reports published by the contemporary press. This popular interest in Schliemann's excavations was amplified when he claimed to have discovered Priam's Treasure in 1873, as summed up by *The Pall Mall*

22 This will be discussed in more detail in the analysis in Chapter 5.

23 Due to the strong amateur tradition, it was much easier for Schliemann to gain popularity in Britain than it was initially on the continent. For a detailed analysis of Schliemann's excavations, cf. Donald F. Easton (2002) *Schliemann's Excavations at Troia 1870-1873*.

Gazette in retrospect:[24] »No discovery of recent date has brought the imagination so directly face to face with the habits and customs of the heroic ages in Greece as have those effected by the persevering enterprise of Dr. Schliemann.« (*The Pall Mall Gazette*, Dec 27, 1877) Innumerable newspapers and magazines reported on and discussed both Schliemann's excavations and his biography:[25]

»In the new world of publicity, others did Schliemann's work for him. Back in Winckelmann's time, to be stirred by his enthusiasm for classical Greece you had to read his books. But now, with Schliemann's own shrewd assistance, every turn of the archaeologist's spade became news. The reading public did not have to wait for heavy tomes to enjoy the adventures of excavation. Newspaper readers held their breath, watching daily for Schliemann's dispatches to *The Times of London*, the *Daily Telegraph*, and the *New York Times*. [...] His portrait by an artist of the *Illustrated London News* was reprinted across the world, making a trademark of Schliemann's broad forehead and heavy moustache, and reporters inventoried his dandyish wardrobe of fifty suits, twenty hats, forty-two pairs of shoes, thirty walking sticks and fifteen riding crops.« (Boorstin 1983: 594)

Although there were also critical voices in regard to his excavations and inferences,[26] Schliemann soon became a popular figure in Britain: »The

---

24 In »Priam's Gold: The Full Story« (1994), Donald F. Easton reconstructs Schliemann's discovery of the so-called Priam's Treasure: »The full story can now be told, and that is what I should like to attempt in this paper.« (Easton 1994: 221)

25 In »Heinrich Schliemann: Leben und Werk im Spiegel der neuen biographischen Forschungen«, Wolfgang Schindler discusses various studies focusing on Schliemann's biography (Schindler 1976: 271-89).

26 This is reflected by David A. Traill's discussion of Schliemann's excavations at Hisarlık: »Schliemann's first excavation campaign at Troy (1871-1873) ended in triumph with the discovery of ›Priam's Treasure‹. (Whether this collection of bronze, silver, and gold objects was a single find, as Schliemann maintained, or numerous small finds he cobbled together is still hotly debated.) In 1877, however, the artist-journalist William Simpson, famous for his drawings of the Crimean War, visited Troy and published a highly critical article in *Fraser's Magazine*. Pointing out that the ruins at Hisarhk [sic] were remarkably unimpressive,

reception accorded to Dr. and Madame Schliemann in this country, testifies the value attached to their discoveries.« (*Athenæum*, Jun 23, 1877: 807) Schliemann promoted his own popularity by having his books published by the prolific and renowned London publishing house Murray[27] and by becoming acquainted with Britain's most noted scientists of the time. Moreover, he regularly sent letters to contemporary British newspapers in which he reported on his recent explorations. Furthermore, he read various papers in Britain and received honorary titles from the Society of Antiquaries, the British Archaeological Association, the Royal Archaeological Institute, the Royal Historical Society, the Royal Institute of British Architects, and the University of Oxford and spent a lot of his time in Britain.[28]

---

he suggested that the modest mud-brick building Schliemann called ›Priam's palace‹ would be better labeled ›Priam's pigsty‹. British antiquarian William Borlase, who visited Troy in 1875, reported that Schliemann's wife Sophia, far from carrying off ›Priam's Treasure‹ from the trench in her shawl, as Schliemann alleged, was not even at Hisarhk [sic] when the treasure was discovered. Schliemann, effectively called a liar in print, must have felt that his reputation was at stake. He resumed work at Troy on September 30, 1878.« (Traill 2002: 54)

27 Cf. Chapter 5.5. In *Mycenæ* Schliemann explicitly thanks his friend John Murray for the collaboration: »Lastly, I here express my warmest gratitude to the learned publisher of this work, my most esteemed friend, Mr. John Murray, as well as to my most excellent learned friend Mr. Philip Smith, for all the kind services they have rendered me and all the valuable assistance they have lent me in carrying out the present work.« (*Mycenæ*: 382) Both Murray and Schliemann were anxious to secure the popular success of Schliemann's books. Schliemann's excavation account *Mycenæ* is a case in point. To stir an interest in the book prior to its publication, a draft version was published in instalments in *The Times* (cf. Traill 1995: 165; cf. Runnels 2002: 14). In addition to that, »[Schliemann's] unfortunate experience with the photographs for the German edition of *Troy and its Remains* and the excellence of the engravings that Murray had produced for the English edition had convinced Schliemann to illustrate *Mycenæ* with engravings« (Traill 1995: 165). For a detailed discussion of Schliemann's individual and international publications, cf. Runnels (2002: 23-58).

28 »Most of June and July Schliemann spent in England, dividing his time between London, where owing to the great success of *Troy and its Remains*, he had be-

To express his gratitude to the British public for their support and interest in his work, Schliemann eventually brought his Trojan Collection to London, where it was open to the public at the South Kensington Museum until 1880: »He was highly honoured in this country, where he was elected a member of many learned societies, and where the good offices of his friend obtained for him an Honorary Fellowship At Queen's College, Oxford, and for some time his Trojan treasures were deposited and exhibited at the South Kensington Museum.« (*The Times*, Dec 29, 1890: 6)

In addition to Schliemann's ability to promote himself and his archaeological excavations among the public, his status as a self-made man was further responsible for his popular reception in Britain (cf. Cobet 2006: 151). It was his status as a self-taught altitudinarian amateur which positioned him firmly in the tradition of the Society of Dilettanti in Britain, whose early members, though noblemen and scholars, were concerned as amateurs in their fields with the study of classical art:[29]

»Antiquarian studies in England attracted a highly motivated, self-taught group of scholars who pooled their knowledge, working cooperatively and subscribing to a belief in the collective importance of the research to which each added. Middle-class members followed their interests on a part-time basis, while those from the elite classes could afford more extended research.« (Allen 1999: 52)

As pointed out in Chapter 3.1, this was part of the reason why amateurs and self-made men in Britain traditionally found a considerable amount of recognition and were also far more accepted by the upper classes, while abroad, where social advancement through amateur interests was almost unheard of and extremely hard, they were often regarded with suspicion:

»He was the quintessential nineteenth-century hero, the *Held der Gründerzeit*, the self-made man par excellence, millionaire and scholar, friend and correspondent of the great, presidents, emperors, prime ministers, kings and queens. If a poor, unedu-

come something of a celebrity, and Brighton, where Sophia preferred to relax with Andromache.« (Traill 1995: 135) For a discussion of Schliemann's academic employees, cf. also Traill (1990).

29 Christiane Zintzen calls Schliemann a dilettante independent Gentleman (cf. Zintzen 1998: 257).

cated, country pastor's son from Mecklenburg [...] did all this, so can you and I. He is the dream figure, the *Ich-Ideal* of every petit bourgeois.« (Calder 1986: 19)[30]

Schliemann thus succeeded in skilfully merging private and public concerns by presenting excavations of collective historical importance on the one hand and having an autobiography that represented the fulfilment of the life dream of the self-made archaeologist on the other. The alleged discovery of ancient Troy can therefore be seen as marking the satisfaction of both individual and collective longings characterised by an amalgamation of fact and fiction. It is against the background of the collective affinity with ancient Greece that culminated in Victorian Hellenism and the general recognition of the self-made man in combination with contemporary discourses that Schliemann's popular reception in Britain becomes clear. This is reflected in A. H. Sayce's preface to Schliemann's 1884 *Troja: Results of the Latest Researches and Discoveries on the Site of Homer's Troy*:

»Why is it that Dr. Schliemann's example has not been followed by some of the rich men of whom England is full? [...] There are few, it is true, who can be expected to emulate him in his profuse generosity, and freely bestow on their mother country the vast and inestimable store of archaeological treasure which it had cost so much to procure; [...] But surely England must contain one or two, at least, who would be willing to help in *recovering the earlier history of our civilization*, and thereby to earn for themselves a place in the grateful annals of science. Dr. Schliemann, indeed, *has created for himself a name that can never be forgotten* even when the memory of the plaudits that have greeted him in the Universities of Germany, or in the oldest University of our own land, shall have passed away.« (Schliemann 1972 [1884]: ix-x, emphasis mine)

30 Although Schliemann eventually became one of the heroes of the German middle-class *Gründerzeit*, German experts were extremely reluctant to take Schliemann seriously (cf. Chapter 5.5 and 5.6), as William M. Calder III points out: »[S]cholarly contemporaries thought Schliemann a thorough rogue, not deserving refutation. [...] The targets are not worth the powder and shot. How many serious scholarly reviews by Germans did Schliemann's books receive? [...] The mandarins detested the self-made man. [...] On the other hand, Schliemann amused the Emperor Wilhelm II and the English were kinder to an amateur.« (Calder 1986: 34-35)

To »recover the earlier history of our civilization« (ibid.) and »create [...] a name that can never be forgotten« (ibid.) is exactly what Schliemann stood for and what made him so popular in Britain.

So far, studies of Heinrich Schliemann have mainly focused on Schliemann's reliability[31] and, in connection with this, on the relevance his discoveries had and still have for archaeology and its development into a scientific discipline.[32] In addition, Schliemann's discoveries, his self-presentation, and his public reception have been read against both his biography[33] and his contemporary cultural background.[34] However, the focus has been almost exclusively on Germany. Recent studies have investigated Schliemann's public image in the media with an emphasis on his self-presentation in the German press.[35] In spite of these various studies on

31 William M. Calder's III midnight lecture »Schliemann on Schliemann: A Study in the Use of Sources« is generally considered the turning point of Schliemann research. Held on the occasion of Schliemann's 150th birthday anniversary on January 6, 1972, in Neubukow, Germany, Calder in his lecture revealed serious inconsistencies in Schliemann's excavation accounts and reporting, initiating a very critical debate on Schliemann and his work. Cf. William M. Calder III (1972, 1986a, 1986b, 1992, and 1998); David A. Traill (1983, 1984, 1985, 1986, 1993, 1995, and 2002); Donald F. Easton (1984, 1992, 1994, 1998, and 2002); Harmut Döhl (1981, 1986); Susan Heuck Allen (1994, 1995a, 1995b, 1998, and 1999); Michael Siebler (1990).

32 Cf. William M. Calder III and Justus Cobet (1990) and Joachim Herrmann (1992).

33 Cf. Wolfgang Schindler (1976, 1990, and 1992); Stefan Goldmann (1992a, 1992b); Tom Crepton and Wilfried Bölke (1990); Hans-Werner Hahn (1990); Helmut Scheuer (1990); Wieland Schmied (1990).

34 Cf. Christiane Zintzen (1998); Justus Cobet (1992, 1997, and 2006); Hans-Joachim Gehrke (2006); Jørgen Mejer (1990); Barbara Patzek (1990); Hanna Leitgeb (2001).

35 Stefanie Samida focuses on Schliemann's public reception and self-presentation in the German press with an emphasis on the renowned daily newspaper *Allgemeine Zeitung* and the satirical magazine *Kladderadatsch*. Cf. Samida (2009, 2010a, 2010c, 2011a, 2011b, 2011c, and 2012). Other texts discussing Schliemann's self-presentation are the works by Wolfgang Schindler (1980);

Schliemann, his figure and his popular reception in Britain have only been touched upon, and no thorough analysis has been offered yet.[36] The concern of the present book is to take a closer look at Schliemann's popular reception in Britain as mirrored by the press and in his popular excavational accounts and analyse it in relation to contemporary nineteenth-century British discourses and the formation of identity.[37]

Brigitte Mannsperger (1992); Kathrin Maurer (2009); David A. Traill (1990); John Vaio (1992).

36 John Vaio (1992) discusses Schliemann's relationship with William Gladstone.

37 For a study of Schliemann's popular posthumous reception, cf. William M. Calder III (1986b).

3.3 EGYPTIAN ARCHAEOLOGY: THE MUMMY IN FICTION

> I met a traveller from an antique land / Who said – Two vast and trunkless legs of stone / Stand in the desert ... Near them, on the sand, / Half sunk, a shattered visage lies, whose frown, / And wrinkled lip, and sneer of cold command / Tell that its sculptor well those passions read / Which yet survive, stamp'd on these lifeless things, / The hand that mocked them, and the heart that fed; / And on the pedestal these words appear: / ›My name is Ozymandias, king of kings: / Look on my works, ye Mighty, and despair!‹ / Nothing beside remains. Round the decay / Of that colossal wreck, boundless and bare / The lone and level sands stretch far away.
> (SHELLEY, »OZYMANDIAS«, 1855: 95)

The first wave of *Egyptomania*[1] reached Europe in the aftermath of Napoleon Bonaparte's invasion of Egypt from 1798 to 1801 and the subsequent publication of the multivolumed and highly illustrated *Description de*

1 *Egyptomania* refers to a prominent style of architecture, art, and dress fashion copied and derived from Egyptian antiquity. While the term in its original sense generally focused on an aesthetic movement restricted to the upper classes, there were also other ways in which ancient Egypt was received and reflected by British culture in the early nineteenth century. Cf. Nikolaus Pevsner (1968); James Stevens Curl (1982) and (1994); Erik Eric (2001); David Jeffreys (2003); Alex Werner (2003). Cf. also Jean-Marcel Humbert, Michael Pantazzi, Christiane Ziegler (eds.) *Egyptomania: Egypt in Western Art, 1730-1930* (1994).

l'Egypte[2] (1809-1813). Despite the significance of this event for the further development of Egyptian archaeology, however, Egyptian antiquities had already been studied by scholars and travellers in previous centuries.[3] Nevertheless, little was known about ancient Egypt,[4] and ancient classical writers such as Herodotus, Strabo, and Diodorus Siculus and the Bible were thus considered the only reliable sources in regard to Egyptian history for centuries. This continued to be the case even when the first scholarly observations of Egyptian antiquities were written and published.

1356 saw the publication of *Travels of Sir John Mandeville*, an account that was fantastic in its second half, while the first half was based on actual travellers' reports. Among others, it discusses ancient Egyptian relics and was particularly popular during the sixteenth and seventeenth centuries. Besides this fictitious book, other factual European publications on observations made in Egypt at the time appeared until the end of the sixteenth century, which were also translated into English.[5] In particular, the end of the sixteenth and the early seventeenth centuries continued to see quite a few English travellers publish their accounts on Egyptian antiquities, among them Sir Walter Raleigh, William Lithgow, and John Sanderson, who had a variety of mummies brought to England:

»Interest in Egypt developed rapidly in the opening decades of the seventeenth century. A new class of traveler soon appeared on the scene, one who would dominate the exploration of Egypt until the end of the eighteenth century. This traveler was usually a well-to-do gentleman or the son of such a gentleman. He had an excellent education in classical languages and literature; the study of Arabic also interested him. He traveled up the Nile to Giza and, in the eighteenth century, farther on to Dashûr, Seqqara, Thebes, and Aswân. He took detailed notes and made careful drawings. The result of his investigations appeared in book form soon after he re-

2 This great pioneering work is also noteworthy for presenting the first scientific description of mummies and includes speculation about the procedures involved in their manufacture (cf. Frost 2008: x).

3 For a detailed survey of the development of Egyptology, cf. John D. Wortham (1971) *The Genesis of British Egyptology (1549-1906)*.

4 This was for the most part due to the fact that the hieroglyphic writing had not yet been deciphered.

5 Cf. Paul and Janet Starkey (1998).

turned home. His interest in antiquities was general, including those of England. Such a traveler gradually turned the trip to Egypt, – formerly either a business journey or an extension of the grand tour – into a serious archaeological expedition.« (Wortham 1971: 16-17)

Many of these books were extremely popular in England at the time, such as George Sandy's *A Relation of a Journey Begun an. Dom.* (1610) and Henry Blount's *A Voyage into the Levant* (1636). Besides these amateur accounts of ancient Egypt, Wortham discusses John Greaves's *Pyramidographia: Or a Description of the Pyramids in Aegypt* (1646) as the »first scientific study of Egyptian pyramids« (ibid.: 19). In addition to the various publications on ancient Egypt, it was also quite common to bring home antiquities as souvenirs and collectibles. Mummies, ascribed certain preternatural forces, were particularly popular and transported to England in huge quantities for all kinds of purposes.[6]

The eighteenth century showed the first more systematic study and documentation of Egyptian antiquities. Men like Richard Pococke, whose two-volume *Description of the East and Some Other Countries* (1743 and 1745) became a standard work of the eighteenth century, and the Dane Frederick Norden, who published his impressive *Voyage d'Egypte et de Nubie* in 1755 and later became one of the charter members of the Egyptian Society in London, fascinated people in Britain with their sketches and illustrations of ancient Egyptian objects and monuments.[7] According to

6 Cf. Carter Lupton (2003) »›Mummymania‹ for the Masses – is Egyptology Cursed by the Mummy's Curse?« The flesh of mummies was sold by chemists all over Europe as a cure-all medicine (cf. Nowel 1982: 224): »[T]here's been a fairly massive traffic in Egyptian mummies for centuries. By the hundred and the thousand mummies were exported, swiped, and smuggled out of Egypt direct to Europe. […] [A]t least since 1100AD doctors had been prescribing mummy for their patients. It was thought to contain the same healing properties as a kind of pitch called mummia, found in Persia. By the sixteenth century, mummy was as commonly prescribed as penicillin is now.« (Larsen 2000: 5)

7 Further excavations and explorations were undertaken in Egypt by Nathaniel Davison, British Consul General at Algiers. Another popular, though controversial, figure in Africa at the time was the British explorer James Bruce. Originally travelling up the Nile with the intention of locating its source, Bruce also

Nowel, Norden's illustrations had a major influence on the development of an Egyptian architectural vogue in Britain, with obelisks, sphinxes, and pyramids starting to appear as eye-catchers in English landscape gardens (cf. Nowel 1982: 233).

As a consequence of these explorations of Egyptian antiquities, the interest in ancient Egyptian history and art also grew in Britain. Detailed copies and descriptions of hieroglyphic writings and ancient paintings, brought home by travellers and explorers, captivated people and thus encouraged further excavations in Egypt. More and more previously hidden tombs and temples were excavated and despoiled by European explorers; anything that was moveable was claimed by the ›discoverers‹ and shipped home. Like the ancient Greek and Roman antiquities brought home from the *Grand Tour*, Egyptian relics became part of many private conglomerations fashionable among the British upper classes during the seventeenth and eighteenth century. One of these aristocratic collectors was Sir Hans Sloane, who, after his death in 1753, bequeathed his extensive collections, including numerous ancient Egyptian items, to the British nation. Together with the King's Library,[8] Sloane's collection formed the first inventory of the British Museum when it opened in 1759.

The widespread interest in Egyptian relics also gave rise to a growing theoretical occupation with ancient Egypt in Britain during the eighteenth century: »Historians wrote of the history and civilisation of ancient Egypt. The embalming practices of the Egyptians attracted the attention of medical doctors, and students of ancient and Oriental languages studied the hieroglyphs.« (Wortham 1971: 37) In 1741 the Egyptian Society of London was formed and regularly attended by antiquaries and explorers, such as Pococke, Norden, and William Stukeley. Comprising both members travelling to Egypt and those devoted to the theoretical study of ancient Egypt, the Egyptian Society encouraged a prolific interchange between these two

stopped to visit the ruins of various archaeological sites in Egypt, by which he was deeply fascinated. In his records, he studies and depicts the different sites and offers a detailed description of the mural paintings in the temples of Karnak and Luxor.

8 The King's Library was assembled by George III and given to the British nation by George IV.

areas. Although the Egyptian Society was disbanded in 1743, their ideas and principles continued in the Society of Antiquaries, founded in 1717.

While the reception of ancient Egypt was mainly reserved to the British upper classes well into the eighteenth century, this changed in the wake of the Napoleonic campaigns. As a member of the Académie des sciences,[9] Napoleon was deeply fascinated by the remains of ancient Egypt and commissioned the first sytematic study of Egyptian antiquities. Between 1798 and 1799 a large troop of interdisciplinary savants were engaged to study, describe, and record antiquities in Egypt. In particular, the illustrations and sketches of Egyptian antiquities by the artist Dominique Vivant Denon, member of the Académie Française, fascinated people all over Europe. Denon's illustrations first appeared in French under the title *Voyages dans la Basse et la Haute Egypte pendant les campagnes du général Bonaparte* (1802)[10] and later became part of the twenty-three volume account of Napoleon's expedition to Egypt *Description de l'Egypte* (1809-1813). Although only a small number of copies were printed of the elaborate and extravagant *Description de l'Egypte*, which made it available only to a social and intellectual elite, its impact on contemporary science and society was immense.[11]

Besides the fascination with (ancient) Egyptian architecture, kindled by Napoleon's expedition and the subsequent publication of *Description de l'Egypte*, the scientific and popular interest in Egyptian history and antiquities grew immensely. In particular, the discovery of the Rosetta Stone by Pierre-François-Xavier Bouchard in July 1799 was epoch-making for the further development of Egyptology and Egyptian archaeology.[12] Although scientists and scholars all over the world instantly recognised its

9 Napoleon Bonaparte later became president of the Académie des sciences (1801-1814).

10 *Voyages dans la Basse et la Haute Egypte pendant les campagnes du général Bonaparte* was also translated into English and published under the title *Travels in Upper and Lower Egypt During the Campaigns of General Bonaparte.*

11 Cf. Andrew Bednarski (2005) *Holding Egypt: Tracing the Reception of the ›Description de l'Égypte‹ in Nineteenth-century Great Britain.*

12 The Rosetta Stone shows the translation of the same passage into three different languages, one of them being classical Greek, the second Demotic, and the third hieroglyphic writing.

great value, it was more than twenty years until the Frenchman Jean-François Champollion (1790-1832) succeeded in decoding the hieroglyphic writing. In contrast to the British physician Thomas Young,[13] Champollion found out that the hieroglyphs comprised not only ideographic but also phonographic signs. Champollion recognised two inscriptions of the names Cleopatra and Ptolemy on the stone, which had already been discovered and recorded by William John Bankes, curator of the British Museum, during the excavation of Ramesses II's temple.[14]

The sensational decryption of the hieroglyphs by Champollion opened up hitherto inconceivable possibilities for scholars and scientists. Firsthand insight into Egyptian history now became possible through the deciphering of ancient inscriptions uncovered by archaeologists. In the following years, increasing numbers of British amateurs and professionals travelled to Egypt to unearth hidden antiquities and explore ancient temples and tombs. With the defeat of the French troops by the British in 1801, France had lost its influence in Egypt, which also meant that it became much easier for the British to ship Egyptian antiquities to their homeland.[15]

---

13 Before Champollion, men like the Swedish diplomat and orientalist Johan David Åkerblad (1763-1819), the French linguist Antoine Isaac Silvestre de Sacy (1758-1838), and the British physician Thomas Young (1773-1829) had studied the Rosetta Stone and presented initial attempts at decoding. In 1815, Young offered a translation of the Demotic writing on the stone and presented a Demotic alphabet, but he failed to translate the hieroglyphic writing. Eight years later, Young published his insights in *Account of the Recent Discoveries in Hieroglyphic Literature and Egyptian Antiquities*.

14 In his famous letter »Lettre à M. Dacier relative à l'alphabet des hiéroglyphes phonétiques« from September 22, 1822, Champollion demonstrated how he had succeeded in deciphering the hieroglyphs on the basis of the name inscriptions of Cleopatra and Ptolemy. In his *Grammaire égyptienne*, which was published posthumously between 1836 and 1841, Champollion later presented and explained the system of hieroglyphic writing.

15 The British even forced the French to hand them over the Rosetta Stone, which along with many other ancient relics was transported to Britain and has been displayed at the British Museum since 1802. The French, however, secretly made a wax copy of the Rosetta Stone, which later made it possible for Champollion to decipher the hieroglyphs.

One of the first professional archaeologists to provide museums and exhibitions in Britain with Egyptian antiquities was John Gardiner Wilkinson (1797-1875). During his long sojourn in Egypt he had several tombs opened and conducted several studies of Egyptian antiquities, which were published in *Topography of Thebes and General Survey of Egypt* (1834) and in *Manners and Customs of the Ancient Egyptians* (1837). Wilkinson also wrote the text for John Murray's guidebook *Hand-Book for Travellers in Egypt* (1847). A popular pendant to Wilkinson's *Manners and Customs of the Ancient Egyptians* was Edward William Lane's (1801-1876) book *Manners and Customs of the Modern Egyptians* (1836), which offered a detailed description of contemporary Egyptian life. Like Lane's translation of *The Thousand and One Nights* (1838-1840), *Manners and Customs of the Modern Egyptians* proved extremely popular and became a bestseller in Britain.

Besides the exhibitions of Egyptian antiquities at the British Museum, ancient Egyptian art was also displayed at the Egyptian Hall at Piccadilly Circus. The Egyptian Hall, which had opened in 1812, was one of the first buildings in London designed in the Egyptian style. Commissioned and owned by William Bullock, it housed various public exhibitions and shows, amongst them Belzoni's Egyptian exhibition of 1821, which stands out as an important event in regard to the popularisation of Egyptian archaeology in Britain:

»In October 1817 G.B. Belzoni discovered the painted relief tomb of Seti I in the Valley of the Kings and during the following year produced a careful record of some of the tomb's wall relief paintings. These formed the basis of a reconstruction of two of the most important of the tomb's rooms, together, with a scale model of the whole tomb, which was put on exhibition at the Egyptian Hall in Piccadilly, London in 1821.« (Pearce 2000: 110)[16]

Belzoni first went to Egypt in 1815 in order to present his hydraulic machine to Muhammad Ali, Wāli of Egypt and Sudan, at the time. When Muhammad Ali did not approve of Belzoni's invention, Belzoni turned

16 For a discussion of Belzoni's early life and later career as an ›archaeologist‹, cf. Judith Pascoe (2006) *The Hummingbird Cabinet: A Rare and Curious History of Romantic Collections*.

towards antiquities and soon became obsessed with the remains of ancient Egypt. In the following years he undertook several expeditions to Egypt, returning to Britain with spectacular archaeological discoveries and reports. Belzoni became most well known through his transportation of the bust of Ramesses II, alias ›Young Memnon‹, to Britain and his discovery of the royal tomb of Seti I, often referred to as ›Belzoni's Tomb‹.[17] The bust of Ramesses II was purchased by and displayed at the British Museum, and facsimiles of Seti I's tomb were exhibited at the Egyptian Hall by him and his wife Sarah in 1820 and 1822, respectively, shortly before Belzoni's death:[18]

»By March 1820 the Belzonis had returned to London, and on 31 March *The Times* informed the world that ›The model of the beautiful tomb discovered by Mr Belzoni in Thebes will be erected as soon as a convenient place shall be found for its reception.‹ This advanced publicity was matched by Belzoni's arrangement with John Murray for the publication of his *Narrative*, which duly appeared in December 1820 (with French, Italian, and German editions) as a handsome quarto volume of text and a separate folio volume of forty-five coloured plates.« (Ibid.: 111)

After his first spectacular archaeological exploration, Belzoni continued to conduct excavations at Edfu, Abu Simbel, Karnak, and Giza. When he returned to Britain in 1819 he published his popular *Narrative of the Operations and Recent Discoveries within the Pyramids, Temples, Tombs and Excavations in Egypt and Nubia* (1820).

In contrast to the predominantly aesthetic interest in ancient Egypt in the wake of the Napoleonic campaign and the publication of *Description de l'Égypte*, Belzoni's discoveries and reports appealed to a broader public and thus marked an important change in regard to the archaeological discourse. While so far Egyptian archaeology had been reserved for the upper classes, Belzoni's exhibition opened it up for a wider public. In particular, his reconstruction of the tomb of Seti I at the Egyptian Hall attracted a popular

17 Belzoni had been engaged by Henry Salt (1780-1827), Egyptologist and British consul to Egypt, to study and secure the bust.

18 There was also a guidebook to the exhibition (cf. Pascoe 2006: 129).

interest in the material remains of ancient Egypt.[19] Furthermore, the exhibitions of his findings offered material proof of Belzoni's success as an explorer and collector and a very direct and material confrontation with what had until that point only been known from printed texts, as Pearce emphasises:

»In this shift of taste, Belzoni's exhibition, and the mummy-unwrapping event which preceded it, were significant. The earlier part of the nineteenth century seems to have perceived a significance in the idea of the ›real‹ thing and its relationship to ›real‹ history, especially where antiquities were concerned, which had not been part of the culture of the later eighteenth century (or before) in which ancient objects acted as decorative or architectural inspiration rather than as a supposed means of access to past realities. Belzoni's Egyptian activities, and the Ancient Egyptian material that he put on display, helped to support this cultural shift, as did his idea of reconstruction of the past as a recreation of that past. But the exhibition of the Egyptian tomb signified more than a positivist assertion of accessibility of the past: its focus was on death presented as mysterious, fascinating and spectacular, and its meaning offered a gothic experience of horror in which individual fragments are dissolved.« (Ibid.: 118)[20]

The public interest aroused by Belzoni is further reflected in Lucy Sarah Atkins Wilson's contemporary book *The Fruits of Enterprize Exhibited in the Travels of Belzoni in Egypt and Nubia. Interspersed with the Observation of a Mother to her Children* (1821), a didactic reworking of Belzoni's *Narrative of the Operation and Recent Discoveries within the Pyramids, Temples, Tombs, and Excavations, in Egypt and Nubia* (1820) directed at children:[21]

---

19 Cf. Susan Pearce (2000) »Giovanni Battista Belzoni's exhibition of the reconstructed tomb of Pharaoh Seti I in 1821«.

20 Cf. also the popular interest in exhibitions in general up to this time (cf. Chapter 2.1).

21 As Frost points out, »[i]t has [...] been suggested that [the writer] Jane Webb may have gotten the inspiration for her magnum opus [, *The Mummy!: Or a Tale of the Twenty-Second Century* (1826),] from a visit to the remarkable exhibition [...] staged by Giovanni Belzoni at the Egyptian Hall [...] in 1821. [...]. [A] book that was definitely inspired by this hugely popular attraction is *The Fruits*

»[Mrs A.]: ›Not so fast, my love. Let us proceed gradually. I have not yet told you that Egypt is divided in Upper, Middle, and Lower; and that it is a country renowned in history, having been once the seat, if not the parent of the sciences. It is not only remarkable for those surprising monuments of antiquity, the famous pyramids, which baffle the researches of the deepest antiquary to fix their origin, but also for many other ›glorious structures‹, astonishing remains of ancient temples, pompous palaces, obelisks, columns, statues, and paintings. Thus is Egypt rendered so interesting; and *it is at the present time peculiarly to us, because a gentleman has lately*, with indefatigable zeal, made many researches in that country, and *his curious discoveries among the pyramids and temples have excited the public attention in no small degree. He has spent many years in this arduous employment*, and is now *amply compensated in knowing that they have not been spent in vain.*‹ […] ›Oh, mamma!‹ exclaimed Emily, ›tell me the name of this gentleman‹ […]. [Mrs. A.] ›[T]he name of the gentleman I mentioned is BELZONI.‹« (*Fruits of Enterprise*: 13, emphasis mine)

Besides the fact that Belzoni is stylised as extremely ambitious and adventurous, his zest for action is referred to here. Similar to Schliemann, Belzoni's personal qualities were combined with archaeological prowess and helped promote his popularity amongst a broader public. In analogy to Schliemann, Belzoni's status as a suitable role model for a wide public in Britain was due as much to his reputation for action and his devotion to hard labour as to his success itself.

While Napoleon's expedition to Egypt was above all an intellectual *venture*, Belzoni's travels to the Near East started off very much as an *adventure* people could identify with. In Belzoni's case, the scientific interest was dominated by a wild romantic dream characterised by an obsession with spectacular discoveries. He was a self-made man who succeeded in making his fortune by means of manual labour, presenting people with relics of a distant time and place. Through his perseverance, courage, strength, and ambition, Belzoni thus did not only embody crucial nine-

---

of Enterprise Exhibited in the Travels of Belzoni in Egypt and Nubia; Interspersed with the Observations of a Mother to Her Children, which was originally published in 1821. Uncredited at the time, the author has recently been identified as Lucy Sarah Wilson.« (Frost 2008: 4) teenth-century moral values, as emphasised in Wilson's *Fruits of Enterprise*, but as a self-made man of action he also appealed to certain dream-fantasies of the working and middle classes. However, Belzoni was not only an enthusiast and a self-made man, but also very much a showman capable of entertaining a wide audience, which becomes clear in regard to the popularity of Seti I's tomb displayed at the Egyptian Hall. People like Belzoni played a key role in the popularisation of archaeology since, by merging an elite scientific discourse with popular culture, they promoted the creation of a popular archaeology in the nineteenth century.

In this context, it is noteworthy that Belzoni's exhibition at the Egyptian Hall had a tomb on display, reflecting the sensational character which dominated contemporary fairs and anticipating the huge success public unrollings of mummies were to have during the following decades:[22]

»Ethnographic freak shows provide a useful context for the sensational Egyptian exhibits and events that were fashionable in England throughout the nineteenth century: the extravagant Valley of the Kings show in 1820; public unwrappings of mummies in the 1830s and 1840s; the Nile panorama at Egyptian Hall; popular lectures on ancient Egyptian culture and history; and museum displays of Egyptian mummies and other artefacts.« (Hurley 2008: 184)

A prominent figure in this context was Belzoni's acquaintance, the medical scientist Thomas Pettigrew, alias ›Mummy Pettigrew‹, who was to become famous for his examination of mummies from the 1820s on.[23] In 1820, Pettigrew and Belzoni »collaborated in the unwrapping of a mummy which Belzoni had brought to England: this was a private occasion, but presumably had its share in generating an interest« (Pearce 2000: 112). One year later,

»Belzoni staged the unwrapping of a young male Egyptian mummy before an audience of distinguished physicians at William Bullock's Egyptian Hall natural history museum in Piccadilly. […] Belzoni's most famous successor was Thomas Pettigrew,

22 Cf. also Nicholas Daly (1994) »That Obscure Object of Desire: Victorian Commodity Culture and Fiction of the Mummy«.

23 Pettigrew's *A History of Egyptian Mummies* (1834) became a bestseller (cf. Sulcer 2007: 153).

a highly respected London physician. Ten years after Belzoni's exhibition, Pettigrew began a series of public unwrappings before audiences packed into such venues as the lecture hall of Charing Cross Hospital, where he was a surgeon and Chair of Anatomy; the Museum of the Royal College of Surgeons; and the auditorium of the Royal Institution. Tickets for these highly publicized ›mummy of the month‹ events, which were sold for one guinea per mummy-side seat and half a guinea for rear seats, were in such demand, accommodation had to be made to appease those whose applications were rejected due to limited space; one lecture theatre was so crowded, neither the Archbishop of Canterbury nor the Bishop of London could gain admission to the show.« (Bridges 2008: 152)

While the first unwrappings of mummies, also aptly referred to as »›scientific‹ striptease« (ibid.: 138), were undertaken privately in front of only a small chosen audience, comprised chiefly of scientists, soon Belzoni, Pettigrew, and other ›showmen‹ conducted the first public unrollings in front of a much wider audience (cf. ibid.: 153): »[W]itnessing the unrolling of mummies became a fashionable pastime amongst antiquaries, dilettanti, and even with the public.« (Dawson 1934: 1)[24] These shows were eventually, quite uncommon for the time, even open to women. What had started off as a private scientific examination of Egyptian mummies had become a popular show by the 1830s, as the ever growing demands for tickets and the public advertisements in the newspapers reveal (cf. ibid.):

»For many late nineteenth-century British critics, the proximity spectators desired and attained to the ancient dead suggested a strange contradiction, for whereas crowds of curious museum-goers flocked to the display of Egyptian mummies in various national galleries and personal collections, concerns for public health and sanitation provoked demands for the segregation of the *modern* dead from the living.« (Bridges 2008: 151)

As emphasised by these various examples, both Belzoni's Egyptian exhibition and Pettigrew's unrolling of mummies spurred the development of a

24 Warren R. Dawson (1934) »Pettigrew's Demonstration upon Mummies. A Chapter in the History of Egyptology«. Cf. also Pearce's article (2002) »Bodies in Exile: Egyptian Mummies in the Early Nineteenth Century and their Cultural Implications« on mummy unwrappings.

popular interest in the subject. Parallel to this rather sensationalist popularisation of ancient Egypt in Britain, a more scientific occupation with the subject continued. In this context, the name of the Scottish Egyptologist Alexander Henry Rhind (1833-1863) is generally associated with a methodological approach to archaeological excavation. Originally a lawyer, Rhind, like many of his contemporaries, first travelled to Egypt to recover from a pulmonary disease; however, he soon became fascinated by Egypt and its antiquities. He started excavating Egyptian ruins in the 1850s and was known for his accurate and methodological approaches as much as for his talent as a writer. One year before his premature death in 1863, Rhind published his archaeological insights and findings, among them the so-called Rhind Papyrus, an ancient Egyptian mathematical treatise, in *Thebes, Its Tombs and Their Tenants, Ancient and Present*.

While earlier archaeologists, such as Richard William Howard Vyse (1784-1853) and Belzoni, had no qualms about using brute force and explosives for their ›excavations‹, their professional successors proceeded more sensibly and systematically. A man who is often paid homage to as the ›father of modern archaeology and Egyptology‹ (cf. Adams 2010: 35) is Sir Flinders Petrie (1854-1942), who spent his early professional years digging in Britain and later travelled to Egypt to apply his archaeological methods to Egyptian antiquities. From the 1880s on Petrie excavated and studied various archaeological sites, *inter alia* at Giza, Tanis, Fayum, and Tel-al-Amarna. His systematic and substantiated approaches to archaeological field studies gained him an international reputation as an archaeologist as well as a knighthood in 1923.

While Petrie revolutionised and promoted archaeological field studies through his new methodologies, the writer Amelia B. Edwards, acclaimed as the ›Queen of Egyptologists‹ (cf. Winslow 1892: 15), championed a sensible handling of Egyptian antiquities *in situ* and in Britain, as emphasised by Adams: »With her appetite for adventure, Edwards might have made a whole life out of simply travelling the world, observing first-class digs, checking off a list of archaeological sites to visit like groceries to buy. Instead, she decided to get involved. She fought for archaeology, for its development, expansion and cause.« (Adams 2010: 39) Together with Petrie she initiated the foundation of the Egyptian Exploration Fund (EEF; now Egyptian Exploration Society) for the conservation of ancient Egyptian relics in 1882: »Blessed by the tireless leadership of Amelia Edwards and the inspired choice of the young Flinders Petrie as its agent, the EEF fanned the Victorian public's enthusiasm over Pharaonic Egypt and doubtless contributed to the interest that would greet mummy stories in the decades to follow.« (Deane 2008: 387) Edwards herself had been travelling in Egypt during the winter of 1873-1874 and was at once fascinated by the country, the culture, and its history. Before returning to Britain, Edwards initiated and participated in an excavation at Abu Simble, where she excavated an ancient sanctuary. On her return to Britain, she compiled the impressions, experiences, and insights she had gained during her travels in Egypt in her famous book *A Thousand Miles up the Nile* (1877), a historical and Egyptological study.

Apart from her archaeological achievements, Edwards also takes on a key role in regard to the popularisation of ancient Egypt in Britain, offering an alternative to Belzoni's brute and sensationalistic approach to the subject. As a prolific and popular writer of novels, non-fiction books, and newspaper articles, Edwards was already a well-known figure in Britain when she left for Egypt. As a consequence, her book *A Thousand Miles up the Nile* was an immediate success in Britain and sparked a wide interest in Egyptian travelogues in general:[25]

»Edwards's best-selling 1877 travelogue *A Thousand Miles up the Nile* can be said to have altered England's relationship to the cultural legacy of Egypt. [...] In addition to her canny use of the picturesque, moreover, Edwards had the advantage of a captive audience for her endeavor. She moved in familiar Victorian literary circles, contributing to Dickens's Christmas collection of ghost stories, and published frequently in the *Saturday Review* and *Household Words*. When she proposed to write about Egypt, then, she did so for the broadest of readership.« (Marx 1998: 56)

Through the way in which she presented and reviewed her archaeological knowledge and experiences in Egypt, Edwards was thus able to transform

25 Other popular examples of travelogues were William Alexander Kinglake's (1809-1891) *Eothen* (1844) and Harriet Martineau's (1802-1876) *Eastern Life, Present and Past* (1848). Barrell refers to the multitude of Egyptian travelogues in general in the first half of the nineteenth century: »In the 1840s and 50s dozens of narratives of tourism in Egypt were published in Britain, as well as the United States, France and Germany.« (Barrell 1991: 101)

specialised knowledge into a form which was accessible both intellectually and financially to a broad public. This made her, together with Belzoni, an important mediator between the specialist and the public discourse, fostering the genesis of a popular archaeology.

Aside from Petrie and Edwards, the three other European archaeologists of high reputation were the two Frenchmen Auguste Ferdinand Mariette (1821-1881) and Gasto Maspero (1846-1916) and the German Karl Richard Lepsius (1810-1884). Mariette was a very productive and successful archaeologist who worked for the Egyptian Service of Antiquities, an institution founded in 1858 for the protection and export control of Egyptian antiquities.[26] After Mariette's death, Maspero, who himself had joined an Egyptian exploration as a linguist, succeeded Mariette as the director-general of the Egyptian Service of Antiquities. While Mariette had been known for his discrimination against non-French archaeologists, Maspero conducted more generous politics in this regard. However, the competition between France, Germany, Italy, and Britain in the search for spectacular discoveries continued far into the twentieth century. In 1922 it reached its climax with the discovery of Tutankhamen's tomb in the Valley of the Kings.[27] After over a decade of intensive search, the British archaeologist Howard Carter (1874-1939), a former pupil of Petrie, eventually discovered the stairway to the royal tomb on November 4, 1922. The expedition was financed by the aristocratic amateur Egyptologist Lord Carnarvon (1866-1923). Tutankhamen's tomb, later known as KV62, was the first Pharaonic tomb which had not been plundered by grave robbers prior to its discovery by Carter. Consequently, both its significance for archaeology and its impact on the public was immense at the time and afterwards. Contemporary British newspapers and magazines regularly informed people at home on the progress of the spectacular excavation and the specialised press fervently discussed the latest findings. The public interest in the excavation was further fuelled by the mysterious series of deaths among archaeologists and other people who had attended the excavation of Tutankhamen's tomb

26 Mariette also initiated the setting up of the Museum of Egyptian Antiquities in Cairo in 1863.

27 This spurred on popular discoveries of mummies in the Nile Valley in the 1870s and 1880s (cf. Marx 1998: 57).

in the aftermath of the excavation.[28] Speculations about a Pharaonic curse, which, according to the warnings found in many ancient Egyptian burial chambers, befell those who disturbed the peace of the dead, were very popular both among amateurs and professionals.

Mummy Fiction

>›What most of us are coming to Egypt for is *mummies*. Egyptian history is too troublesome, anyhow, for a normal man to grasp. Give me mummies! There's something *in* them.‹
>(*IT HAPPENED IN EGYPT*: 113)

It is within the context of the popular nineteenth- and early twentieth-century fascination for Egyptian archaeology that we have to read the prominent presence of ancient Egypt in literature at the time. While a huge variety of contemporary texts dealing with ancient Egypt were produced, both fictional and factual, ranging from children's literature to historical novels and translations of ancient Egyptian narratives,[29] Egyptian archaeology is most prominently featured in the so-called ›mummy narrative‹, thereby reflecting the popular obsession with Egyptian mummies[30] in the nineteenth and at the beginning of the twentieth century:

---

28 Cf. Colla Elliott (2007) *Conflicted Antiquities: Egyptology, Egyptomania, Egyptian Modernity*.

29 For a comprehensive bibliography of ancient Egypt in fiction cf. Noreen Doyle's »Ancient Egypt in Fiction« http://www.noreen-doyle.com/Egyptomania Org//aef/Egyptfiction.html. For an overview of novels, short stories, films, etc., cf. Brian J. Frost (2008) *The Essential Guide to Mummy Literature*, Richard Dalby's article »The Mummy«, Susan D. Cowie and Tom Johnson (2002) *The Mummy in Fact, Fiction and Film*.

30 Cf. L. Green (1992) »Mummy Mania: The Victorian Fascination with Ancient Egypt's Mortal Remains«.

»British homes and public galleries became the often-overcrowded repositories of ancient Egyptian artefacts, from stone obelisks to colossal statues to fragments of stone obelisks and colossal statues. But no other vestige of oriental antiquity captured the British imagination like the mummy.« (Bridges 2008: 139)[31]

The popular fascination with mummy fiction was continuously kindled by further archaeological findings in Egypt and their discussion in the contemporary British press:

»Interest was fuelled by Sir Gaston Masparo and Emile Brugsch Bey's discovery, in 1881, of a cache of mummified kings in the tomb of the Pinejem family. This tomb, surrounded by the romantic story of its discovery and robbery by the native Abd er Rassul family, and their subsequent torture and eventual succumbing to reveal its whereabouts, proved to contain the bodies of several key figures from the Nineteenth Dynasty, including Seti I and his son, Ramesses II, the latter being the supposed pharaoh of Exodus and the Israelites' departure from Egypt. The English press went wild with the discoveries; the most informed report of the time was probably that by Amelia Edwards, full of exquisite line drawings, in the *Illustrated London News*, 4 February 1882.« (Pearson 2000: 220)

In the present book the mummy will be analysed in its function as a popular *pars pro toto* of Egyptian archaeology in general.

Although there were a few texts featuring mummies earlier in the century, such as Jane C. Webb Loudon's popular science-fiction romance *The Mummy! A Tale of the Twenty-Second Century* (1827), mummy fiction was most prominent at the end of the nineteenth and the beginning of the twentieth centuries, which Daly sums up as follows:

»By the end of the nineteenth century the vogue for unrolling of mummies was long over, and the mummy's power as spectacle was confined to the museum. At the same time, the appearance of mummy fiction suggests that the mummy retains its

31 Cf. also Hurley: »The mummy is a recurring figure in British imperial Gothic fiction at the fin de siècle, when interest in Egyptian and other ancient cultures was being fueled by museum exhibits, popular lectures on art, archaeology, and history, and the increasing availability of ›exotic‹ artefacts for private collections.« (Hurley 2008: 181)

importance in the cultural imaginary of late Victorian and Edwardian Britain.« (Daly 1994: 25)

Accordingly, the mummy narratives compiled for my text corpus comprise short stories and novels published between 1870 and 1914.

Mummy texts have been discussed by various critics, with a particular emphasis on turn-of-the-century fiction. Daly (1994, 1999) argues that the representation of the mummy in fiction mirrors contemporary changes in commodity culture, resulting in »inverted relations of objects and people within an *imperial* global economy« (Daly 1999: 27).[32] As a result, the foreign objects, such as mummies, gain an »uncanny power« (ibid.: 90) over the consumer. In order to compensate for this loss of control, »[m]ummy stories offer a fantasy of the restoration of human mastery over the world of things through their gender logic; married to the commodity, the (masculine) consumer once again enjoys the upper hand over the feminized object, though this fantasy, too, is not without its dangers for the consuming subject« (ibid.: 27). Stephen Arata (1990) and Patrick Brantlinger (1990) read the figure of the haunting mummy in the western hemisphere in the context of ›reverse invasion‹, while David Seed (1998) conceives the mummy as the embodiment of ancient evil forces that erupt in the present in the form of the primitive and often feminine. Both Robert Edwards (1998) and Susan Pearce (2002) explore the perception of the mummy by British contemporaries as both ›alien‹ and ›familiar‹, which turns the encounter with the mummy, such as at the occasion of public mummy unrollings, into »a narcissistic experience detached from any obligation to form a relationship« (Pearce 2002: 70). In a similar way, Kelly Hurley (2008) conceives of the mummy as the »uncanny double of the late-Victorian subject in a process both fearsome and pleasurable« (Hurley 2008: 182). Along these lines, Aviva Briefel (2008) argues that Victorians were both »[f]ascinated and repelled by the mummy and its hand« (Briefel 2008: 264), since for her the mummy's hand functions as a constant reminder of »the manual craftsmanship that produced it« (ibid.: 266), that is, the ancient Egyptian embalmer and »the [mummy] hand's own ability to make« (ibid.). Breadley Deane (2008) sees the figure of the mummy as representing the complex relationship of the British Empire and Egypt in

32 Cf. Chapter 1.

»that it is [...] the eroticized suspense of mummy fiction that betrays its entanglement within the complicated aspiration of Victorian imperialism and, most particularly, within the political riddle posed by Egypt, that unacknowledged but crucial corner of the empire whose independence was forever hovering just over the horizon« (Deane 2008: 385). On the basis of Deane's assumption, Karen Macfarlane (2010) in her recent article suggests that the figure of the mummy »is the quintessential monster of imperial gothic«. For her, »[t]he sudden interest in a figure that some would describe as a fundamentally flawed monster« (Macfarlane 2010: 5) reflects »a profound anxiety about the epistemological underpinnings of the imperial project« (ibid.) in that mummies are frequently represented in connection »with lost and powerful knowledge: they are priestesses, sorceresses, and witch-queens« (ibid.: 6). Concentrating on Bram Stoker's texts, David Glover (1992, 1995, 1996) and Carol Senf (1997) explore the figure of the mummy in terms of ancient versus modern knowledge, in particular in regard to science and the occult. Meilee Bridges (2008) focuses on the mummy curse in mummy fiction and connects it to the »genre's relationship to nineteenth- and early twentieth-century popular and scholarly Egyptology« (Bridges 2008: 139). According to Bridges, »the critical significance of the mummy's curse narrative derives from the serious question it raises about the often-deleterious treatment of the dead in the name of acquiring knowledge about an ancient civilization« (ibid.).

In the following, I want to argue that previous approaches to mummy fiction can essentially be understood as based on major Victorian myths reflecting contemporary sociocultural issues, presented in my introduction as the (1) *Frankenstein Discourse*, (2) *Dracula Discourse*, (3) *Jekyll and Hyde Discourse*, and (4) *Sherlock Holmes Discourse*.[33] Furthermore, I claim that all the hitherto presented approaches to the subject can be traced back to the assumption of an initial alienation of the individual subject rooted in the lasting sociocultural changes of Victorian and Edwardian Britain. This isolation and loneliness is represented by the eponymous

33 Although parallels between mummy fiction and certain aspects of the different discourses have been pointed out by others, no thorough analysis has been offered yet. *Frankenstein Discourse*: Hopkins (2003); *Dracula Discourse*: Edwards (1998) and Glover (1992, 1995, and 1996); *Holmes Discourse*: Daly 1999: 98-102, Korte (2000a), Patzek (1993, 1999), and Neuhaus (1999).

characters of the four discourses introduced, who, as mentioned above, are all – sooner or later – outcasts of society: Victor Frankenstein and his creature, Count Dracula, Dr Henry Jekyll and Edward Hyde, and Sherlock Holmes.[34] It is against the background of these discourses that the figure of the mummy assumes a central position in regard to the negotiation of identity and alterity within the archaeological space. As pointed out by other critics, the mummy is the embodiment of various oppositions, oscillating between the familiar and the unfamiliar.[35] Most prominent in this regard is the fact that the mummy apparently undermines the laws of nature: In spite of being a dead body, it is not subject to processes of decay (cf. Pearce 2002: 58). Paradoxically, then, the dead mummy seems very much alive. In this sense the mummy represents both death, associated with the ›other‹, and life, associated with the self: »The borderline between Ourselves and Others is weakened, and the meaning of life and death that were associated with those two domains becomes somewhat confused. This is where mummies, skulls and skeletons become our fetishes in seeking for meaning.« (Wieczorkiewicz 2005: 68) This ambivalence is enforced by the fact that the mummy is on the one hand associated with a different time and space, while on the other it is by virtue of its humanness very close to the here and now. The mummy thus represents both the past and the present, home and abroad. It is this multilayered paradoxical tension embodied by the mummy which made it so popular at the time.

This is prominently reflected by the popularity of Belzoni and Pettigrew's exhibitions and shows, respectively. It is crucial to note in this context that both Belzoni's exhibition of Seti I's tomb and Pettigrew's unrollings of mummies focused on the ancient Egyptian death cult and thus drew attention to aspects of life, death, and immortality. However, while Belzoni's exhibition lacked the central figure, that is, the mummy of Seti I, and thus by providing the tomb as a *pars pro toto* only alluded to death, Pettigrew's demonstrations on mummies did not disappoint their audience in this respect. The fact that in Pettigrew's case we are dealing with dead human bodies, stripped naked and displayed in front of an audience, gives the whole event an atmosphere charged with voyeurism and sensationalism

34 Note that all these characters are *searching* for meaning and identity within a society of which they are essentially not part.

35 Cf. Edwards (1998); Pearce (2002); Hurley (2008).

which is reminiscent of that of the present-day ›Body Worlds‹ exhibition of the German Gunther von Hagens and confirms Wieczorkiewicz statement that mummies are »the organisational core of many museum exhibits« (Wieczorkiewicz 2005: 52). As Bridges points out:

»[u]nlike other objects of Egyptian antiquity, mummies invite their spectators to contemplate their dead bodies as visual manifestations of the grotesque and the horrible rather than as conduits to knowledge about ancient Egyptian life and culture. In this sense, exposing the ancient embalmed dead caused observers to gawk rather than to think – museum displays translate mummies from objects of contemplation or knowledge to objects of perverse and mesmerizing spectacles.« (Bridges 2008: 149)

Even though the context is different, the popularity of both the mummified body à la Pettigrew and the plastinated body à la von Hagens suggests a universal and timeless human fascination, which could be summed up as sensationalism, prominently represented in Victorian culture, as underlined by Brantlinger:

»In an age that was preoccupied with bodies as spectacles that signified everything from criminal behaviour, psychological disorder, moral standing and racial categorizations the body of the mummy functions as a signifier that mediates between imperial fantasies of control and immortality and fin de siècle fears of regression, invasion, atavism and dissolution.« (Brantlinger 1990: 230)

It is this sensationalism, however, as characterised by a voyeuristic pleasure in another person's suffering or death distinguished by a feeling of simultaneous alienation and identification with the subject/object in question, which is essentially disturbing:

»But mummies subvert this basic categorization, offering the possibility that the rational classification of subject and object can break down. Mummies are dead bodies and are, as such, objects, in much the same way as our own bodies are objects to us, to be pierced, tattooed or sunburned. But mummies are also the physical presence of what were once living people, with life stories of their own, who viewed their own bodies as we do ours. This ambivalence is part of the power of the unwrapping experience, where examination becomes violation. As the subject-nature of the mummy develops, so the self of the viewer becomes objectified within the mummy's gaze – a reversal which lies at the heart of the succeeding mummy fiction, a process well traced by Daly. Mummy unwrappings force the viewer to acknowledge the lure of fantasy in which exile from one's own self as subject creates a self as acted-upon object by a dead object turned back into personified subject.« (Pearce 2002: 65)

This simultaneity of alienation and identification is true of both the mummy unrolling event and the ›Body Worlds‹ exhibition, since in both cases the displayed object is the familiar human body, which, however, is alienated through its particular artificial preparation (cf. ibid.: 71).[36] As a consequence, the displayed body is perceived by the observer as being both the same as and different from one's own body. As pointed out by Pearce and Pollock in regard to Freud and the threat associated with the past represented by the mummy and archaeology in general, it is this tension between the known in what is seemingly unknown to us that makes people perceive something as uncanny (cf. Pearce 2002: 67):

»In his essay on ›the Uncanny‹ [*Das Unheimliche*], 1919, Freud defines that curious thing, the uncanny, lying between concept and domain, as an effect to two possible conditions: the return of what was once familiar but later repressed which, he argues, concerns castration anxiety above all, and the revival of infantile beliefs, later surmounted by adult understanding, such as led one to believe in the omnipotence of one's wishful thoughts in the face of powerlessness or fear.« (Pollock 2007: 75)

Thus, one reason which renders the preserved human body so interesting to people can be seen in a voyeuristic curiosity about an abnormal state of the human body which can be looked at by the observer with a certain sensationalist *schadenfreude,* with the recognition that what happened to the individual on display could have happened to oneself, but *did not*. Another crucial aspect for the fascination people feel with these bodies is certainly rooted in the fact that the preserved human body personifies the paradox of

36 Significant in this context is the fact that von Hagen's exhibition showing *modern* Western bodies has been received as highly controversial by the public, while the mummy unrolling displaying bodies from another time and culture was generally popularly accepted at the time.

immortality in death; it is »the living embodiment of the dead« (Bergstein 2009: 2): »The mummy is an incarnate oxymoron: a decaying body preserved from decay. It speaks to the afterlife, to the prospect of immortality and the transcendence of the body.« (Hurley 2008: 192) In this way, while the observers are immediately confronted with death, they also get a glimpse of physical immortality. These bodies are dead, while at the same time physically they *seem* very much alive. It seems only natural, then, that in fiction these preserved bodies behave exactly according to what one secretly expects and fears while looking at them: They wake up from their deathly slumber and come to life. In fact, this is nothing more than the accomplishment of an unnatural process which was started by the preservation of the human body.

Considering the seemingly timeless human fascination with preserved human bodies as exemplified by the popular exhibitions of Pettigrew and von Hagens, it makes sense that the mummy, the very epitome of the living death, should have become so popular in fiction from the last few decades of the nineteenth century to the present (cf. Daly 1994: 25). By coming to life, the mummy turns from a passive object into a subject with agency, which means a loss of control for the hitherto omnipotent voyeurs, thereby posing a sudden threat to them:

> »Some stories deal with hostile mummies who revenge themselves on those who have disturbed their original resting places. Others exploit the erotic possibilities of the mummy. Still others vacillate between these categories. However, all feature mummies who come to life or otherwise attain agency in modern Europe.« (Ibid.: 36)

In regard to the popularity mummies enjoyed in turn-of-the-nineteenth-century fiction, it can be argued that this fear of the loss of power and control over the mummy mirrors general fears of instability in nineteenth-century British society. In contrast to the mummy discourse of the first half of the nineteenth century, which seems to depend very much on contemporary events such as Pettigrew's mummy unrollings, towards the end of the century the mummy discourse had turned into a complex and multi-layered discourse, as pointed out by Daly:

»If commodities were on the move, so too were mummies: toward the end of this period mummies first begin to stalk the pages of popular fiction. The mummy, and to a lesser extent the collector and the detective, were figures around which another such theory of the new relations of subjects and objects finds articulation. Indeed, it could even be argued that mummy fictions tell us more about Marx's model for explaining the inverted relationship of subjects and objects in a nascent consumer culture as Marx can tell us about such fictions. [...] The mummy is the type of object which becomes a commodity simply because it becomes desirable for consumers, and is thereby drawn into economic exchange. [...] What makes mummies valuable, though, is not the cost of their production but the fact that people *want* them. [...] Similarity, in the stories of Everett, Haggard, Conan Doyle, and others, the mummy, as a markedly foreign body within the British economy, articulates precisely the imperial dimension of the nineteenth-century British economy that Marx underplays.« (Ibid.: 27)

In addition to this intrinsic fascination with the mummy, the public reception of the mummy thus underlines in what way the popularity of ancient Egypt was dependent on *how* people got in touch with Egyptian archaeology – and connected to that, *who* got in touch with it – and second, on *what* aspects of it they were confronted with. This is emphasised by the different receptions of the spectacular exhibitions and show events à la Belzoni and Pettigrew as opposed to a multivolumed and exclusive work in the Napoleonic style.

The following analysis of mummy fiction focuses on four major texts of the time, namely Theo Douglas's *Iras: A Mystery* (1896), Guy Boothby's *Pharos, the Egyptian* (1898), Bram Stoker's *The Jewel of Seven Stars* (1903), and Richard Austin Freeman's *The Eye of Osiris* (1911), to show the ways in which they mirror contemporary social and cultural concerns. To support the findings in these texts, however, I will also include other texts in which mummies appear[37] to show how they too reflect different contemporary discourses.

37 Jane C. Webb Loudon (1827) *The Mummy! Or, A Tale of the Twenty-Second Century*; *Fruits of Enterprise in the Adventures of Belzoni in Egypt and Nubia;* Lucy Sarah (Atkins) Wilson (1821) *Fruits of Enterprise in the Adventures of Belzoni in Egypt and Nubia; With an Account of his Discoveries in the Pyramids, among the Ruins of Cities, and in the Ancient Tomb*; Grant Allen (1878) »My New Year's Eve among the Mummies«; Grant Allen (1884) »The Miraculous Explorer«; Arthur Conan Doyle (1890) »The Ring of Thoth«; Arthur Conan Doyle (1892) »Lot No. 249«; Guy Boothby (1894) »A Professor of Egyptology«; Ernest R. Suffling (1896) »The Strange Discovery of Dr. Nosidy«; George Griffith (1903) »The Lost Elixir«; George Griffith (1906) *The Mummy and Miss Nitocris: A Phantasy of the Fourth Dimension*; Algernon Blackwood (1908) *Nemesis of Fire*; Lilian Bagnall (1910) »In the Tombs of the Kings«; Henry Rider Haggard (1912-1913) »Smith and the Pharaohs«; E. and H. Heron (1898) »The Story of Baelbrow«; Sax Rohmer (1903) »The Mysterious Mummy«; Cutliffe C. J. Hyne (1904) »The Mummy of Thompson-Pratt«; Clive Pemberton (1906) »The Bulb«; Fergus Hume (1908) *The Green Mummy*; Sax Rohmer (1913) »The Case of the Headless Mummies«; Sax Rohmer (1914) »The Cat«; Alice Muriel and C. N. Williamson (1914) *It Happened in Egypt.*

3.4 Zimbabwean Archaeology: The Empire as a Space of Negation and Construction

> Not in the waste beyond the swamps and sand, / The fever-haunted forest and lagoon, / Mysterious Kor thy walls forsaken stand, / Thy lonely towers beneath the lonely moon, / Not there doth Ayesha linger, rune by rune / Spelling strange scriptures of a people banned. / The world is disenchanted; over soon / Shall Europe send her spies through all the land. / Nay, not in Kor, but in whatever spot, / In town or field, or by the insatiate sea, / Men brood on buried loves, and unforgot, Or break themselves on some divine decree, / Or would o'erleap the limits of their lot, / There, in the tombs and deathless, dwelleth *She*!
> (LANG, »SHE«, 1889: 93)[1]

On April 1, 1892, *The Journal of the Manchester Geographical Society* published the following article on recent archaeological explorations carried out in Rhodesia, present-day Zimbabwe, by the British archaeologist Theodore Bent:

»Through the kindness of Mr. Theodore Bent we are able to inspect a valuable and interesting collection of photographs of the ruins of Zimbabwe in Mashonaland, as well as of the interesting remains of stone carvings, religious symbols, pottery, tools, and weapons which were found by excavations. The remains of Zimbabwe were

1 The Scottish anthropologist and writer Andrew Lang dedicated his poem »She« to his friend Henry Rider Haggard, after Haggard had dedicated his African adventure romance *She: A History of Adventure* (1887) to Lang.

visited by Mr. Bent, accompanied by his wife, last year, for the purpose of archaeological research, with much success. His enquiries resulted in discoveries of a most interesting character, and prove beyond doubt that long ago a settlement of people existed in South Africa who had reached a high standard of civilisation. From the remains found, Mr. Bent believes that the originators of Zimbabwe were colonists, who visited the district in search of gold, coming most probably from Arabia. They were evidently settlers in a hostile land, as the ruins prove that they took every care to protect themselves from invasion, their settlements being placed in positions affording great natural strength, and being, moreover, fortified with much care.« (*The Journal of the Manchester Geographical Society*, Apr 1, 1892: 188-189)

The following analysis will focus on the reception and interpretation of Zimbabwean archaeology as represented by the ancient stone ruins generally known as Great Zimbabwe in Mashonaland, which were ›rediscovered‹ by European explorers in the second half of the nineteenth century south of present-day Harare.[2]

Since the first mention of the Zimbabwe ruins in accounts of Portuguese settlers in the sixteenth century, people in both Africa and Europe had been deliberating the question of their origin and creators, as reflected in Henry Rider Haggard's introductory note to his romance *Elissa: Or, the Doom of Zimbabwe* (1899): »The world is full of ruins, but few of them have an origin so utterly lost in mystery as those of Zimbabwe in South Central Africa. Who built them? What purpose did they serve? These are questions that must have perplexed many generations, and many different races of men.« (*Elissa*: vii) Although Great Zimbabwe was buried in oblivion for a few centuries after being first discovered by Portuguese settlers, the interpretation of the ruins from their first mention until far into the twentieth century has been characterised by severe racism. This is reflected in the journal of the German African explorer Karl Mauch, in which Mauch refers to the builders of the Zimbabwe ruins upon hearing about

2 Great Zimbabwe was ›rediscovered‹ by the American big game hunter Adam Renders in 1867. Renders showed the German explorer and geographer Karl Mauch the ruins in 1871. For a summary of the history of the different archaeological interpretations of Great Zimbabwe, cf. Martin Hall (1995) and (1996); Laurence P. Kirwan (1938); Anthony Chennls (2007); Merrick Posnansky (1982); Bruce Trigger (1998); Daniel Tangri (1990).

them from indigenous people: »There was one among these people who came forth with still more important news, namely of the presence of quite large ruins which *could never have been built by blacks.*« (Mauch 1969: 139, emphasis mine) To deny a black origin for Great Zimbabwe was common far into the twentieth century.[3] Accordingly, various popular alternatives were proposed in regard to the origin of the ruins. In a speech delivered at the Royal Geographical Society in London on February 22, 1892, Bent attributed the origin of the ruins to Arabian settlers:[4]

»The ruins were not in any way connected with any known African race; and they doubtless formed a garrison for the protection of a gold-producing race in remote antiquity. There was little room for doubt, he thought, that the builders and workers of the Great Zimbabwe came from the Arabian Peninsula, and belonged to a pre-Mohammedan period.« (*Leisure Hour*, Apr 1892: 429)

The assumption that Great Zimbabwe must have been built by foreigners simply because the Western colonisers could not – or rather were not willing to – think of any African ethnic group to attribute the ruins to, shows the subtle and implicit racism, which besides the overt racism reflected in the reception of Great Zimbabwe, had been characteristic of the interpretation of the ruins. Only after Zimbabwean independence was achieved in 1980 was it publically accepted that the origin of Great Zimbabwe lies in the Shona culture, as Chennells points out:[5]

---

3 Cf. below.

4 The assumption that the ancient settlers were either of Arabian or Phoenician origin was widespread. On the same occasion, Bent also displayed artefacts found at Great Zimbabwe: »On February 22nd Mr. Theodore Bent will give the Society an account of his researches in the ruined cities of Mashonaland. The many objects of art and handiwork belonging to the ancient inhabitants of these strange buildings, which Mr. Bent found by excavation in and about the ruins, will be on exhibition the same evening, after the reading of the paper.« (*Proceedings Royal Geographic Society* 1892: 120)

5 In spite of archaeological evidence proving a black origin of Great Zimbabwe provided by archaeologists such as David Randall-MacIver (1905/1906) and Gertrude Caton-Thompson (1929) at the beginning of the twentieth century, the official interpretation of the ruins attributed their origin to white settlers until as

»Nearly a century of professional archaeological work has reconstructed in some detail the history of Great Zimbabwe. The old city was not an outpost of some ancient and exotic empire but an African city, built by the Shona, who are still the dominant people of the southern Zambezian plateau, and depended on African economic networks, African politics, and African religion for its existence and status.« (Chennells 2007: 1-2)

While racism dominated the interpretation of Great Zimbabwe for centuries, in the realm of the British Empire the interest in and effort of claiming a white history for Great Zimbabwe was particularly strong:

»The British Association for the Advancement of Science sponsored excavations by Randall-MacIver in 1904 and by Gertrude Caton Thompson in 1930 which clearly established for Great Zimbabwe both a medieval date and that it had been constructed by an indigenous African population. Nevertheless, the writings of amateurs like Hall (1905) and Neal at the time and in more recent years by Bruwer (1965) and Gayre (1972) kept alive, unfortunately not only among fringe romantics, the notion that what was advanced in Africa's past was the work of outside invaders, merchants or metal-workers, variously derived from Phoenician, South Arabia, Israel, India and Indonesia. This was an idea that was extremely attractive to a White community that at no time amounted to even 10 per cent of the total population of Rhodesia. It served to justify their deprecation of Black capabilities and past achievements and their occupation of more than half the best agriculture land. The preservation of a White civilization ultimately retarded the discovery of Black civilization and hampered the progress of archaeology. As late as 1971, a Rhodesian Inspector of Monuments was forced to resign his position because he was unwilling to interpret the significance of Zimbabwe in terms dictated by the government of the day.« (Posnansky 1983: 347)

Accordingly, the popular interest in the Zimbabwe ruins reached its climax during the high phase of imperialism towards the end of the nineteenth century. One of the leading figures in that process was Rhodes, who dreamt of an Anglo-Saxon super empire under the supremacy of Britain.[6]

---

late as the independence of present-day Zimbabwe in 1980 (cf. Woodham 1989: 25).

6 Cf. Chapter 2.2.

In addition to being used as proof for an allegedly white African past, the ruins were also commonly associated with the biblical gold city Ophir, the source of King Solomon's wealth, ever since their first ›rediscovery‹.[7] The popularity of this belief is mirrored in an article on the latest discoveries in South Africa in the *Illustrated London News*:

»Strange stories have been told of late about the Ophir of Solomon having been discovered. The recently-opened diamond-mines of South Africa led to explorations further north, which resulted in the revelation of extensive gold-mines. Mr. Hartley, the lion-hunter, and Mr. Mauch, the German explorer, went further, and made known a more northern auriferous district. It is in the last discovered gold-field that the real Ophir is supposed to have been seen.« (*Illustrated London News*, Jan 11, 1873: 46)[8]

---

7 »The mention of Solomon [in Walmsely's *The Ruined Cities of Zululand* (1869)] introduces a new element into the argument. One of the notions influencing those who believed that outsiders had colonized southern Africa was a tradition that the region had been the site of the Biblical Ophir, a belief that can be traced back at least to Portuguese explorers of the sixteenth century.« (Tangri 1990: 294) Cf. also Scott T. Carroll (1988) and Lindy Stiebel (2001) for the development of the Solomonic myth in connection to the Zimbabwe ruins.

8 »This theory about Ophir became extremely popular in southern Africa. The Portuguese themselves launched expeditions into the hinterland to find the site, and the connection between southeast Africa and Ophir was accepted by many influential later authors. By the nineteenth century the theory was acceptable enough to influence explorers into believing they had found Ophir.« (Tangri 1990: 294) The German Carl Peters, who went on an expedition in order to prove that Great Zimbabwe had been built by white settlers, played a crucial role in the popularisation of the Ophir myth: »A representative of Reuter's Agency who met Dr. Carl Peters on his arrival from Africa yesterday has obtained details regarding the discovery of Ophir. As a result of his investigations on the subject, the explorer was convinced that he knew the position of Ophir. Dr. Peters at the time kept his information secret, but claims now that his original theory was correct, and that he has actually discovered Ophir.« (*The Derby Mercury*, Dec 27, 1899)

The first British archaeologist commissioned by Rhodes to investigate Great Zimbabwe was Bent, who, amongst others, popularised the interpretation of the Zimbabwe ruins as the remains of an ancient white settlement:

»Bent's expedition had been sponsored by Cecil Rhodes, himself quite convinced that Great Zimbabwe was of a lost civilization. The artefacts found by Mauch and Bent encouraged the formation, in 1895, of Rhodesia Ancient Ruins Ltd., a company with exclusive rights to work all ruins except Great Zimbabwe itself. The results of the orgy of pillaging that followed were brought together by a local journalist, Richard Nicklin Hall, in a periodization that had Sabaens (2000-1100 B.C.) followed by Phoenicians and a later ›decadent period‹ of miscegenation between the original builders of the walls and the local population.« (Hall 1996: 109)

The Zimbabwe ruins thus became instrumentalised within the imperial discourse as a means of legitimising white supremacy in Zimbabwe and Africa in general. In Britain, Bent's archaeological descriptions of Great Zimbabwe were followed with interest and reported on in contemporary newspapers and magazines such as *The Times*:

»According to despatches received here, Mr. Theodore Bent, who is exploring the ruins and historical remains in Mashonaland, has discovered some images and pottery at the Zimbabye ruins which are supposed to be of Phœnician workmanship. The news of the discovery has awakened great interest, and further particulars are eagerly awaited in the hope that the relics, even if not identified as the handiwork of the discoverers of the land of Ophir, will enable archæologists to decide when and by whom the Zimbabye ruins were originally built.« (*The Times*, Jul 23, 1891: 5)[9]

In 1891 Bent published his book *The Ruined Cities of Mashonaland*, an account of his excavations and exploration of the ruins. It was widely received and discussed in both Britain and South Africa: »The glamour and

9 Cf. Rebecca Stott (1989) for the mediation of the colonial image of Africa in the contemporary British press: »Popular magazines of the period, such as the *London Illustrated News*, catered for public demand with years of continual reporting of savage African life using visual images supplied by travelling artists.« (Stott 1989: 69)

fascination posed by the discovery of ancient ruins in southeast Africa captivated a sizable, contemporary literary audience in Europe. Living in a secure Victorian world, the British public experienced vicarious pleasure reading the latest exploits of adventurers in the African interior.« (Carroll 2004: 236) In the preface to the second edition of *The Ruined Cities of Mashonaland*, Bent formulates this as follows:

»No one of the many reviewers of my work has criticised adversely my archæological standpoint with regard to these South African remains: on the contrary, I continue to have letters on the subject from all sides which make me more than ever convinced that the authors of these ruins were a northern race coming from Arabia – a race which spread more extensively over the world than we have at present any conception of, a race closely akin to the Phœnician and the Egyptian, strongly commercial, and eventually developing into the more civilised races of the ancient world.« (Bent 1896: xvii-xviii)

What Bent sums up here in regard to the ruins points towards popular diffusionist theories assuming the origin of Western civilisation in ancient northern races settling in Africa. Furthermore, Bent also helped re-popularise the idea that Great Zimbabwe might be the remains of the legendary Ophir. In addition, the reception of the Zimbabwe ruins in popular texts by Bent and other explorers of the time was frequently characterised by discourses of degeneration and doom.

The three assumptions prominent in Bent's accounts on Great Zimbabwe, first, its white origin, second, its connection to the biblical Ophir, and third, it being the remains of a once powerful and flourishing high culture that was extinguished by doom and degeneration, are paradigmatic for the popular discourse on Zimbabwean archaeology in Britain at the time. This is mirrored in a variety of contemporary fictional texts dealing with the topic. Between 1880 and 1912 nine novels were published by British authors in which the Zimbabwe ruins played a prominent role. Four of these novels alone – *King Solomon's Mines* (1885), *She. A History of Adventure* (1887), *Elissa: Or, the Doom of Zimbabwe* (1899), and *Benita: An African Romance* (1906) – were published by the prolific writer Henry

Rider Haggard.[10] Due to their great popularity, Haggard's *She* and *Elissa* will form the focus of my analysis, which will show in what way they reflect dominant and recurring discourses in connection to the Zimbabwe ruins as represented in both fictional and factual texts of the time. Other fictional texts on Great Zimbabwe, published at the same time as Haggard's novels, will further inform the analysis.[11]

Central to all these texts is the fact that the Zimbabwe ruins were conceived as disturbing in the first place, since they seemed to contradict the main imperial assumption that Africa did not have a civilised history *before* its colonisation by white settlers. This is emphasised in Haggard's preface to Alexander Wilmot's book *Monomotapa (Rhodesia): Its Monuments, and Its History from the Most Ancient Times to the Present Century* (1896):[12]

»Southern and South Central Africa has been named *the country without a past.* Till within recent years its untravelled expanses were supposed from the beginning *to have harboured nothing but wild beats and black men* almost as wild, who for ages without number had pursued *their path of destruction* as they rolled southward from the human reservoir of the north, each wave of them submerging that which preceded it. Within the last thirty or forty years, however, rumours arose that this was not true, or at least was not all the truth. Baines, and other travellers now dead, reported the existence of great ruins in the territories known as Matabele and Mashona Lands, and on the banks of tributaries of the Zambesi River, which from

10 The popularity of Rider Haggard's novels is reflected by the high sale figures of his bestselling romances *King Solomon's Mines* (cf. Scheick 1991: 19; Tangri 1990: 295) and *She* (Pocock 1993: 63).

11 In addition to the novels by Haggard already mentioned, my text corpus comprises the following texts: Hugh Mulleneux Walmsley (1869) *The Ruined Cities of Zulu Land*; Ernest Glanville (1891) *The Fossicker. A Romance of Mashonaland*; A. A. Anderson and A. Wall (1891) *A Romance of N'Shabé. Being a Record of Startling Adventures in South Central Africa*; Edward Markwick (1896) *The City of Gold*; Iver McIver (1910) *An Imperial Adventure*; Gertrude Page (1910) *The Rhodesian*; Rider Haggard (1906) *Benita*.

12 Alexander Wilmot was commissioned by Cecil Rhodes after Bent to investigate Great Zimbabwe and find out more about its origin and former inhabitants. Wilmot also published the multivolumed *The History of Our Own Times in South Africa* (1899).

their construction must have been built by a race of civilised men; and in 1871 Herr Mauch re-discovered the fortress-temple of Zimbabwe, that now, as in the time of the early Portuguese, was said to be nothing less than the site of one of the ancient Ophirs.« (Wilmot 2005: xiii-xiv, emphasis mine)

Characteristically, Haggard here refers to South Central Africa, the former Rhodesia and present-day Zimbabwe, as a country »without a past« (ibid.). The fact that he is pointing out that apart from »wild beasts and black men almost as wild« (ibid.) there was nothing to be found in that region further alludes to Haggard's understanding of what ›having a past‹ actually means. For him as for Western society at the time in general, ›having a past‹ implied that a country had to be able to produce some indications of a civilised history, which, according to colonial thinking, had long not been the case for South Central Africa. »Wild beasts« (ibid.) and allegedly »wild [...] black men« (ibid.) do not count in this respect, since they did not produce any culture. Instead of being productive, progressive, and civilised – as the three main attributes associated with culture – these »wild [...] black men« (ibid.) caused nothing but destruction, explicitly stated by Haggard when he refers to the »path of destruction« (ibid.) these people pursued. By establishing this right at the beginning of his preface, Haggard makes it quite clear in regard to his further description that the ruins discovered in this area could not have been built by black men, since they clearly bear the marks of civilisation and thus »must have been built by a race of civilised men« (ibid.), that is to say, *white* men.

At the same time, Haggard's claim that South Central Africa was hitherto considered to not have had a past at all can be read as standing for the alleged blank space colonists associated with the African continent in general and its past in particular: »By summoning up the image of a *tabula rasa*, the vast, empty, pristine landscape of Africa was made receptive to the European imperial drive.« (Eaden van 2004: 25) According to this view, Africa only became civilised through the process of its colonialisation and the exploitation of its ›wilderness‹[13] and past by white settlers. As

13 The use of the term ›wilderness‹ to describe various landscapes in which the European explorers found themselves is here acknowledged as no neutral term. Wilderness usually denotes space uninhabited by humans, which is how the ex-

Haggard points out in regard to Africa's history it was only in »the last thirty or forty years« (Wilmot 2005: xiii-xiv) that the idea of an »uncivilised black past« (ibid.) had started to crumble. Significantly, this is the time span in which the Zimbabwe ruins were ›rediscovered‹ and depicted by white explorers, such as Mauch and Baines, to whom Haggard explicitly refers.[14] These people and their reports on as well as their descriptions and sketches of the ruins were responsible for the fact that not only the discovery of the ruins but also their *past* was soon associated with white people. This is underlined by the dominant interpretation of the ruins as being the remains of white settlers. Thus, the very fact that the ruins were ›rediscovered‹, explored, and interpreted by white colonialists turned them into an instrument for creating a pseudo *white* African identity. This was only possible on the grounds of the aforementioned Western concept of Africa as a blank, uncivilised space prior to its colonisation. In the aftermath of Mauch's ›rediscovery‹ of the ruins, Subsaharan Africa consequently became a space of construction and negation: »The quest to find a ›rational‹ explanation for the existence of the ruins mirrors imperial discourse's attempts to find a rationality for its own operation, but the quest is also to mystify these ruins, to inscribe them as a vacant site of indeterminacy.« (Chrisman 1990: 50) While the Western colonialists, represented by prominent figures like Cecil Rhodes, were anxious to construct a white past and thus identity for Africa in order to justify and legitimate their imperial-

---

plorers often appraised the spaces they found themselves in, thereby eradicating the inhabitants and their claim to the land.

14 John Thomas Baines (1820-1875) was a British artist and explorer who accompanied David Livingstone on his expedition along the Zambesi in 1858 to paint the Zimbabwean landscape. His book *The Gold Regions of South Eastern Africa* was published posthumously in 1877 (cf. Chennells 2007: 3-4). For a discussion of Baines, Haggard, and Great Zimbabwe, cf. Lindy Stiebel (2001). The relevance of Great Zimbabwe for the British Empire discourse is further emphasised by Stiebel: »Not only are the ancient ›white‹ ruins in Haggard's Africa a historical precedent for Britain's own presence in Africa, they also could act as a demonstration of the inevitable transience of empires, including Britain's own, together with a reminder of civilised Britain's intimate relationship with a savage Africa in which the classical world, humankind's cradle, had also once had its roots.« (Stiebel 2001: 130)

ist behaviour, the same people negated Africa's black history and thus denied the black Africans any right of self-identification with their country and its past.

In this context it is crucial to consider how these ideas were generated, communicated, and spread both in Britain and its empire. Haggard's introductory note to *Elissa* can be seen as a case in point in this respect:

»The researches of Mr. Wilmot prove to us indeed that in the Middle Ages Zimbabwe or Zimboe was the seat of a barbarous empire, whose ruler was named the Emperor of Monomotapa, also that for some years the Jesuits ministered in a Christian church built beneath the shadow of its ancient towers. But of the original purpose of those towers, and of the race that reared them, the inhabitants of mediæval Monomotapa, it is probable, knew less even than we know to-day. The labours and skilled observation of the late Mr. Theodore Bent, whose death is so great a loss to all interested in such matters, have shown almost beyond question that Zimbabwe was once an inland Phœnician city, or at the least a city whose inhabitants were of a race which practiced Phœnician customs and worshipped the Phœnician deities. Beyond this all is conjecture.« (*Elissa*: vii)

As in many other fictional and factual texts of the time, Haggard's note is characterised by intertextuality to a great extent.[15] After rendering an intro-

15 Cf. Haggard's two non-fictional texts on Great Zimbabwe: first, his introduction to Wilmot's Monomotapa, and second, his article »The Real ›King Solomon's Mines‹«, published in Cassell's Magazine in 1907: »Haggard wrote two non-fiction popular accounts of Great Zimbabwe. In both of these he stressed the similarity between the Phoenician and British empires. As he duly noted, recent work indicated that Great Zimbabwe was ›the heritage of the Anglo-Saxon race‹.« (Tangri 1990: 296) The intertextuality and merging of fact and fiction is further emphasised in *The Rhodesian*: »›I am not prepared to agree offhand that Zimbabwe is probably the ancient Havilah of the Scriptures, but I see no very good reason why it should not be. On the other hand, the ancient workings and fortifications and temples may have been the work of Phœnicians or Mongols several thousand years ago. Certainly against Mr. McIver's theory, that the Temple was the work of Bantus a few hundred years ago, I think we may put the fact that an admirable drainage system has been unearthed; – drainage systems of any kind being more or less unknown to black races of a low order. In the

ductory sentence on the ruins in general and posing the popular question as to their origin, Haggard refers to Wilmot and Bent as two prominent figures associated with the respective research and exploration of the ruins. In doing so, Haggard successfully positions his fictional text in the contemporary scientific debate. In other words, he establishes a connection between his fictional text and alleged scientific facts, which helps create a legitimating backdrop to his made-up story. The fact that Haggard himself then wrote a preface to Wilmot's *Monomotapa* is significant, since it illustrates how closely the reception and interpretation of the Zimbabwe ruins was characterised by intertextual references at the time. In regard to the creation of an alleged white African past, this is crucial in that it shows how different discourses, including fictional, scientific, and political ones, were conjoined and thus established a tightly interwoven web that developed into a constructed white history. At the same time, the close interrelation between fact and fiction in regard to the interpretation of the ruins resulted in the blending of these two realms. This also meant that out of the many different rumours, stories, and interpretations in regard to the Zimbabwe ruins there soon developed a popular story about the ruins and their origin. This popular story gradually became solidified as the history of Zimbabwe and also Africa, which corresponds to Gehrke's concept of intentional history. The process of constructing an intentional history for Africa that was taken for *real* by the contemporary reader was further fuelled by the publication of illustrated African romances as books and in magazines and newspapers, as Pascal Fischer shows in his analysis of the *Graphic* edition of *She* (1886/1887): »By shaping an impression of Africa as mysterious, fear-inspiring and hideous, Haggard's novel was fabricating a *reality* for the Victorian reader. The visual dimension of the novel plays a central role for the depth as well as the duration of that impression.« (Fischer 2007: 276, emphasis mine)[16]

---

meantime, we can but await fresh clues, which may put us upon the track of proofs, and hope that the day is not very far distant when much of the mystery will be cleared.‹« (*The Rhodesian*: 127, emphasis mine)

16 Scott T. Carroll supports the crucial role of the press in regard to the popularisation of the Solomonic myth in connection to Great Zimbabwe: »So the popularity of the Solomonic theory in the nineteenth century, to a large degree, was the

This is further underlined by Wilmot's own introduction to *Monomotapa*, where he refers to the public interest in the ruins, the colonialisation of South Africa, and the legendary Ophir of King Solomon:

»The great Zimbabwe ruins have been specially brought to public notice almost simultaneously with the conquest of the country in which they are situated, and a natural desire has arisen to read the riddle of this South African sphinx. It is desirable also to discover all possible traces of the history of the wonderful country of Monomotapa, the Ophir of King Solomon, and the land which is marked ›rich in gold‹ in maps of the sixteenth and seventeenth centuries.« (Wilmot 2005: vii)

Wilmot's introductory comment on the ruins is further significant since it points towards central aspects of the contemporary reception of Great Zimbabwe: first of all its popularity, that is, the »public notice« (ibid.), second the Empire, that is, »the conquest of the country« (ibid.), and finally the Bible, that is, »the Ophir of King Solomon« (ibid.). In this respect, Haggard's reference to Wilmot in *Elissa* is crucial as it emphasises Haggard's own opinion on the subject and further contextualises *Elissa* as a fictional text in the contemporary scientific debate. This is also confirmed by the conclusions Wilmot presents in his own book:

»First, the builders of the Zimbabwes in South-east Africa and of the Nauraghes in Sardinia were Nature worshippers of the early Phœnician cult, when stone worship was one of the leading features of that religion. Second, the arguments of MM. Perrot and Chipiez point to the fact that the Nauraghe builders came from Libya, and we shall see in due course that the intermixture of the African and Arabian peoples was of peculiar and intimate character. Third, we may venture to attribute a very remote antiquity (the Bronze Age) to both classes of buildings: certainly we can scarcely be wrong in concluding that the oldest of the Zimbabwes of South-eastern Africa were erected before the ninth century B.C.« (Ibid.: 36-37)[17]

---

result of the press encompassing both the novelist and the journalist« (Carroll 1988: 238).

17 Georges Perrot (1932-1914) was a French archaeologist, and Charles Chipiez (1835-1901) was a French architect, historian, and Egyptologist. Perrot and Chipiez published the results of their architectural studies in Greece, Turkey,

Wilmot here infers that the Zimbabwe ruins are of Phoenician origin, that there had been an »intermixture of African and Arabian peoples« (ibid.), and that the ruins could be attributed to a »remote antiquity« (ibid.). In regard to Haggard's *Elissa* this is illuminating, as Haggard identifies the inhabitants of Zimboe as Phoenician and sets the story in a very remote past. Furthermore, Haggard renders the theme of intermixture between different peoples as a central theme in his novel. He further alludes to the intermixture between different peoples when he, in his introductory note to *Elissa* following Wilmot, states that in »the Middle Ages Zimbabwe or Zimboe was the seat of a *barbarous empire*« (*Elissa*: vii, emphasis mine). Implicitly, Haggard's reference to the »barbarous empire« (ibid.) can be read as an allusion to the fact that the highly civilised city of Zimboe was eventually destroyed by ›barbaric‹ tribes, which in *Elissa* he cites as a reason for its degeneration and final downfall.

In addition, it is significant that Haggard mentions Bent in his introductory note to *Elissa* as one of the leading investigators of the ruins at the time. Although Bent himself did not commit to the popular thesis that the ruins were the remains of the biblical Ophir, he did not dismiss it either in his study *Mashonaland*:

»As to the vexed question of the land of Ophir, I do not feel that it is necessary to go into the argument for or against here. Mashonaland may have been the land of Ophir or it may not; it may have been the land of Punt or it may not; Ophir and Punt may be identical, and both situated here, or they may be both elsewhere. There is not enough evidence, as far as I can see, to build up any theory on these points which will satisfy the more critical investigation to which subjects of this kind are submitted in the present day. All that we can satisfactorily establish is that from this country the ancient Arabians got a great deal of gold; but as gold was in common use in prehistoric times, and lavishly used many centuries before our era, there is no doubt that the supply must have been enormous, and must have been obtained from more places than one [...]. The study of Arabian and Phœnician enterprise outside the Red Sea is only now in its infancy – we have only as yet enough evidence to prove its extent, and that the ruins in Mashonaland owe their origin to it.« (Bent 1896: 228-229)

Iran, and Egypt in their book *Histoire de l'art dans l'Antiquité, Égypte, Assyrie, Perse, Asie mineure, Grèce, Etrurie, Rome* (1882).

Significantly, although Bent refused to definitively identify the Zimbabwe ruins as the remains of the legendary Ophir or Punt, he did not deny the possibility and discussed it extensively.[18] This meant that even though Bent did not come to any conclusion in regard to the »vexed question of the land of Ophir« (ibid.), he still called the reader's attention to the question and left it to her to decide for herself what to believe. However, the very fact that he mentions Ophir and Punt in relation to the ruins is significant in regard to the popularisation of Great Zimbabwe, since as delineated above, the line between fact and fiction became increasingly blurred in the process of this popularisation. As Carroll notes:

»This notion so stirred the readers' imagination and at the same time aligned itself with their romantic appreciation for African exploration that they soon forgot that it was fiction. [...] By the end of the nineteenth century, the Solomonic origin of the Great Zimbabwe had been accepted *carte blanche*. Fiction and popular notion had been mistaken for fact. Undaunted by the question of truth, the Victorians seemed the more eager to adopt Solomon as a precedent to attempt to justify their imperialistic designs in Africa, if not simply to assuage their conscience. Rhodes and company would exploit Africa after the legendary manner of the Hebrew king (and with twice his splendor).« (Carroll 1988: 236-237)

In other words, because the popular interpretation of the Zimbabwe ruins was characterised by a blending of fact and fiction anyway, forming an intentional history, statements such as those made by Bent fostered popular theories on the white origin of the ruins, as emphasised in Hugh Mulleneux Walmsley's novel *The Ruined Cities of Zulu Land*:[19]

---

18 Martin Hall comes to a similar conclusion in his assessment of Bent's work: »Bent's [*The Ruined Cities of Mashonaland*] initiated the ambivalent combination of collision and collusion with popular mythology that was to characterize subsequent archaeological work at Great Zimbabwe. [...] But although Bent dismissed King Solomon and the Queen Sheba from his stage, he retained and strengthened the structure of the mythology that had sustained them for so many years.« (Hall 1995: 188-189)

19 Tangri points out that »[i]n sum, Walmsley's book reflects two major views about southern African antiquities: that local black people were too primitive to

»Here, then, was a confirmation of all his theories. Here, the vast ruins among the gold fields of King Solomon; here the source of the Sabe, or Golden River, down whose stream the boats of bygone days floated gold, cedar-wood, and precious stones. An Englishman's first impulse at once seized on Hughes, and yielding to it, the two exchanged a vigorous shake of hand.« (*Ruined Cities*, Vol. I: 166)

In regard to the popular reception of Great Zimbabwe, both Haggard's preface to *Monomotapa* and his introductory note to *Elissa* are paradigmatic. First, these texts underline the disturbing quality of the ruins for the colonial discourse as such, in that they stand for a highly advanced African past centuries *before* the continent's colonisation by white settlers. Second, both texts show how the white imperialists dealt with this unsettling recognition by negating an indigenous history for Great Zimbabwe and constructing a white history by spinning their own story around the ruins.

My analysis of Zimbabwean archaeology consequently focuses on the reception of Great Zimbabwe as a confounder for the imperial discourse in fictional texts of the time: »The discovery of ancient stone-walled cities and gold mines in Africa posed a problem for these were unknown in comparable European Iron Age sites – hence the theory that other, European races must have built them in some far distant age.« (Stiebel 2001: 128) This popular practice of dealing with the ruins can be described as a process of assimilation. The colonisers' initial confrontation with the ruins through their reappearance in the popular discourse in the form of reports, rumours, and stories after their rediscovery by Mauch in 1871 has to be seen as a deeply disturbing and traumatic experience causing feelings of threat and insecurity. All of a sudden, the ruins emerged from the alleged ›darkness‹ of the African continent, hitherto conceived as a space without a history, and stood out as disproof of the colonisers' original conception of Africa as a blank space. In an attempt to retrench the disturbing reality of the ruins discovered, the colonisers identified with them on the grounds of a constructed collective and individual affiliation with the ›other‹, that is, the ruins. Rather than having the desired comforting effect on the colonisers, however, this identification with the ›other‹ in the form of the ruins culmin-

have produced them, and that Ophir may have lain in southern Africa« (Tangri 1990: 294).

ates in an over-identification with them and all they represented, eventually resulting in fears of degeneration and doom:

»All the recent work on what is called colonial discourse and on the relation of the West and the East has started from the binary distinction, however much it may come to be complicated, between the western self and its exotic other or oriental other, in which the other is conceived in terms of the projection of whatever in western psyche is an object of disgust and terror: the East as the embodiment of the western fears.« (Barrell 1991: 99)

This projection constitutes the final step in the process of assimilation, which ironically leaves the colonisers with an even deeper feeling of threat and insecurity than was the case before the assimilation process.

PART II: Popular Receptions of Archaeology

Popular Receptions of Archaeology

In the following analysis, I focus on the different representations of Greek, Egyptian, and Zimbabwean archaeology as popularly reflected in the reception of the self-made archaeologist Heinrich Schliemann, the figure of the mummy, and the ruin complex Great Zimbabwe in the context of nineteenth- and early twentieth-century negotiations of individual and collective identities at a time of profound sociocultural changes in Britain. In Chapter 4 the archaeological space is introduced as an indefinite realm of ambivalence that had an equally fascinating and repelling effect on British contemporaries. Chapter 5 focuses on the (re)creation and performance of individual and collective identities by the example of Schliemann's excavations, which, evoking ancient Greece as the *most familiar strangeness*, can be seen as operating as a prosthetic memory for the masses in Britain. In Chapter 6 the emphasis is on the inversion and subversion of identities in the reception of Egyptian archaeology as what I refer to as the *less familiar strangeness*. Chapter 7 analyses the negotiation of the more alien side of Egyptian archaeology and the Zimbabwe ruins as the *most unfamiliar strangeness*. Chapter 6 and 7 trace a gradual shift prominent in contemporary perceptions of the past from fears of reverse colonisation to a radical negation of the past materialising in the process of constructing an intentional history. After each of the four main chapters, I present a summary of the interim findings to finally discuss the central points of my analysis in the conclusion of the book.

4. Archaeology as a Space of Ambivalence

›I speak to you now with my feet upon the threshold of the other world.‹
(*RING OF THOTH*: 152)

Archæological exploration is not free from its own risk.
(*ACADEMY*, NOV 18, 1882: 361)

In the present chapter, I introduce the archaeological space as represented in popular fictional and factual texts as the sphere of the temporal and spatial ›other‹, forming a counter-discourse to contemporary master discourses. I show how this space was experienced by contemporaries as indefinite and ambivalent due to its lack of discursive structure, oscillating between the familiar and the unfamiliar, attraction and disgust, desire and fear, essentially representing Said's *Orientalism*. I further demonstrate that the archaeological space the Western explorers are confronted with in the search for the Zimbabwe ruins might in its unfamiliarity be read as standing for a dual ›otherness‹: On the one hand it represents the spatial and temporal ›other‹, and on the other it symbolises the psychological ›other‹, the subconscious of the Western explorers. Against this background, archaeological exploration thus becomes the exploration of the unfamiliar side of both the self and the ›other‹, mirroring what I refer to as the *Jekyll and Hyde Discourse*. At the same time it underlines how the archaeological search and the search for identity are closely intertwined. Analysing the representation of archaeology in Zimbabwean fiction and reports on Schliemann's excavations, the first chapter illustrates how the ambivalent experience of the Western explorers in regard to the archaeological space materialises itself in the anthropomorphisation and feminisation of the archaeological landscape. The second chapter delineates how in Bram Stoker's *Jewel of Seven Stars* the characters are confronted with the ambivalence of the archaeological space *in* Britain in the form of mysterious nightly attacks and a diffuse smell caused by and attributed to an Egyptian mummy, respectively. In the third chapter I demonstrate in regard to Zimbabwean fiction how the Western explorer applied what I term an ›inversed empiricism‹ represented by the (geographic) map to the archaeological space in the attempt to import contemporary British discourses, representing the spatial and temporal known, into the unknown. It is this endeavour to structure and categorise the archaeological space which marks the attempt to create meaning and eventually identity within that ambivalent space, as I will show in Chapter 5 in regard to the reception of Heinrich Schliemann.

4.1 Penetrating the Darkness: Experiencing the Unknown

> But still it was a blessing that [the country] bestowed, a potent charm, in spite of, nay, because of, the peril, – that magic charm of penetrating the unknown, of gaining privileged access to the privacies of the closely-veiled, of enjoying the rare confidences of jealously-withdrawn.
> (*Imperial Adventure*: 25)

> It was an ambitious project, and one which promised a good meed of excitement and danger – excitement from the chase of big game, and danger from the unknown tribes whose territories would be invaded, for the first time in all probability, by a white man.
> (*N'Shabé*: 1)

In his article »Schliemann's ›Dream of Troy‹«, David Traill concludes, referring to Schliemann's lifelong[1] ambition to excavate ancient Troy, that not only was »Schliemann's childhood dream of excavating Troy [...] a *romantic fabrication* but that the motives for this fabrication were less innocent than might have been supposed« (Traill 1985: 23, emphasis mine). This »romantic fabrication« (ibid.) might be seen as paradigmatic for the archaeological venture in both fictional and factual texts at the time in general in that it might be read as being motivated by the Lacanian desire to regain wholeness as experienced by the individual in the realm of the Real prior to the entrance into the symbolic state. When one understands this

1 As various critics have shown, Schliemann's story of his alleged ›lifelong‹ dream of excavating Troy was made up by him only after he had discovered what he claimed to be the ancient remains of Troy.

Lacanian desire as the driving force of Storey's concept of identity formation, the archaeological remains in question become sites of individual and collective memory representing both the past, anchored in the concepts of Halbwachsian memory, and the future never to be attained, that is, the desire to regain wholeness. In this context, the ›romantic fabrication‹ combines both the familiar in the form of the here and now as the origin, that is, reality, of the dream and the ›other‹ in the form of a phantasmatic projection onto the spatial and temporal unknown, that is, fantasy. The archaeological quest thus becomes the departure into the realm of the ›other‹ in an attempt to (re)discover a more complete self in terms of Storey's concept of identity formation, one which contemporary Victorians felt that they had lost in the process of industrialisation and urbanisation. The journey into the past through the archaeological sites consequently symbolises the return to the ›roots‹ (cf. Storey 2008: 81-86) in an attempt to regain completeness of the self in the future. The confrontation with this original and unknown part of the self is what I identitfy as the *Jekyll and Hyde Discourse*.

The close link between archaeological exploration and the exploration of the Western self is particularly prominent in Haggard's *She*, where an African expedition becomes the confrontation with a multidimensional ›other‹.[2] From the beginning of the novel the story is introduced as an »extraordinary history« (*She*: 11) of an African exploration presented as an embedded narrative by an anonymous editor, who is later revealed to be a friend of the narrator, the Cambridge professor Horace Holly. Quite typically for these kinds of texts, the reader is promptly informed by the fictional editor of the story's mysterious character:[3] »In giving the world the record of what, looked at as an adventure only, is I suppose one of the *most wonderful and mysterious experiences ever undergone by mortal men*, I feel it incumbent on me to explain what my exact connection with it is.« (Ibid., emphasis mine) What follows is the description of how the eccentric Holly came to be the warden of Leo Vincey, the orphaned son of a friend, and how, according to the instructions of Leo's late father, on Leo's twenty-fifth birthday they open a chest containing an antique potsherd, the Sherd of Amenartas, and a letter by Leo's father. Through the sherd and the letter the men learn of Vincey's line of descent, which reaches back to

2 Cf. also Malley 1996a and 1997.

3 Note that many of the mummy texts are introduced in a similar way.

ancient Egypt, and a mysterious white queen living in the African ›wilderness‹. Even though Holly, Leo, and their butler Job at first entertain the characteristic doubts as to whether the story told by Vincey and the potsherd about Leo's ancestors and the mysterious immortal white queen can be trusted,[4] they eventually decide to leave Britain for Africa to »begin [their] search for the ruined city and the Fire of Life« (ibid.: 56), which is supposed to be the secret of the queen's immortality. In the further course of the novel, the realisation of the »quest« (ibid.: 27) into the unknown takes place, with which Vincey had challenged Holly shortly before Vincey's death: »The circumstances were almost uncanny, so much so that, though I am by no means nervous, or apt to be alarmed at anything that may seem to cross the bounds of the natural, I grew afraid, and began to wish I had had nothing to do with it.« (Ibid.) Holly's fear and his awareness that their venture means to »cross the bounds of the natural« (ibid.) is crucial in this context as it marks the trespassing from the space of the ordinary and familiar into that of the extraordinary and unfamiliar characteristic of the archaeological discourse. That the urge of the Western explorer to trespass on and explore the African landscape is more than just an imperial venture is pointed out by Etherington, who reads this encounter as a symbolical exploration of the temporal, cultural and psychological ›other‹ underlying modern civilisation:

»Haggard's adventurers proceed from present-day Europe towards an encounter with the past in Africa. *The past they encounter is their own past.* Centuries, even millennia, are stripped away in the course of each quest. Significantly, Haggard's Africa teems with the ruins of white civilizations, quite unlike real Africa, which, apart from the Nile Valley and Zimbabwe, is singularly bare of monumental ruins. Under Africa's spell, English gentlemen regress and become Vikings; Cambridge rowers become ancient Greeks.« (Etherington 1978: 84, emphasis mine)[5]

---

4 Holly comments on Vincey's letter as follows: »›I believe that the whole thing is the most unmitigated rubbish. I know that there are curious things and forces in nature which we rarely meet with, and, when we do meet them, cannot understand.‹« (*She*: 54) Holly's doubts in regard to the reliability of Vincey and his story can be seen as characteristic of the archaeological romance in general.

5 Cf. also Malley (1996a: 175; 1997: 275-97). »In a recent essay on *She* and archaeology, Shawn Malley argues that archaeology is essentially hermeneutic

The journey into Africa and the search for the ancient ruins thus becomes the journey not only into the spatial and temporal but also into the psychological and cultural ›other‹.[6] In other words, the penetration into the unknown becomes the penetration into the ›darkness‹ of human civilisation and subconsciousness at the same time and thus stands for both an individual and collective self-exploration: »The white man must explore and penetrate this foreign territory, but he must also resist it or be threatened with

[...]. It deals with a search for origins, and constructs an evolutionary sequence which establishes material continuity to the development of civilization. For him, this is also the pattern of *She*. Malley associates Haggard with the anthropology of his friend Andrew Lang; his lines are familiar ones. As Rebecca Stott notes, while suggesting an anthropological reading of Haggard, ›The anthropologists were to tailor and adapt evolutionary theory to a more accessible and less troubling shape for the general public. Anthropology both played to (and played out) the widespread alarms caused by the ›discovery‹ and common ancestor of man and the ape alleviated such fears by confirming at every point the evolutionary superiority of the white races.‹« (Pearson 2000: 221-222)

6 The metaphorical connection between the past in general, archaeology in particular, and the human psyche has already been emphasised by Sigmund Freud (cf. Freud's analysis of the novel *Gradiva: A Pompeian Fancy* by the German author Wilhelm Jensen, published in *Delusion and Dream in Wilhelm Jensen's* Gradiva (1907)) and discussed by various critics (cf. Pollock 2006: 2). As Pollock points out, psychoanalysis corresponds to archaeology, since »Freud's analytical theatre, full of objects and casts, stand [sic] for the shattered, incomplete and repressed histories, no longer available in their original unity or vitality. Instead, each item is marked by both oblivion and anamnesis, exemplifying in material form the shards of memory and fantasies that analytical sessions will conjure up in the transferential presence/present of the analyst with whose partnership, some hermeneutic sense of these discontinuous fragments may be rewoven into a tissue of shifting, subjective meaning. In that space of induced reverie and daydream, the adult is invited to fall back into an active relationship with what Freud considered the subject's infantile ›prehistory‹ and to excavate his/her subjectivity in its sedimented layers that are also interleaved temporal strata, resurfacing across the immediate plea of language and the ›speaking‹, fantasizing body« (ibid.: 5). Cf. also Kuspit (1989); Bowdler (1996); Pollock (2007a), (2007b); Orrells (2012).

absorption into otherness: cultural otherness and sexual otherness.« (Stott 1989: 77)

Unlike Shawn Malley, who sees this as specific of Haggard's texts (cf. Malley 1996: 175), I argue that this applies to the majority of texts dealing with archaeology, and in particular with the Zimbabwe ruins. Africa, the fathomless continent, with its ruins that first had to be discovered before they could be explored, functioned more than any other continent as a mirror and projection screen for the collective and individual Western psyche and culture, as noted by Brantlinger: »So the Dark Continent turned into a mirror, on one level reflecting what the Victorians wanted to see – heroic and saintly self-images – but on another, casting the ghostly shadows of guilt and regression.« (Brantlinger 1985: 198) More than Egyptian and Greek antiquities, which for various reasons were usually easier to access, Zimbabwean antiquities were often conceived as mysterious and portrayed as dark and fathomless, thus functioning as the *most unfamiliar strangeness*.[7] In this context it is the ›other‹ that has to be explored and civilised, while the self has to be rediscovered, completed, and cultivated, as underlined by the following comment by Holly from *She*:

»›[A] country like Africa […] is sure to be full of the relics of long dead and forgotten civilisations. Nobody knows the age of the Egyptian civilisation, and very likely it had offshots. Then there were the Babylonians and the Phœnicians, and the Persians and all manner of people, all more or less civilised, to say nothing of the Jews whom everybody ›wants‹ nowadays. It is possible that they, or any one of them, may have had colonies or trading stations about here. Remember those buried Persian cities that the consul showed us at Kilwa.‹« (*She*: 70)

Inspired by the ruins the men encounter upon landing on the East African shore, Holly here comments on what they see. Significantly, it is in this unknown territory[8] that Holly, after having survived the first adventure of their expedition, discovers the remains of a »long dead and forgotten civil-

7 Egyptian and Greek antiquities were for the most part to be found quite easily, whereas the exact location of the Zimbabwe ruins, although there existed various myths in regard to this, remained unknown until Mauch rediscovered the ruins in 1871.

8 Note that no clear information is provided about where exactly the men are.

isation« (*She*: 70), which he loosely connects to various ancient, albeit notably materially ›developed‹ civilisations, namely the Egyptians, Babylonians, Phœnicians, Persians, and Jews. This search for and anxiety to (re)discover ancient civilisations that are seen as ancestors of Western civilisations in the unknown ›wilderness‹ of Africa underscores the idea that the ›other‹ that is being explored is somehow part of one's own, and more complete, self (cf. Etherington 1978: 84).[9] The simultaneous danger associated with this exploration of the unknown is underscored by a passage from Walmsley's *The Ruined Cities of Zululand* (1869), which was one of the first British texts to deal with the Zimbabwe ruins, with the following exchange:

»›Yes! will you join me in the search?‹ replied the missionary, eagerly, pausing for a while as the others looked moodily into the embers without replying; and then continuing ›[…] There are dangers and difficulty in the path, but it is one which has never been trodden by European foot. Up to the present moment all efforts made to penetrate the country have failed, and the old temples and palaces of a once glorious race, if indeed they do exist, serve as a den for beasts of prey, or a refuge for the hardly less savage Kaffir.‹« (*Ruined Cities* Vol. I: 49)

Here, the danger of »penetrat[ing]« the unknown, which hitherto »has never been trodden by European foot« (ibid.), can first of all be read as pointing towards very concrete threads such as hostile indigenous peoples, wild animals, diseases, and so forth. In addition to that, however, it can also be read as alluding to the less obvious but nevertheless existential dangers of being confronted first with an unknown side of the self *and* second with the incompleteness of the self of the symbolic state in the Lacanian sense. On a more abstract level, this disturbing association of the ruins is supported by the fact that »the temples and palaces of a once glorious race« (ibid.) are now inhabited by wild animals. This clearly evokes latent fears of degeneration, which alludes to an ultimate identification of the present incomplete self with the ruins as opposed to a former self characterised by perfect wholeness of the state of the Real in the Lacanian sense. The ruins are thus

9 In *She* this is clearly demonstrated by the sherd of Amenartas, which traces Leo's family history back through time and space and thus manages to create a continuity between ancient Egypt and nineteenth-century Britain.

no more than a reminder of the incomplete self of the beholding subject, while they evoke irretrievable past grandeur symbolised by »the temples and palaces of a once glorious race« (ibid.).[10] Additionally, the trajectory of exploration often parallels in form and expression the sexualised notions of conquest. This reification of the land to be explored in feminine terms displays the dialectic of incomplete subjects. To better foreground such language, I cast the net further in the next section and draw on other sites of feminisation in the archaeological process of the time.

The perception of archaeological sites as sites promising a potential retrieval of wholeness is further supported by their frequent association with the female body and more precisely with both that of the mother *and* the lover, as illustrated in regard to Schliemann's archaeological discoveries. Reporting on Sophia and Heinrich Schliemann's travels in Asia Minor in 1877, *The Times* published the following account:

»But going with Mrs. Schliemann, from Nauplia to Mycenæ, they found it impossible to pass Tiryns [...] without stopping a week to explore it, its huge Cyclopean walls deemed by the Greeks themselves as the work of the demons, and more stupendous than the Pyramids of Egypt, bound them with a spell the more resistless from the fact that the pickaxe of no explorer had ever touched its virgin soil.« (*The Times*, Mar 23, 1877: 10)

The archaeological landscape depicted here is essentially characterised by ambivalence, imbuing the whole scene with a sense of mystery that makes it impossible for the Western explorer to pass by. This ambivalence is created by the juxtaposition of the »huge [...] walls« (ibid.), associated with »demons« (ibid.) and thus suggesting an uncanny power, with the »virgin soil« (ibid.), evoking an innocent vulnerability. It is exactly this ambivalent meaning of power and vulnerability conveyed by the archaeological site that deeply fascinates the explorer. The very fact that the site is said to cast a »spell« (ibid.) on the explorer, in combination with the reference to its »virgin soil« (ibid.), both anthropomorphises and sexualises it. This is further underscored by the Schliemanns' reaction to the site, as »they found it impossible to pass Tiryns [...] without stopping a week to explore it« (ibid.). The sexual connotation of the alluring female making it impossible

10 Cf. Chapter 7.6.

for the (male) beholder to pass by without ›exploring‹ her evoked here is further enforced by the information that the archaeological site has »bound them with a spell *the more resistless from the fact that the pickaxe of no explorer had ever touched its virgin soil*« (ibid., emphasis mine).

The ancient site thus becomes the luring virgin that casts her spell on the explorer, waiting to be taken by him for the first time, while the Western archaeological pickaxe becomes the penetrating phallus, conquering spheres that have never been touched before. This is further underlined by Schliemann's repeatedly proclaimed conviction that archaeological sites were *waiting for him to explore them*, as pointed out by Döhl: »Schliemann sometimes had a very peculiar way of expressing himself. He wrote several times that a particular place urgently expected him, Schliemann, with hoe and spade. Hoe and spade for Schliemann were the instruments with which to reveal history.« (Döhl 1986: 103)[11]

The consequence of the archaeologist's penetration of the virgin soil is that after months of labour (pain) the archaeologist is finally able to bring to light what has hitherto been contained by Mother Earth. The metaphorical parallels to the act of giving birth are further underlined by the repeated use of the phrase *to bring to light* in connection with Schliemann's excavations.[12] The archaeological excavation here becomes a creational act by breaking out the ancient relics *to expose them to the light of day*, and the archaeological spade functions as a means of establishing meaning and identity by revealing past realities to the daylight in an attempt to regain wholeness where before only fragmentation or nothing at all existed: »Unlike the ruins of the classical period, which were mostly visible at the surface, prehistoric remains are totally hidden beneath the earth. Thus, the archaeologist must determine whether remains are in fact concealed at any given spot.« (Ibid.: 105)

The association of the archaeological site with the female body is also prominent in texts dealing with the Zimbabwe ruins, where, as in Haggard's *She*, it is frequently considered the quest of the male explorer to

---

11 Cf. also Franz Georg Maier (1992) *Von Winckelmann zu Schliemann – Archäologie als Eroberungswissenschaft des 19. Jahrhunderts*.

12 Cf. *Athenæum*, Mar 10, 1877: 327; *Athenæum*, Aug 21, 1880: 245; *Glasgow Herald*, Dec 15, 1876.

conquer the feminised African continent and its past to spread Western (male) civilisation:

»In such a narrative the landscape plays the role of the virginal female body waiting to be taken. What is more interesting is that the female body-to-be taken in this narrative (the land) is based upon Western mythical and stereotypical models of femininity which are often deeply problematic and contradictory in themselves.« (Stott 2001: 151)[13]

What Stott points out here in regard to the African landscape can also be confirmed for the African past in the form of the Zimbabwe ruins in that these are often symbolically equated with the female virgin waiting to be mated: »The normal gendering processes of the colonialist imaginary often proceed on an implied equation between feminisation and subjugation, casting the land as a feminine space to be ›husbanded‹ by the incoming colonist: Raleigh named his colony Virginia.« (Hopkins 1996: 8) At the same time, however, Africa is not solely a virgin. As its ancient ruins representing its old age suggest, Africa also symbolises a mother archetype frequently assuming the function of a founding myth for the Western explorer. This at first sight contradictory dual nature of Africa, and particularly Zimbabwe with its ancient ruins, as a mother *and* virgin lover, representing both present and past, is paradigmatic for the perception of the archaeological landscape in the nineteenth century.

In *She*, this dual perception of archaeology is made explicit by the figure of Ayesha,[14] who rules the ancient necropolis of Kôr:

»There were, however, [Ustane] informed us, mounds of masonry and many pillars near the place where *She* lived, which was called Kôr, and which the wise said had once been houses wherein men lived, and it was suggested that they were descended from these men. No one, however, dared go near these great ruins, because they were haunted: they only looked on them from a distance. Other similar ruins were to be seen, she had heard, in various parts of the country, that is, wherever one of the

13 Cf. also Sandra Gilbert (1983) »Rider Haggard's Heart of Darkness« and Catherine L. Innes (1994) »Virgin Territories and Motherlands: Colonial and Nationalist Representations of Africa and Ireland«.

14 Ayesha is also referred to as ›She‹ in the novel.

mountains rose above the level of the swamp. Also the caves in which they lived had been hollowed out of the rocks by men, perhaps the same who built the cities. [...] They had a Queen, however. *She* was their Queen, but she was very rarely seen, perhaps once in two or three years, when she came forth to pass sentence on some offenders, and when seen was muffled up in a big cloak, so that nobody could look upon her face.« (*She*: 95-96)

As Holly and his companions are informed here by Ustane, one of Ayesha's subjects, the powerful Queen Ayesha inhabits the ancient ruins, a fact which immediately links her with the past and old age. This is later enforced when Ayesha talks to Holly about her past, alluding to her old age:

»›Oh, tell me of the philosophy of that Hebrew Messiah, *who came after me*, and whom thou sayest doth now rule Rome, and Greece, and Egypt, and the barbarians beyond‹ [...]. I had recovered myself a little by now, and, feeling bitterly ashamed of the weakness into which I had been betrayed, I did my best to expound to her the doctrines of Christianity, to which, however, with the single exception of our conception of Heaven and Hell, I found that she paid but faint attention, her interest being all directed towards the Man who taught them.« (Ibid.: 194-195, emphasis mine)

In this passage, which is only one of several in the novel referring to Ayesha's old age, Ayesha is associated with a past that lies *before* what is traditionally conceived as the origin of contemporary Western civilisation, as the reference to the beginning of Christianity suggests. In addition, material proof for Ayesha's immense old age is introduced when Holly perceives Ayesha's footprints in the solid rock of her cave: »I do not think that anything that I had heard or seen brought home to my limited understanding so clear a sense of this being's overwhelming antiquity as that hard rock hollowed out by her soft white feet.« (*She*: 234) Besides the repeated emphasis of her old age, however, Ayesha is also presented in terms of a beautiful *young* woman:

»She lifted her white and rounded arms – never had I seen such arms before – and slowly, very slowly, withdrew some fastening beneath her hair. Then all of a sudden the long, corpse-like wrappings fell from her to the ground, and my eyes travelled up her form, now only robed in a garb of clinging white that did but serve to show its perfect and imperial shape, instinct with a life that was more than life, and with a certain serpent-like grace that was more than human. [...] I gazed above [her arms] at her face, and – I do not exaggerate – shrank back blinded and amazed. I have heard of the beauty of celestial beings, now I saw it.« (Ibid.: 158)

The way Holly describes Ayesha's appearance, which he eventually sums up as the »beauty of celestial beings« (ibid.), makes explicit that Ayesha does not only represent antiquity, that is, both the past and old age, but also the beauty of a virgin.[15] In analogy to the African landscape and the ruins, Ayesha thus embodies both old age *and* youth, mother *and* lover at the same time. This ambivalent role turns her into the woman seemingly to be consumed and the woman that creates at the same time,[16] which is further supported by what Young has pointed out in regard to Haggard's romances in general: Haggard's »narratives frequently involve a journey towards a womb-like space where the protagonists feel engulfed and threatened with annihilation« (Young 2005: 25). The »womb-like space« (ibid.) referred to here inevitably evokes association with the archaeological site as both the creative, fecund mother and the consuming lover.

Taking this into consideration, Ayesha can be seen in this context as a representation of archaeology in general, in that the ancient archaeological sites are frequently associated with the female and more precisely with the creational mother figure, the virgin lover, or both. The (male) archaeologist encounters the ruin as something that is immensely old – representing the mother – but that has at the same time never been touched (by Western

15 Although Ayesha's evil character is already alluded to in this passage with the serpentine description, part of her also represents the beauty and innocence of a young woman. Sandra Gilbert points out that Aysha »[u]like the women earlier Victorian writers had idealzed or excoriated, [...] was an odd but significant blend of the two types – an angelically chaste woman with monstrous powers, a monstrously passionate woman with angelical charms. [...] She was in certain important ways an entirely New Woman: the all-knowing, all-powerful ruler of a matriarchal society« (Gilbert 1983: 125).

16 However, the reader also learns that Ayesha is not a virgin, as she has seduced various men in the past.

men) before – thus representing the virgin lover.[17] This ambivalence is once more summed up by Holly's description of Ayesha's face: »Though the face before me was that of a young woman of certainly not more than thirty years, in perfect health, and the first flush of ripened beauty, yet it had stamped upon it a look of unutterable experience, and of deep acquaintance with grief and passion.« (*She*: 159)

This dual nature of the archaeological site also indicates that in a very similar way to the archaeologist's relationship to the mummy the ancient African ruins are both objectified *and* subjectified through the confrontation with the archaeologist/explorer (cf. Chapter 6.3). On the one hand they symbolise the *passive* female that is waiting to be taken by the explorer, while on the other hand they stand for the *active* and creational mother. As in most of the mummy texts in which the mummy first has to be resuscitated by modern men, the ancient sites have to be ›reanimated‹ by the explorer/archaeologist before they can unfold their powers as sites of memory.

The creational power of the ancient archaeological sites is paradigmatically underscored in McIver's *An Imperial Adventure* (1910):

»In new lands as yet but half settled, where Nature can still work untrammelled by brick and mortar and iron rails, her skies unsullied by smoke, she exercises an incalculable influence over all who come within reach of her. Though they may have no conscious knowledge of her, yet they are what she chooses to make them, and by each fervid noon, each sunset and sunrise, and by each placid night, she fashions their souls as she fashions the hills and the rivers. The dwellers in an artificial civilisation may be as strangers to her maternal bosom, but those who make their abiding-place at her tent doors are indeed as her sons, drawing their life directly from her into whose lap they shall empty it again in a few short years, projecting no thought of which she is not the inspirer, confidante, and guide, subdued to her commands, akin to the soil, to the hills and vales about them which she has moulded. Nowhere is this more manifest than in those lands south of that ancient river on whose waters the vessels of Solomon may once have floated. It is she, here Africa the Sphinx, who

17 This is emphasised by the many texts in which the ancient cities are inhabited by powerful white women, symbolising both the ancient first mother, functioning as a founding myth, and the virginal lover (*City of Gold*: Nazir ul Mālaho; *A Romance of N'Shabé*: Queen of N'Shabé and Princess Etellé; *She*: Ayesha).

has made the Boer what he is, who will fashion the Englishman in like manner when he strays into her purlieus.« (*Imperial Adventure*: 83)

Here the archaeological landscape functions as a site of identity creation. As a synecdoche of Africa, Mashonaland is personified as the mother of the Western explorer, offering maternal shelter and nutrition in the most pure and original way. It becomes the source and the driving life-force for Western explorers, the beginning and the end concurrently, a place of origin and creation as well as dissolution, as underlined by the fact that the explorers are »drawing their life directly from her into whose lap they shall empty it again in a few short years« (ibid.).[18] The African landscape is thus attributed an organic and dynamic power that is both soothing and threatening at the same time.[19] The reference to »the ancient rivers on whose waters the vessels of Solomon may once have floated« (ibid.) and the metaphor of »Africa the Sphinx« (ibid.) further bring into play time as a determining factor. The old age of Africa is set into sharp contrast with the short life of the Western explorers by the allusion to their mortality, as they – unlike the ›immortal‹ continent – are bound to die in a »few short years« (ibid.). The fathomless age of Mashonaland and its ruins, with the power resulting from this old age, is further implied in Anderson and Wall's *A Romance of*

18 Cf. also: »And the silence, the silence unruffled in its oceanic profundity, was no longer a fell harpy merciless and fatal destruction, no longer a bird of prey feeding upon his heart. It was a couchant lion with a woman's face that brooded eternally and unalterably upon the dark mysteries of the earth and this strange continent. It gathered itself together and gazed at him through the wide, questioning, melancholy eyes of Sheba, – of Sheba, Ophirean Queen, of Africa herself. It was the spirit of Africa that met him here face to face; it was Africa, why, it was his mother Africa, who had brought him forth and bound him about with the infrangible bonds of love, upon whose heart he now lay! It was his mother, his mother!« (*Imperial Adventure*: 194)

19 Cf. also Hinz's reading of the name Kôr in this context: »The name of the ›fallen city‹ finally is Kôr, which whether read backwards or forwards, phonetically or figuratively connotes the fixed centre, the axis mundi. The journey to Kôr, therefore, takes the form of the basic mythological movement – *the return to the beginning*.« (Hinz 1972: 420, emphasis mine)

N'Shabé (1891), where a group of explorers is confronted with the African landscape:

»These lonely solitudes, which were grown forests before the Pyramids were built or ancient Nineveh existed, exercise a peculiar influence over the mind. While calling back memories of our own past, they tend to soothe the harsh asperities of life and charm the brain into peaceful repose. The disagreements of yesterday vanished before the lulling power of these enchanted forests, and only a sensation of rest remained. This is Nature's opium – health-giving, not soul-destroying.« (*N'Shabé*: 145)

Significantly, as in *An Imperial Adventure*, where Africa is referred to as the »Sphinx« (*Imperial*: 83), Mashonaland is here again depicted in relation to the ancient monuments of Egypt and Nineveh. This underlines how Africa and particularly Mashonaland is closely associated with the interpretations of the past in general, even if its ancient monuments are not explicitly referred to. The sense of past evoked by Mashonaland, however, is not restricted to the African landscape but becomes universal, as the memories evoked of »our own past« (*N'Shabé*: 145) make explicit. Instead of having an upsetting influence on the Western mind, the landscape here soothes it and has a therapeutic »health-giving« (ibid.) as opposed to a »soul-destroying« (ibid.) effect on the beholder. The African landscape can thus also have a healing power over its observer by resolving »the disagreements of yesterday« (ibid.) through its drug-like power. It creates the illusion that wholeness can be regained and that the »disagreements of yesterday« (ibid.), evoking the traumatic cessation of the state of the Real, can be smoothed out.

As underlined by these examples through their triple function as sites of creation (mother), sites of consummation (lover), and sites of memory (mother archetype), evoking both a collective and an individual past, the archaeological sites become sites of identity formation representing what Storey conceives as the determining parameters of the latter. Through their symbolic function as the creational mother archetype they represent the Lacanian wholeness and the desire to return to the state of the Real, while their association with the female lover refers to sexual/romantic love as a substitute for that state of wholeness never to be attained. Furthermore, the archaeological sites' status as sites of memory underscores the function of memory conceived by Storey as the ›root‹ of identity formation. In other words, these different connotations of the archaeological sites emphasise their potential to become sites of identity formation by seemingly providing what cannot be found elsewhere. The archaeological sites on the one hand allegedly provide the founding myth for the Western explorer as the mother archetype, while on the other they fuel the Lacanian desire to regain wholeness as the mother *and* the lover figure. As sites of memory, creation, and consummation they thus pose as projection screens for the cravings of the Western mind.

It is this inherent ambivalence, seemingly promising the fulfilment of the Western explorer's longings, that imbues the ruins and their surroundings with a deep sense of mystery and renders them so attractive to the Western explorer, as illustrated in Gertrude Page's *The Rhodesian* (1912) when the protagonist is looking down on the ruins: Major Peter Carew »climbed on up the winding pathway, enfolded with mystery and romance concerning the feet that trod it in the far-off centuries, and made his way between the mighty natural boulders out on to the high platform, where eyes, all those long centuries ago, must have looked out even as his, across the lovely land« (*Rhodesian*: 12). Here the »mystery« (ibid.) and »romance« (ibid.) surrounding the ruins and their power over the beholder are explicitly referred to.[20] Significantly, this sense of mystery frequently

20 This is further underlined by the missionary Wyzinski's comment on the ruins in *The Ruined Cities of Zulu Land*: »›Yes, the extent of them is a mystery at present. We can but affirm their existence. What deep dead silence hangs over the spot. Let us go on.‹« (*Ruined Cities*, Vol. I: 170) Furthermore, the Zimbabwe ruins and Mashonaland are repeatedly referred to as mysterious in *An Imperial Adventure*, *The City of Gold*, *Benita*, and *The Rhodesian*: »[Sugdon] was intent upon the undecipherable ruin and the many amazing secrets it might conceal. He had ransacked many such another through the length and breadth of that wild land, partly from an eager interest in their origin, wrapped in fables of an unchronicled past, partly because of the gold they sheltered or to which they were supposed to give a clue.« (*Imperial Adventure*: 196), »Surely the mysterious city of his day-dreams must have something more than a fanciful existence to render possible so marked an act of self-abnegation!« (*City of Gold*: 85), »Possibly they might lead me whither my thoughts and aspirations had long turned; to that mysterious region in the very heart of Africa where a strange race

excites a very intense emotional reaction on the part of the beholder, as emphasised by the British woman Meryl's statement in *The Rhodesian*:

»›It is extraordinary how Zimbabwe gets one's heart. I have never seen anything anywhere that appealed to me quite like those old walls, with their untold story and their patience of the ages. The Sphinx in Egypt may be older, but we know how it came to be there and who built it. One of Zimbabwe's fascinations seems to be the absence of all knowledge about it, of all why and whereof.‹« (*Rhodesian*: 161)[21]

Meryl's strong feelings for the Zimbabwe ruins underscore the Western beholder's personal connection to the ruins, which becomes possible through »the absence of all knowledge about« (ibid.) them, that is, a lack of clear meaning. The personal connection felt to the ruins is further underlined by the fact that frequently the mystery surrounding the ruins induces feelings of melancholy and nostalgia:

»Yet as she looked around upon the empty desolation her heart grew sad with a nameless sorrow; that old, old ache, and old, old tiredness, for the utter futility of

---

of white men was reputed to guardian against contact with the outside world.« (Ibid.: 24), »Those mighty ancient walls built by hands unknown, which had seen so much history and so much death; the silent, triple ring of patient, solemn men, the last descendants of a cultured race, the crouching figure to be communing with his god – it was all very strange, very well worth the seeing to one who had wearied of the monotony of civilization.« (*Benita*: 80), »It is one of Rhodesia's mysteries, and one also of its fascinations; those mysteries and fascinations which so far have effectually baffled all efforts to find the clue and read the closed book. Who was it came for gold in those old, old days? Who was it built the line of forts to Solfala on the coast to guard the route along which the gold was undoubtedly carried, and of which remains may still be seen at regular intervals the whole distance? Where was the gold taken to from Solfala, and by whom?« (*Rhodesian*: 122)

21 A very similar feeling in regard to the ruins is conveyed in *The Ruined Cities of Zulu Land*: »A deep interest attached itself to the scene, for between him and those mountains the Zambesi must run, and somewhere among the forest must lie the ruined cities of Zulu land, if indeed they existed at all, save in the excited imagination of the missionary.« (*Ruined Cities*, Vol. I: 146)

work and of striving, that sometimes seems to fill the human heart, when in a depressed mood it looks upon the ruins of something that has once had strength and greatness.« (Ibid.: 116)[22]

The melancholy and nostalgia characterising Meryl's perception is significant, as it first suggests an identification with the archaeological sites on the part of the beholder and second points towards a sentimental mourning for something that once had been there in the past but has been lost in the present, causing an essential *feeling of lack*. These feelings of melancholy and nostalgia frequently evoked by archaeological remains can thus once more be read as pointing toward their ambivalent function in regard to the Western self as sites where on the one hand incompleteness painfully prevails and on the other completeness seems to be retrievable.

22 This and the following are only a few examples of archaeological landscapes evoking feelings of melancholy: »All this was to be to those who dared, [Sannie] said to herself, but not to her, – not to her who yearned after that land in all its mystery and loneliness who loved it passing well even on hearsay.« (*Imperial Adventure*: 25) And: »Such melancholy thoughts would occasionally obtrude themselves upon my notice and I imagined that all travellers who have, as it were, become absorbed into a grand, majestic, and mysterious region must suffer from them. The very immensity of the scenery, with the horizon fading far away into bluish, cone-shaped hills, the brilliancy of the floral coverings, and the deathlike stillness which pervades nature, are each and all provocative of sadness.« (*N'Shabé*: 24)

4.2 The Texture of the Past: Dreams and the Subliminal

> Mrs. A. – ›Many persons could not withstand the suffocating effect it produces; and the enterprising traveller is also annoyed with the immense quantity of fine dust, and the effluvia arising from the mummies.‹
> (*Fruits of Enterprise*: 105)

While in the passages discussed so far the emphasis was on encounters with an individual and collective past in the realm of the geographical ›other‹, there are a number of texts, in particular mummy narratives, in which the past as the ›other‹ emerges at home in Britain. A case in point is Bram Stoker's *Jewel of Seven Stars*, in which dreams and the subliminal in association with the past play a prominent role and thus evoke the *Dracula Discourse* with its emphasis on liminal psychological states, as pointed out by Byron: »Moving into the world of dreams, somnambulism, hypnotism and telepathy, Stoker engages here with many of the current debates in mental physiology.« (Byron 2007: 55) This chapter looks at how archaeology and the past as the unfamiliar were perceived as closely related to dreams, sensuality, and the subconscious in the novel, creating a counter-discourse to the contemporary civil discourse that infiltrates Britain and its private spheres, thereby challenging the Western rational self.

Characteristically, *Jewel* starts with Malcolm Ross's[1] dream in *medias res*, which is indirectly linked to archaeology when the daughter of a famous Egyptologist seeking for help later disturbs Ross's sleep:

»It all seemed so real that I could hardly imagine that it had ever occurred before; and yet each episode came, not as a fresh step in the logic of things, but as something expected. It is in such wise that memory plays its pranks for good or ill; for

---

1 The lawyer Malcolm Ross is the homodiegetic narrator of the novel.

pleasure or pain; for weal or woe. It is thus that life is bitter-sweet, and that which has been done becomes eternal.« (*Jewel*: 7)

This introductory passage, in which Ross is reflecting on one of his dreams in which he processes a recent encounter with Margaret Trelawny, the Egyptologist's daughter he is about to fall in love with, anticipates the general themes of the novel, which in the first part »drift[s] along in a slow, hypnotic pace that reflects the characters' tendency to fall into trances and surreal waking dreams« (Deane 2008: 404). Consequently, Ross's comment that »[i]t all *seemed* so real« (*Jewel*: 7, emphasis mine) can be read as paradigmatic for the whole story that is to follow, since it implies the ambivalence and the blurring of boundaries between dream and reality, fact and fiction. As revealed by the combination of the words »seemed« (ibid.) and »real« (ibid.), the stability of reality as such is being questioned and attention is turned to the unreliability of subjective perception in general.[2] In the novel this becomes particularly clear once Ross has come to Margaret's aid and enters the Trelawnys's house, a sphere where the borderline between reality and fantasy but also the familiar and unfamiliar becomes increasingly vague. This is further emphasised by the fact that Ross refers to »memory [as] play[ing] its pranks for good or ill« (ibid.), which implicitly also suggests that his whole narration presented to the reader in retrospect is subjected to the same processes. Thus, to follow Ross's story and accompany him on his nocturnal visit to the Trelawnys's house means to some extent to leave behind everyday reality and enter into a liminal space characterised by fantasy, dreams, memory, and the past.

Against this background it is crucial to note that Ross is woken from his dream and summoned by Margaret to her father's house in the middle of the night: »All at once the gates of Sleep were thrown wide open, and my waking ears took in the cause of the disturbing sounds. Waking existence is prosaic enough – there was somebody knocking and ringing at some one's street door.« (Ibid.: 8) Rather than leaving the realm of sleep, then, Ross here seems to open it up for his »prosaic« (ibid.) »waking existence« (ibid.), as suggested by the formulation »the gates of Sleep were thrown

2 With reference to Berger/Luckmann (1966), this could also be read as pointing towards the socially constructed character of reality. Furthermore, it anticipates the subjectivity of human perception in general emphasised by modernism.

wide open« (ibid.). However, surely enough Ross wakes up in the further course of events and makes his way to Kensington Place Road rather than leaving it behind, and the dreamy state in which he first introduces himself at the beginning of the novel prepares the reader for the dreamlike atmosphere at the Trelawny house and the surreal events that take place at this scene.

Doctor Winchester and Superintendent Dolan, who are already present when Ross arrives at the Trelawnys's house and represent the ›prosaic‹ reality as doctor and detective, seem in this context rather to function as a temporary intrusion of reality into the realm of a phantasmagoric parallel world. As Edwards points out, »[t]he uncanny is associated with the past in general, and is opposed to the prosaic, which is associated with the present in general« (Edwards 1998: 108). This evokes the *Dracula Discourse* with its emphasis on the psychological ›other‹ in the form of dreams, fantasies, and semi-conscious states. As the further events at the house show, Doctor Winchester and Superintendent Dolan sooner or later become overwhelmed and even absorbed by the strange atmosphere of the house to the extent that their rational patterns of explanations fail. Doctor Winchester eventually contemplates:

»›Two lines: Fact and – Fancy! In the first there is this whole thing: attacks; attempts at robbery and murder; stupefying; organised catalepsy which points to either criminal hypnotism and thought suggestion, or some simple form of poisoning unclassified yet in our toxicology. *In the other there is some influence at work which is not classified in any book that I know – outside the pages of romance.*‹« (Ibid.: 70, emphasis mine)

Although Doctor Winchester names a number of things that might theoretically explain the attacks on Mr Trelawny and his subsequent stupor, he seems particularly drawn to the line of what he calls »fancy« (ibid.). Winchester admits this himself after he has given an extensive analysis of the factual side of the situation:

»›The other horn of the dilemma is a different affair altogether; and if we once enter on it we must leave everything in the shape of science and experience behind us. I confess that it has its fascination for me; though at every new thought I find myself romancing in a way that makes me pull up suddenly and look facts resolutely in the face.‹« (Ibid.: 72)

However, the question whether or not the group agrees with Winchester and decides to »leave everything in the shape of science and experience behind« (ibid.) at this point of events is irrelevant, since, whether unwittingly or not, they have already done so.

This is underscored by Winchester's further comment in which he ponders on the peculiar scent coming from Mr Trelawny's study: »›I sometimes wonder whether the influence or emanation from the sick-room at times affects me as it did the others – the Detective, for instance.‹« (Ibid.: 72) The »›emanation from the sick-room‹« (ibid.), which is a recurring motive in the novel, is the smell attributed to the innumerable Egyptian objects in the house and in particular in Mr Trelawny's study: »Impalpable yet seemingly sentient and inescapable, the scent of the past threatens to ivade the physical bodies of the present, enervating them with its soporific effects and causing the members of Trelawny's household to become ›lost in‹ inescapable dreams of ancient Egypt.« (Bridge 2008: 146)[3] Significantly, Winchester names this smell as a possible source for a peculiar, subtle, mind-expanding influence on him and the others which seems to affect people once they have spent a certain time in the sickroom. Although the influence and effect of the smell as well as its character remain diffuse, its potency is self-evident right from the beginning of the novel when Winchester tries to give an initial diagnosis of Mr Trelawny's strange stupor:

»›Of course, there is ordinarily in this room so much of *mummy smell* that it is difficult to be certain about anything having a delicate aroma. I dare say that you have noticed the *peculiar Egyptian scents*, bitumen, nard, aromatic gums and spices, and so forth. It is quite possible that in this room, amongst the curios and hidden by stronger scents, is some substance or liquid which may have the effect we see.‹« (Ibid.: 21, emphasis mine)

No matter which of the many ancient curios in the study the smell actually comes from, the fact that it is referred to as a collective of a variety of »pe-

3 Cf. also the danger associated with the foul smell attributed to the Count in *Dracula* (*Dracula*: 250-251, 300).

culiar Egyptian scents« (ibid.) is significant. This is particularly true if we consider that the scents mentioned here all refer to ingredients essentially used in ancient Egyptian embalmment, which thus can probably be ascribed to mummies. This assumption is further underscored by the »mummy smell« (ibid.) mentioned by Winchester as »ordinarily« (ibid.) dominating the room. Considering this, one can assume that the main source of the collective of »peculiar Egyptian scents« (ibid.) emanates from the mummies in Mr Trelawny's study, more precisely from one particular mummy, the mummy of the ancient Egyptian Tera, which somehow seems to be connected to the nightly attacks on Mr Trelawny. This turns the collective Egyptian smell into a *pars pro toto* of the mummy, being itself a *pars pro toto* of ancient Egypt, which in turn is a *pars pro toto* of the past as much as the innumerable Egyptian curios Mr Trelawny has assembled in his house. This means that on a symbolic level we are dealing with a convoluted structure of semantic references, which finally all turn out to be facets of a semantic whole, namely the past.[4] This conflates the »peculiar Egyptian scents« (ibid.) with the ›smell of the past‹, which impalpably infiltrates the present (cf. Bridges 145-146).[5]

Significantly, the *extra*ordinary power of the past in the form of the »peculiar Egyptian scents« (ibid.) seems to have a particular influence on

4 The convoluted structure of the semantic field can also be seen as representing the multilayers of archaeological strata. Cf. also Malley (1997), who notes in regard to the potsherd in Haggard's *She*: »The sherd represents an enclosed text, a nested narrative that interconnects past with present for the reader. The motif of boxes within boxes, of stories within stories, of history buried under history, takes many forms as the party ventures toward the sherd's provenance: a stratified narrative reflects the archaeological content.« (Malley 1997: 278-280)

5 Although this peculiar Egyptian smell plays a particularly prominent role in *Jewel* (cf. *Jewel*: 21, 35, 36, 45, 58, 72, 73), the power and ›otherness‹ associated with it in general is underlined by the various references in other mummy texts of the time. Other texts in which the Egyptian smell is referred to are Rider Haggard's »Smith and the Pharaohs« (1912-1913), Fergus Hume's *The Green Mummy* (1908), E. and H. Heron's »The Story of Baelbrow« (1898), Cutliffe Hyne's »The Mummy of Thompson-Pratt« (1904), Sax Rohmer's »The Case of the Headless Mummies« (1913), and Clive Pemberton's »The Bulb« (1906).

the characters' minds, as underlined by Ross's perception during the night-watch at Mr Trelawny's sickbed:

»[Nurse Kennedy] had a quality of common-sense that seemed to pervade everything around her, as though it were some kind of emanation. Up to that moment I had been building fancies around the sick man; so that finally all about him, including myself, had become involved in them, or enmeshed, or saturated, or… But now that she had come, he relapsed into his proper perspective as a patient; the room was a sick-room, and the shadows lost their fearsome quality. *The only thing which it could not altogether abrogate was the strange Egyptian smell.* You may put a mummy in a glass case and hermetically seal it so that no corroding air can get within; but all the same it will exhale its odour. One might think that four or five thousand years would exhaust the olfactory qualities of anything; *but experience teaches us that these smells remain, and that their secrets are unknown to us. To-day they are as much mysteries* as they were when the embalmers put the body in the bath of natron… All at once I sat up. I had become lost in an absorbing reverie. The Egyptian smell had seemed to get on my nerves – on my memory – on my very will.« (Ibid.: 35-36, emphasis mine)

In this passage nurse Kennedy, with her »quality of common-sense« (ibid.), appears to Ross, who has »become lost in an absorbing reverie« (ibid.), as a saviour who prevents him from slipping away from the present into a mysterious and unknown past.[6] The fact that Kennedy's »quality of common-sense« (ibid.) is compared to »some kind of *emanation*« (ibid.) that »pervade[s] everything around her« (ibid.) links her quality of common sense to the semantic field of smell. Considering the fact that Ross perceives Kennedy's appearance as a relief, since she causes him to see the things in their »proper perspective« (ibid.) again, her »emanation« (ibid.) »of common-sense« (ibid.) can be read as the antipode of the »strange Egyptian smell« (ibid.). This is further emphasised by the mystery and permanence Ross attributes to the Egyptian smell and finally by the power Ross feels it has on his »nerves« (ibid.), »memory« (ibid.), and »will« (ibid.). Thus, the »emanation« (ibid.) of Kennedy's »common-sense« (ibid.) becomes emblematic of the present, while the »strange Egyptian smell« (ibid.) primarily

6 Nurse Kennedy's role thus corresponds to Mina Murray's role in *Dracula* as the rational and matter-of-fact woman in the novel.

associated with mummies, as made explicit by Ross's reflection on the peculiar character of the Egyptian smell, represents the past.

Interesting in this context is the effect the ›smell‹ of the past versus that of the present has on Ross, or rather the way in which these different ›smells‹ are perceived by him. While before the appearance of Kennedy Ross had been controlled by what he refers to as »fancies around the sick man« (ibid.), these »shadows [lose] their fearsome quality« (ibid.) once she has entered the room. The very fact that Ross in his concluding sentence makes the »Egyptian smell« (ibid.) responsible for his reveries prior to Kennedy's entry clearly suggests that this smell of the past appeals to Ross's subconsciousness, that is, his nerves, while it seems to benumb his rationality, that is, his will. This benumbing but subtle power of the »Egyptian smell« (ibid.) is so strong that Ross can even perceive it after nurse Kennedy has entered the room with her »emanation« (ibid.) of »common sense« (ibid.). In other words, even though Kennedy, who in her function as a nurse represents common sense and modern science, is able to relieve Ross of the grip of the past, she cannot disperse it from the present, as the lingering of the penetrating Egyptian smell suggests.

This is also what Ross realises when he finally links his own experience to Mr Trelawny's mysterious stupor:

> »At that moment I had a thought which was like an inspiration. If I was influenced in such a manner by the smell, might it not be that the sick man, who lived half his life or more in the atmosphere, had gradually and by slow but sure process taken into his system something which had permeated him to such a degree that it had a new power derived from quantity – or strength – or… I was becoming lost again in reverie. This would not do. I must take such precaution that I could remain awake, or free from such entrancing thought.« (Ibid.: 35-36)

That Ross suddenly considers the Egyptian smell as a possible explanation for Mr Trelawny's state is interesting since Mr Trelawny's stupor, which he and the others suspect in some way to be connected with the ancient Egyptian relics at the house, resembles that of a mummy. This is particularly true if we consider that in the case of the mummy Tera – as in that of many other mummies in fiction – we are actually not dealing with a dead body but with a person sleeping a deep sleep, that is, a stupor, who under certain circumstances might be restored to life.[7] This is also the case with Mr Trelawny, who under mysterious circumstances slips into his strange stupor: »›This disease, or condition, or whatever it may be called, from which Mr Trelawny is suffering, is in some way connected with Egypt.‹« (Ibid.: 106) As formulated by Corbeck here, he sees the cause of Mr Trelawny's stupor as somehow linked to the Egyptian past. That Corbeck's assumption seems to be correct is further underscored by the strange trance-like condition into which Ross falls after having spent some time at the sickbed. Eventually, all characters at the house become aware of the powerful and mysterious influence of the Egyptian past on the present, which seems to be strongest in Mr Trelawny's study.

Consequently, all of the characters try to protect themselves as best as they can against the strange forces of the past, as becomes clear from what Sergeant Daw and the nurse express verbally and nonverbally, respectively: »›I sleep lightly and I shall have my pistols with me. I won't feel so heavy-headed when I get out of this mummy smell‹« (ibid.: 37) and »›I noticed that [the nurse] had a vinaigrette in her lap. Doubtless she, too, had felt some of the influence which had so affected me‹« (ibid.). Apart from noticing the precaution taken by the detective and the nurse against the influence of the past, Ross also refers to his own safety measures in the form of his respirator: »Again I found myself thinking of the Egyptian smell; and I remember that I felt a delicious satisfaction that I did not experience it as I had done. The respirator was doing its work.« (Ibid.) Interestingly, both in Daw and Ross's comments on their personal security, the Egyptian/mummy smell is mentioned as the prominent threat to the characters' safety in the room.

It seems important to ask why the Egyptian smell is actually so threatening and in which way it affects people. At first glance, the Egyptian smell seems to be not only responsible for Mr Trelawny's stupor but also in some way connected with the sanguinary attacks on Mr Trelawny, in that both the Egyptian smell and the nocturnal incidents in Mr Trelawny's study appear to originate in a remote past. Furthermore, as much as the assaults on Mr Trelawny are wild and brutal, the Egyptian smell, though not as explicitly aggressive, is also connoted negatively in that it is repeatedly referred to as ›strange‹ and ›peculiar‹. This turns both the smell and the

7 At the end of the novel, Tera *is* woken up from her sleep.

attacker into envoys of a *strange* and *wild* ancient Egyptian past infiltrating modern Britain, reflecting attitudes of Orientalism and fears of reverse colonisation both prominent in the *Dracula Discourse* (cf. Arata 1990). Significantly, although the nocturnal assaults are directed solely against Mr Trelawny, the smell seems to affect everyone who spends a certain time in the sickroom. Ross and Daw's references to the oppressive smell emphasise this, but it becomes most drastically apparent when Kennedy, too, suddenly falls into a stupor:

»She was, to all intents and purposes, turned into stone. There was no special expression on her face – no fear, no horror; nothing such as might be expected of one in such a condition. Her open eyes showed neither wonder nor interest. She was simply a negative existence, warm, breathing, placid; but absolutely unconscious of the world around her.« (*Jewel*: 39)

The »cataleptic trance« (ibid.) which has befallen the nurse during her nightwatch, though not as extreme, essentially seems to be of the same kind as Mr Trelawny's stupor.[8] Significantly, she is referred to as a »negative existence, warm, breathing, placid« (ibid.) and »absolutely unconscious of the world around her« (ibid.), which suggests that she is completely bereaved of her waking conscious and her will.[9] This corresponds to what Ross experiences when he describes the Egyptian smell as seeming »to get on [his] nerves – on [his] memory – on [his] very will« (ibid.: 36). Sergeant Daw, who has entered the room shortly before Ross becomes aware of the nurse's condition, seems similarly affected by the atmosphere in the room:

»In the middle of the room Sergeant Daw, in his shirt and trousers and stocking feet, was putting fresh cartridges into his revolver in a dazed and mechanical kind of way. His eyes were red and heavy, and he seemed only half awake, and less than half conscious of what was going on around him.« (Ibid.: 40)

8 This emphasises the danger of half-conscious states, such as sleep, which is also prominent in the *Dracula Discourse*.

9 This again alludes to the *Dracula Discourse* in that it corresponds to the somnambulistic state Lucy falls into under the power of the Count.

What is central to all these cases in which the forces of the past have their effect on the characters is that these forces – apart from the additional physical assaults on Mr Trelawny – essentially affect an individual's mind and will: »One property of the past suggested [...] is that it affects the unconscious parts of the mind, and that its influence is not always susceptible to voluntary control.« (Edwards 1998: 101) In addition to Ross's explicit reference that he feels his »will« (ibid.: 36) affected by the Egyptian smell, this is further underscored by the nurse's »negative existence« (ibid.: 39) as much as by the fact that Daw charges his pistol in a »dazed and mechanical kind of way [...] and [...] seem[s] [...] less than half conscious« (ibid.: 40) of what he is doing.

The fact that the forces of the past predominantly aim at a character's mind and will as the centre of rationality and control is particularly instructive in regard to their emblematic function for nineteenth-century Britain, with its emphasis on science, technology, and common sense in general. Accordingly, the ancient Egyptian smell can be read as an archaic assault of the spatial and temporal ›other‹ on modern rationality: »Egypt is treated as if it were some ominous state of mind, irresistibly taking hold of the lives of all who come into contact with it.« (Glover 1996: xiii) Significantly, this implicitly reflects the Orientalist view on the East as dominated by sensualism as opposed to the Western rationality and common sense (cf. Said 1978: 26). In general, the threat perceived as coming from the archaic past in the form of the Egyptian smell thus mirrors the widespread anxiety of possible weak points of an empire hitherto characterised by rationality and progress. The invisibility of the Egyptian smell befalling the Western mind can further be read as symbolising the *Dracula Discourse*, with its underlying fear of invading forces that can neither be grasped nor controlled by contemporary society. In particular around the turn of the century, many contemporaries saw these dangers in decadence and antecedents of modernism with their emphasis on the senses.

In the novel, this fear is intensified by the fact that the respirator provided by Ross and the detective's pistol fail as emblems of modern technology to thoroughly protect the characters:

»I had on my respirator and knew that I breathed freely. The Nurse sat in her chair with her back towards me. [...] The sick man lay as still as dead. It was rather like a picture of a scene than reality; all were still and silent; and the stillness and silence were continuous. Outside, in the distance I could hear the sounds of a city, the occasional roll of wheels, the shout of a reveller, the far-away echo of whistles and the rumbling of trains. The light was very, very low; […]. The room, for all its darkness, was full of shadows. It seemed in my whirling thoughts as though all the real things had become shadows – shadows which moved, for they passed the dim outline of the high windows. Shadows which had sentience. […] I sat as one entranced. At last I felt, as in nightmare, that this was sleep, and that in the passing of its portals my will had gone.« (*Jewel*: 38)

This passage is crucial since it demonstrates how modern technology as represented by the respirator is outplayed by the subtle influences from the past that slowly expand into the room and also into Ross and the nurse's consciousness, until Ross eventually feels that his own »will ha[s] gone« (ibid.).[10] Interestingly, Ross feels as »one entranced« (ibid.) before he perceives his condition as sleep to which he has unconsciously abandoned himself. This emphasises the subtle influence of the past on the mind and its affinity with dreams and the subconscious. Thus, similar to Dracula, who imperceptibly intrudes into people's private spheres to affect and manipulate their minds, the mummy smell, as the temporal and spatial ›other‹, invades and occupies people's homes and bodies. As a consequence, »[t]he domestic space in these fictions is increasingly experienced as foreign; the present is increasingly infiltrated by what it has designated as archaic« (Daly 1994: 34).[11]

In *Jewel* as in *Dracula*, the atmosphere is intensified by the fact that the scene takes place in the middle of the night, with repeated reference to the »stillness« (*Jewel*: 38), »silence« (ibid.), and »darkness« (ibid.) of the room, as well as by the isolation of the house, enforced by the sounds of the distant city. This is characteristic of how the influence of the past in archaeological fiction, and particularly in mummy fiction, tends to be very strong at night, which emphasises the isolating function of archaeological artefacts on individuals but also reflects a belief that modern people's susceptibility to these influences of the past is heightened at that time of day. In that

10 This again alludes to the *Dracula Discourse*, where modern technology and science are unable to eliminate the Count.

11 This already marks the inversion of the familiar and unfamiliar, the self and the ›other‹, particularly prominent in mummy fiction (cf. Chapter 3).

context, night as the realm of sleep and dreams again represents a liminal sphere in which the past emerges beyond the present and stimulates people's psychological ›other‹.

Ross's perception of the »shadows which [have] sentience« (ibid.) evoked by the strong presence of the past in the stillness and isolation of Trelawny's study can thus also be read as an image of the human psyche as it is emphasised by the references to the subconscious, sleep, and dreams in this and other scenes of the novel. To be isolated from the present, its technology, and society and confronted with the past signifies danger. This is, for instance, prominently underlined in Guy Boothby's novel *Pharos*, where the protagonist Forrester is symbolically raped by forces of the past in the isolation of an Egyptian pyramid (cf. Chapter 6.4), or in Theo Douglas's *Iras*, where the archaeologist Lavenham and his wife are eventually hunted down by the past, embodied by an evil ancient Egyptian priest in the solitude of the Scottish Highlands (cf. Chapter 7.2). As has been shown in regard to *Jewel*, the presence of the past in the form of the ancient Egyptian relics and the Egyptian smell creates a liminal sphere characterised by the dominance of subconscious and sensual experiences once the characters are alone and isolated.[12] In these situations, the characters generally lose control of themselves and experience their surroundings as strange and alienated, as is underlined by Ross's reference to the familiar room Ross perceives as dark and »full of shadows« (ibid.).[13] Consequently, it is in those moments that the characters are most vulnerable and helpless, as the nocturnal attacks coming from the ancient Egyptian objects and mummies in the form of both physical and mental assaults suggest.

12 Note that even though in the passage quoted above Ross is not alone in the room but in the company of Nurse Kennedy, the fact that they do not communicate and Ross's perception seems completely detached from the nurse's creates the impression of isolation and loneliness.

13 The alienation of the familiar in the present of the past is also typical of archaeological fiction.

4.3 MAPPING THE PAST

›I distinctly remember his showing me a statement about some hidden treasure in Mashonaland […]. The paper, however, contained no definite instructions, without which a hunt for treasure in a wild land would be a wild-goose chase. It refers, however, to a key to be found on Table Mountain.‹
(*FOSSICKER*: 56)

While so far the archaeological space has been described in its ambivalence, having both a fascinating and a threatening effect on the Western archaeologists and explorers, the present chapter focuses on the relevance attributed to maps in regard to the archaeological sites. For this purpose, the focus will be on fictional texts dealing with the Zimbabwe ruins, in which the geographic map becomes the major instrument in the process of appropriating the past in the present. I argue that the map assumes the function of an ›inverse empiricism‹ in an attempt to create structure and meaning in the spatial and temporal ›other‹ as a means of securing and protecting the Western self in this otherwise indefinite discursive space. This is emphasised by the many texts introducing maps as a means of exploring and reclaiming the land in search for the Zimbabwe ruins.

The importance of the map in regard to the colonial discourse in general has been discussed by various critics in connection with the African landscape, in particular with an emphasis on its sexual symbolism and misogyny in Haggard's texts:[1]

1 In Haggard's *She*, the map is ascribed a crucial role right at the beginning of the novel. Cf. William J. Scheick (1991) »Adolescent Pornography and Imperialism in Haggard's *King Solomon's Mines*«; Rebecca Stott (1989) »The Dark Continent: Africa as Female Body in Haggard's Adventure Fiction«; and Jeanne van Eaden (2004) »The Colonial Gaze: Imperialism, Myths, and South African Popular Culture«.

»In addition to reproducing the female body, it is reasonable to assume that the map features a significative illustration of Haggard's entire sexual discourse and especially of his misogyny. In an interesting study of Haggard's landscape, Lindy Stiebel pushes this hypothesis even further, suggesting that the map may very well be a visual version of the writer's bodyscape ideology, born of the friction between sexual repression within Victorian society and the unconfined libertine phantasies about the colonised female.« (Elio Di Piazza 2007: 91)

In a broader sense the map can be seen as the materialisation of the coloniser's internal image of the landscape, thus functioning as the representation of his conception of both the African landscape and its past:

»In a very general sense, landscape description in literary works is connected to the observer's cognitive standpoint – i.e. the prospective line along which the contours, dimension, colourings, and spatial relations among the landscape's various components are elaborated as signifiers. […] The dynamic expansion into the dominion of the Beyond, in this case, takes us to an ambivalent ekphrasis that reduces landscape's elements to a-temporal structures (albeit historically established ones), to be seen through the deforming lens of fore knowledge and presuppositions. […] This is particularly true of Haggard's novels where the relationship between the vicinity of the landscape object (obvious) and that of knowledge (intended) is regulated by the ideological subcodes conceived in the golden period of British colonial expansion.« (Ibid.: 88)

It is the map's representative function of the implicit, subtle, and unsaid that makes it a key factor in this context, since it externalises what Di Piazza refers to as the »ideological subcodes« (ibid.) underlying the perception of the landscape: »In terms of cultural geography, maps can be held subject to textual interpretation as they are culturally encoded mental constructs.« (van Eaden 2004: 29) At the same time, in the case of the Zimbabwe ruins the map often represents what I want to call an ›inversed empiricism‹ in that it represents something that in most cases has not been *experienced* first hand – at least not by the Western explorer – and thus cannot be known yet. The maps are usually old and frequently based on stories told about the ruins rather than representing reality. This makes them *prescriptive* rather than *descriptive*, indicating the Western explorers' urge to import structure into the archaeological space as the spatial and temporal unknown. Once more this emphasises the power of the »ideological subcodes« (ibid.) referred to by Di Piazza, as the image of the ruins, landscape, and past represented on maps is not determined by reality but by the explorers' *a priori* mental conception of it. The measuring, categorising, and appropriating of the landscape and the past therefore is also always a narrative, a fictional act that makes its reality both mentally and physically accessible for the Western explorer. Characteristically, in many Zimbabwe texts the map functions as a guide not only to the landscape but also to the past, which once more implies that the explorer carries his own mental image of the past with him when accessing the unknown territory.[2] In regard to the archaeological space, which I conceive as an initially more or less blank discursive field, the map thus becomes an attempt by the Westerner to project and transfer the structure and knowledge of the Western present onto and into that space of the ›other‹, that is, the past and Africa, in an effort to create structure and eventually meaning.

The most popular example of a map relating to the Zimbabwe ruins is the ancient map introduced in Haggard's *King Solomon's Mines* as once created by the Portuguese José da Silvestra in 1590 and handed down over generations before finally falling into the hands of the British Captain Good:[3]

»›Ay, but the document‹, said Sir Henry, in a tone of deep interest. ›Yes, the document; what was in it?‹ added the captain. ›Well, gentlemen, if you like I will tell you. I have never showed it to anybody yet except my dear wife, who is dead, and she thought it was all nonsense, and a drunken old Portuguese trader who translated it for me, and had forgotten all about it next morning. The original rag is at my home in Durban, together with poor Dom José's translation, but I have the English rendering in my pocket-book, and a fac-simile [sic] of the map, if it can be called a map. Here it is.‹« (*Solomon*: 54-55)

2 Maps appear in *The Ruined Cities of Zulu Land* (1828), *King Solomon's Mines* (1885), *She* (1887), *The Fossicker. A Romance of Mashonaland* (1891), *A Romance of N'Shabé* (1891), and *John Prester* (1910).

3 The first edition of *King Solomon's Mines* included a fold-out frontispiece showing a photograph-like image of the map.

As in many other texts referring to maps of the ruins, what follows is not only a detailed description but also a visual reproduction of the map. On a symbolical level this parallels the idea that the map represents the coloniser's internal image of the landscape and the past in that the maps published in these books illustrate what is otherwise only expressed in a less accessible form in the written text. Furthermore, it is significant that Captain Good mentions the fact that his late wife considered the story and what was shown on the map as »nonsense« (ibid.). Frequently, the maps introduced in fictional texts dealing with the ruins only make sense to those who know *how to read* them and, in analogy to the archaeologist, are thus able to discover the meaning behind otherwise meaningless signifiers. In this context, the deciphering and decoding of what is shown on these maps already becomes the process of exploring the unknown both in a spatial and temporal sense:

»It is one of Rhodesia's mysteries, and one also of its fascinations; those mysteries and fascinations which so far have effectually baffled all efforts to find the clue and read the closed book. Who was it came for gold in those old, old days? Who was it built the lines of forts to Solfala on the coast to guard the route along which the gold was undoubtedly carried, and of which remains may still be seen at regular intervals the whole distance? Where was the gold taken to from Solfala and by whom?« (*Rhodesian*: 122)

The mystery surrounding the Zimbabwe ruins is conceived in terms of a »closed book« (ibid.) that is yet waiting to be read by someone who understands its signifiers, in this case the archaeologist who assumes the role of the decoder of the past.[4] At the same time, the fact that it is reserved to a few chosen people to first access and then decipher these maps emphasises the dominant role of the Western explorer in regard to the African land-

4 Again, the many questions in the passage from *The Rhodesian* point towards the blank space associated with Great Zimbabwe and Africa in general that needs to be filled with meaning, as emphasised by the following passage from Haggard's *Benita* (1906): »›Stones cannot speak, the spirits are silent, and we have forgotten.‹« (*Benita*: 75)

scape and past:[5] »Maps are a fundamental part of the metaphorical organization and representation of space, and since they are part of the discourse of colonialism, they are automatically implicated in power relation.« (van Eaden 2004: 29) This attitude is reflected in the preface to *A Romance of N'Shabé*:

»After having revised our book, we are vain enough to form a hope that some parts of it may be interesting to ›children of a larger growth‹, for the story covers some of those regions about which there will, sooner or later, be fierce bickerings amongst certain European nations. *The question of the identification of the locality of the land of Sheba will also exercise maturer intellects*, and they may come to believe, as we do, that when Josephus called the Queen of Sheba, the Queen of ›Egypt and Ethiopia‹, he was much nearer the mark than the huge host of modern investigators.« (*N'Shabé*: vi, emphasis mine)

As suggested by the reference to the »European nations« (ibid.) in combination with the »maturer intellects« (ibid.), the location and interpretation of the Zimbabwe ruins is reserved to the Western world, since it requires the learned ability to decode the signs of the past. In *King Solomon's Mines* this is further emphasised by Captain Good's description of the events and stories surrounding the ancient map he possesses:

»*But here and there you meet a man who takes the trouble to collect traditions from the natives and tries to make out a little piece of history of this dark land.* It was such a man as this who first told me the legend of Solomon's Mines, now a matter of nearly thirty years ago. […] His name was Evans, and he was killed next year […]. I was telling Evans one night, I remember, of some wonderful workings I had found whilst hunting koodoo and eland in what is now the Lydenburg district in Transvaal. I see they have come across the workings again lately in prospecting for gold, but I knew of them years ago. There is a great wide waggon road cut out of the solid rock, and leading to the mouth of the working or gallery. Inside the mouth of this gallery are stacks of gold quartz piled up ready for crushing, which shows that the workers, whoever they were, must have left in a hurry, and about twenty paces in the gallery

5 Usually it is the Western explorer who manages to make sense of stories told by indigenous peoples – and thus imbue them with meaning – and decipher maps and the like.

is built across, and a beautiful bit of masonry it is. ›Ay‹, said Evans, ›but I will tell you a queerer thing than that‹, and he went on to tell me how he had found in the far interior a ruined city, which he believed to be the Ophir of the Bible, and, by the way, other more learned men have said the same long since poor Evans' time. I was, I remember, listening open-eared to all these wonders, for I was young at the time, and this story of an ancient civilisation and of the treasure which those old Jewish or Phœnician adventurers used to extract from a country long since lapsed into the darkest barbarism took a great hold upon my imagination, when suddenly he said to me, ›Lad, did you ever hear of the Suliman Mountains up to the north-west of the Mashukulumbwe country?‹ I told him I never had. ›Ah, well‹, he said, ›that was where Solomon really had his mines, his diamond mines, I mean.‹ ›How do you know that?‹ I asked. ›Know it; why what is ›Suliman‹ but a corruption of Solomon! and, besides, an old Isanusi (witch doctor) up in the Manica country told me all about it.« (Solomon: 51-52, emphasis mine)

Here, the first sentence, in which Good introduces the hunter Evans as someone who »tries to make out a little piece of history of this dark land« (ibid.), is crucial in this context, since it points towards the allegedly blank archaeological space. Moreover, it emphasises the constructed and arbitrary character of history being influenced by the perceiving subject in the present and thus by his/her »ideological subcodes« (Di Piazza 2007: 88), referred to by Di Piazza in connection to the exploration of the African continent in general. Evans's urge to create a ›history‹ for Africa conceived as the ›dark land‹ can thus be read as representing the Westerners' imperial drive to appropriate the temporal and spatial ›other‹ according and in relation to their present.[6] The fact that Good refers to Evans as a man who tries to ›make a history‹ for Zimbabwe emphasises the whole process of *creating* meaning – history and continuity – for an otherwise meaningless space.

The alleged relevance of the Western colonisers for the Zimbabwe ruins and their history is further emphasised in *The Rhodesian*: »But in the meantime the Valley of Ruins no longer lies alone and unheeded in the sunlight; and no longer do the hills look down upon rich plains left solely to the idle pleasure of careless black people. The forerunners of to-day's great

6 This points to Gehrke's intentional history, which emphasises the social function of history at a certain point of time rather than whether a specific history is ›true‹ or not.

civilising army have marched into the valley.« (*Rhodesian*: 5) Here, once more the need for someone from *outside* to interpret the ruins is underlined, since, according to the British, the indigenous population lacks the interest and ability to adequately deal with the ruins. This once more calls to mind the metaphor of Africa as the ›dark continent‹ that is being enlightened by the Western colonisers: »English adventurers apparently saw themselves as moving from the civilised to the savage, enlightening the natives as they went.« (Hopkins 1996: 4)

At the same time, the process of collecting and processing data from indigenous people to create a coherent picture/story of the past is reminiscent of archaeological work as such, which is very much concerned with piecing together individual fragments to recreate a past materiality, as has been shown in the previous chapters. Once one has constructed a story/picture of the past by falling back on different sources – such as legend, rumours, antiquities – the aim is to confirm it by linking this story to a spatial and temporal reality. This also represents that fact that, in analogy to what I have termed ›inversed empiricism‹, the archaeological search for the ruins is strongly characterised by an *a priori* assumption of the ruins: Rather than the interpretation of the ruins being based on the archaeological remains, the interpretation is created *prior* to the actual encounter with the ruins. This can, for instance, be observed in *A Romance of N'Shabé*, where one of the local people asks the explorer Montfort after the latter's arrival at the city: »›How did you discover that there was anything antique or venerable about my nation?‹ ›By piecing together two very distinct rumours.‹« (*N'Shabé*: 167) In order to successfully link the fictional accounts – stories, etc. – to reality, that is, the ruins, however, the structure of the story/picture of the Western explorers is characteristically documented on paper in the form of maps:

»*The whole country lay spread like a map before him.* Far away to the east lay the face of a friend, for the blue line of the ocean was distinctly visible. [...] To the north, stretched at the foot of the mountains, lay the plains they were to traverse, *the unknown land of promise.* It looked one dense forest, broken at rare intervals with open, and intersected by two rivers, one appearing a considerable sheet of water, while in the far distance, a range of lofty mountains loomed, dim, blue, and ghost-like in their outline.« (*Ruined Cities* Vol. I: 145-146, emphasis mine)

Here the depiction of the country as seen through Captain Hughes's eyes is significant, since it is actually referred to as »spreading like a map before him« (ibid.).[7] It is the Western explorer's ›colonial gaze‹ (cf. van Eaden 2004) that is reflected here, conceiving the unknown landscape as a map and suggesting that mentally the country and its past has already been explored and categorised, that is, colonised, by the Western mind:

»The colonial expansion of the Victorian age is often personified in the character Mary Louise Pratt calls the seeing man, ›whose imperial eyes … look[ed] out and possess[ed]‹ Africa and Asia, and who came into being in the early part of the century as England expanded its mission from maritime trade to colonial annexation.« (Marx 1998: 52)

The cultural function of the colonial explorer as the agent to process and appropriate the African landscape is pointed out by Stiebel in regard to people in Britain: »For those at ›home‹ in England in the nineteenth century, explorers and writers in southern Africa had above all to represent a localised landscape which helped their audience move from a vague sense of remote space to its contextualisation in a recognisable place on the map of their imagination.« (Stiebel 2001: 123)[8] Furthermore, Walmsley's reference in the above quote to the »unknown land of promise« (*Ruined Cities* Vol. I: 145-146) refers to the biblical idea of the Western explorers as the chosen people predetermined to explore the land and all its riches: »This land of Prester John is a simulacrum to Europe: a dreamed-of world beyond the dark, barbaric territories with their semi-human savages.« (Hall 1995: 183).[9]

It is thus only by means of maps serving as a guide and manual to both the country and its past that the latter can actually be tracked down. The

7 Cf. also »The country of the Batonga lay mapped before them.« (*Ruined Cities* Vol. I: 70)

8 Frequently, the power of the ›colonial gaze‹ is also reflected in what Stiebel refers to as an ›imaginary aerial perspective‹: »Haggard, whose future fame was to result from his adventure stories set in Africa, described African landscapes which depict similar vistas of untamed wilderness, seen from an imaginary aerial perspective.« (Stiebel 2001: 126)

9 Cf. also John Buchan's eponymous novel *John Prester*.

importance associated with maps is emphasised in the following dialogue between the British explorers Dick and Smedley in regard to a map of Ophir they own:

»›Seriously, Dick – weren't you too confidential with that man Foster?‹ ›Well, I don't know; I certainly did tell him that we were bound for Ophir, and as he was so deeply interested in the ancient history of Mashonaland, I gave him the latitude of that fabulous mine. The longitude I had clean forgotten.‹ ›You are indeed a flat!‹ ›Nonsense! *What's he to Ophir, or Ophir to him, without the plan*?‹« (*Fossicker*: 2-3, emphasis mine)

The plan to Ophir becomes the decisive instrument for approaching the spatial and temporal unknown; without it this place remains nothing but a myth, as is further suggested by Foster's comment on the plan later in the book:

»›I did meet a man in Kimberley who had been up in the Mashona country. It was understood that he had discovered a store of gold near one of those old ruins to be found up there, but he kept his secret. At my suggestion, he drew up a plan of the place, with description of the route to be followed, but *he lost the paper. He disappeared* – died from drink, I think – *and of course his secret died with him.*‹« (Ibid.: 13-14, emphasis mine)

Here the loss of the paper and the disappearance of a person are equated with the loss of knowledge and orientation regarding the treasure: »[T]here is only one thing left, and that is to find whether there is any truth in this story of the treasure, and of the existence of the Fossicker.« (Ibid.: 154) Considering the function of the map as a guide to the spatial and temporal ›other‹, a space essentially lacking structure and meaning, the disappearance of a person in connection to the loss of the map can be read on a symbolic level as standing for the failure to sustain identity in the realm of the ›other‹ once the manual, representing the Western discursive structures, has been lost. That the map represents the Western colonial attitude towards the past and its landscape is emphasised by the eponymous character of the book, who is introduced as the only living person who knows the location of the treasure and thus, as long as there is no map, functions as the key to the past: »Smedley took up the quill which had nearly cost him his life, and drew out therefrom a roll of paper. He unrolled it eagerly, stretched it out, and looked for a plan of the locality where the treasure had been deposited. There was no plan, merely a line: ›The Fossicker knows the treasure.‹« (Ibid.: 103)

As the different examples of texts dealing with the Zimbabwe ruins have shown in regard to maps, the process of reading the map becomes the process of reading the landscape and eventually the past in an attempt to finally define individual and collective identity *and* alterity. Significantly, this process is very much characterised by a *prescriptive* approach that emphasises how by mapping and reading the past is interpreted and categorised according to the ideological context of the colonial mind in the present. The map thus symbolises the *a priori* cognitive conception of the African landscape and past, which I have referred to as ›inversed empiricism‹, satisfying the nineteenth-century's obsession with empiricism. It functions as the representation of what van Eaden refers to as ›colonial gaze‹ in regard to the present *and*, in addition to that, as I have argued, to the past.

Interim Findings: Archaeology as a Space of Ambivalence

In the present chapter, the archaeological discourse has been introduced as a space essentially marked as the realm of the temporal and spatial ›other‹. As the various examples have shown, the realm of the ›other‹ evoked by archaeological remains in the broadest sense can be encountered abroad or at home in Britain. In both cases, the archaeological space is characterised by an ambivalence present in the negotiation between the familiar and the unfamiliar. Significantly, in all the cases looked at, the confrontation with archaeology as the temporal and spatial ›other‹ is strongly linked to the confrontation of the Western characters with their own subconscious self as the hitherto unfamiliar, evoking the *Jekyll and Hyde Discourse*. This emphasises the function of the archaeological space as an initially indefinite space in which underlying contemporary longings, needs, anxieties, and threats are negotiated. Furthermore, archaeological relics, in the form of ruins or mummies, prominently work as sites of memory confronting the beholder with their own incomplete subconscious self, while at the same time they seem to offer the possibility for completion. In terms of Storey's idea of identity formation, archaeological relics as sites of memory thus symbolise the alleged ›roots‹ in terms of memories as well as the ›routes‹ in that they seem to promise future fulfilment of the Lacanian desire for wholeness. In this sense, the confrontation with the archaeological remains has both a creative and a threatening potential to it. On the one hand it functions as the confirmation of identity and on the other it reflects the instability of identity through the reflection of its incompleteness and the futility of the search for completion. As an attempt to deal with this disturbing ambivalence, the explorers turned to maps as a means of importing structure into the archaeological space as the spatial and temporal ›other‹ to thus appropriate it for the modern Western imperial mind.

5. Heinrich Schliemann's Troy as the *most familiar strangeness*[1]

> So literally do present opinions of all shades bear the mark of Dr. Schliemann's personality, so completely have his character and the manner of his life he led invaded our whole consideration of the heroic age in Greece, that the biography prefixed to the present book is one of its most instructive chapters.
> (*SPEAKER*, FEB 5, 1892: 175)

In the previous chapter, the archaeological discourse was introduced as a space characterised by indefiniteness and ambivalence. Following the assumption that the archaeological search can be read as the symbolic desire to retrieve Lacanian wholeness, the present chapter focuses on the popular reception of Heinrich Schliemann as an example of how the archaeological space was used as a site of individual and collective identity formation in contemporary cultural texts and practices. On the basis of what I have termed the *Frankenstein Discourse*, I demonstrate how on a symbolical level Schliemann, in analogy to Victor Frankenstein, created identity arti-

1 I would like to thank Professor Barbara Patzek for her helpful references in regard to Heinrich Schliemann.

ficially by piecing together various heterogeneous fragments.[2] Incorporating both an individual and collective past and present, Schliemann thus constructed a coherent *his*tory[3] that was popularly received by his British contemporaries. I argue that British society adopted Schliemann's *his*tory, which was both highly subjective and androcentric, as a prosthetic memory to create meaning and identity for themselves at a time of instability. I further show that what was decisive for Schliemann's popularity in Britain was first of all the object of his excavations, that is, ancient Greece as the *most familiar strangeness*, and second Schliemann's personality, which, in the tradition of the British Dilettanti movement, represented the self-made man and was particularly suitable as a Victorian role model. Both encouraged the identification of contemporaries with Schliemann's *his*tory.

2 These ›heterogeneous fragments‹ can be individual and collective memories, stories, history, images, artefacts, etc.

3 In the following I use the term *his*tory both to refer to a self-created story that eventually solidifies into a ›history‹ and, following feminist approaches of the late 1960s, to refer to a male-dominated and determined his-story as opposed to her-story (cf. Scott 1999).

5.1 A Case in Point: Heinrich Schliemann as a Victorian Role Model

Setting, as we instinctively do, the body above the mind, our sympathy goes out to the brave Homeric tales across the intervening gulf of age and race.
(*Quarterly Review*, Jul, 1889: 109)

The story of Schliemann's life reads like a classic rags-to-riches story.[1] He spent his childhood as the son of a poor clergyman who could not afford higher education for the young Schliemann. Heinrich started working as a grocer's shop boy when he was fourteen years old, left his hometown Ankershagen for Hamburg and Amsterdam, and eventually succeeded in entering the world of international commerce after years of hard work. Alongside his job in the grocer's business, Schliemann began to study a variety of languages, until he was finally fluent in sixteen modern foreign languages (cf. Cobet 2007: 35). His talent for languages and dogged ascension in society laid the foundation for the mythological status his life was to acquire. As the literary magazine *Belgravia* put it:

»Languages were the young clerk's passion. He believed that his fortune lay in their mastery, that with them he could storm the heavens. […] He wrote from dictation, and paid great attention to his chirography, which was sadly at fault. […] Though constantly running about the streets, he always had his grammar and dictionary under his arm. Necessity invented for him a method which greatly facilitated his studies.« (*Belgravia*, Jul, 1877: 78)

It was through this self-study, further years of hard work, commercial skills, and luck that Schliemann made his fortune in Saint Petersburg and

1 For a detailed description and analysis of Schliemann's career and excavations, cf. David Traill (1995) *Schliemann of Troy: Treasure and Deceit*, David Traill (1993d) *Excavating Schliemann*, and William M. Calder III (1996) »A New Picture of Schliemann«.

Sacramento, which allowed him to quit the commercial world in order to realise his lifelong dream and apply himself to the study of archaeology. In this field he continued his approved method of self-study in Saint Petersburg and Paris. Schliemann travelled the world before eventually visiting the Troad in 1868, where he spontaneously decided to excavate the city of ancient Troy. In 1870 Schliemann started his first excavations at Hisarlık without a *firman*,[2] following the opinion of the British archaeologist Frank Calvert in assuming that Hisarlık was the site of ancient Troy:

»It was Calvert's excavations that Schliemann saw when he visited Hisarlik on 10 August. On 15 August [1868], when Schliemann paid him a call in the Dardanelles, Calvert realized that this amateur archaeologist had the enthusiasm, drive and money to complete the task he had begun. He told Schliemann of his view that Hisarlik was the site of Troy. Though Schliemann had heard mention of this theory in Paris, he seems to have forgotten about it. Now, for the first time, he was forced to take it seriously. [...] Schliemann has reduced Calvert's role to merely confirming his own view of the identity of Hisarlik with the site of Troy.« (Traill 1995: 56-57)[3]

It was in the years 1872 and 1873 that Schliemann first found his way into the public eye by claiming to have discovered ancient remains at Hisarlık that definitely proved it to be the Homeric Troy, causing *The Times* to enthusiastically proclaim:

»Troy has, indeed, risen again from the ground which has buried it for more centuries than the Iliad has books; and hardly had its resurrection been proclaimed to the world when the signals of strife were in the sky, and there began a new War of Troy, destined, probably, to last at least as long as that in which Hector and Achilles strove. Dr. Schliemann commenced his excavations at the Mound of Hissarlik in the Troad – the spot attested by a tradition extending Greece as the site of Heroic Ilium

2 A *firman* is an official excavational permit issued by the Turkish government.

3 Cf. also Arentzen: »Schliemann was not the real inventor of the identification of Hisarlik as Troy, and in *Ithaka, der Peloponnes und Troja*, he tells the world about those who had come to this conclusion before.« (Arentzen 2001: 171) Schliemann submitted *Ithaka, der Peloponnes und Troja* (1869) as his dissertation to the University of Rostock, from which he subsequently received his doctorate.

– in April, 1870. […] The Homeric Troy, it has been argued, was a city in the clouds. But, given these facts, – the constant tradition that such a city as Troy once existed, the discovery of a town on the site to which this tradition points, and the correspondence of these remains with the town called Troy in the Iliad, – the ›ethereal‹ view has at least probabilities against it, and, in any case, is irrelevant to the question as between Hissarlik and another site.« (*The Times*, May 31, 1875: 7)

By claiming to have discovered the remains of ancient Troy, Schliemann not only brought to light what for centuries had only existed as a collective myth and dream, thus fulfilling the wishes of a larger public, but also realised the dream of his own life in two respects. He achieved this first by purporting to have realised a childhood dream in excavating Troy and second by appearing with it before an admiring public.

It was the combination of Schliemann's ambition, devotion, and energy in the form of hard and continuous work that finally not only made possible the realisation of his personal aims in life but also brought him recognition from a public which constantly marvelled at both his achievements and the comforting trajectory of his success:

»The story of his life would have been well worth telling, even if his unselfish enthusiasm had been disappointed at Mycenæ and Ilium as it was at Ithaca. It is the story of a self-made man, who started from something like indigence, and acquired rapidly a considerable fortune; but also of a self-educated man, who pursued knowledge concurrently with fortune, and prized both chiefly as means for realizing an early dream of liberal ambition, and who has realized it while yet well on the better side of the grand climacteric. Mere industry and resolution and sagacity even when conjoined with fervour do not compass all this without some aid from exceptional endowments; and one of these certainly was a marvellous facility in acquiring language.« (*Athenæum*, Dec 4, 1880: 749)

With regard to contemporary British society, it is both Schliemann's personal achievements and his public fame that make him particularly suitable as a Victorian role model, as underlined by the reference to his story being that of a self-made man (cf. ibid.), a feature which was emphasised even posthumously, as an obituary by the Irish classicist John Pentland Mahaffy in 1891 notes:[4]

»Even from the notices in the press of Europe it can be clearly gathered that Schliemann occupied a unique position, and had attained an interest and an importance in the public mind of Europe quite extraordinary beyond the sphere of politics. When we add that he raised himself from obscurity by many years of toil, that he did not come before the world till he had made his fortune in business, at the age of forty-six, we add to the distinction of the man.« (*Athenæum*, Jan 3, 1891: 27)

That Schliemann was considered the embodiment of the self-made man and the perfect role model for nineteenth-century Britain is further made explicit in Dickens's popular periodical *All the Year Round*: »Time alone will show to what point Dr. Schliemann will now direct his amazing energy and indomitable patience and perseverance. His life reads the moral to *our* copy-book phrase: ›Labor omnia vincit‹.« (*All the Year Round*, Jun 1874: 205, emphasis mine) In this comment, Schliemann's life is again mentioned as an example that »reads the moral to our copy-book phrase: ›Labor omnia vincit‹« (ibid.), which refers to nineteenth-century British society in two respects: first due to the fact that as the author of the article Dickens speaks of »*our* copy-book phrase« (ibid., emphasis mine) and thus alludes to the cultural background of the time, and second because the phrase ›labor omnia vincit‹ can be read as one of the guiding principles of Victorian society in general. This is also made explicit by a statement on Schliemann Gladstone made before the Society of Antiquaries in 1875: »›We owe a debt to [Schliemann] for his devotion and for his example – a debt which could never be cancelled even if he were to fail in the proof of any of his conclusions‹ (Hear, hear).« (*The Times*, Jun 26, 1875: 9)

---

4 The popularity of Schliemann's autobiographical accounts corresponds to a general Victorian fasciation with autobiographical writing: »Throughout the Victorian period the autobiographical impulse flourished more variously than ever before. The keeping of journals and diaries as memorials to the immediate past, a practice already brought to a considerable perfection by the Romantics, became a common pursuit, to some indeed almost a responsibility, an occasion for self-accounting and communion.« (Buckley 1967: 98) Cf. also Linda H. Peterson (1986, 2001).

Furthermore, it is particularly interesting in this context to find this statement in *All the Year Round*, as it points towards striking parallels between Schliemann's life and career and Dickens's fictional heroes in novels such as *Hard Times* and *Great Expectations*. Like the protagonists in these novels, Schliemann gained fame and fortune through hard work and perseverance, which helped him climb the social ladder from a poor clergyman's son to a gentleman of the world.[5] Similar to Dickens's novels, Schliemann's life as narrated in popular texts demonstrates with great assurance that people who employ the right moral values of perseverance and ambition have the chance to succeed in life, despite the poor conditions in which they grow up: »He only came in contact with the lower classes of society, and as he was forced to work from five in the morning until eleven at night, it is scarcely astonishing that he rapidly forgot the little he had learnt in his youth. Yet for all that he did not lose the desire to study« (*All the Year Round*, Jun 1874: 200), enthused *All the Year Round*. That this idea was very popular at the time is further mirrored by the popularity of Dickens's novels and periodicals on the one hand and the innumerable contemporary magazine and newspaper articles which presented Schliemann as the embodiment of this type of role model on the other. In the following, the different facets of Schliemann as a role model for Victorian society will be further illustrated in relation to the nineteenth-century magazine and newspaper landscape. In this context, I will show how certain aspects of Schliemann's life and character were continuously recycled in different articles and essays of the time.

In *Peter Parley's Annual*,[6] a periodical for children with colourful illustrations published at Christmas-time in London from 1840-1892, the author in his essay »Dr. Henry Schliemann« comments on Schliemann as follows:

---

5 This emphasis on success through hard work is also underscored in Traill's article »Schliemann on Dickens« (1991), where Traill comments on Schliemann's visit to Dickens's public reading of *A Christmas Carol* in New York on January 3, 1868. Schliemann's life can also be read in terms of an educational novel (cf. Helmut Scheuer 1990).

6 The editor of *Peter Parley's Annual* was William Martin, writing under the pen name of Peter Parley.

»If you have [heard of Schliemann], you will feel interested to hear something about the story of his life; and if you have not, you will not on that account find *the story* any the less *edifying and instructive.* Dr. Schliemann's life is a splendid illustration of the saying of Virgil that ›Labor conquers everything‹ (*labor omnia vincit* [sic]). Only let a *boy or girl have* some one *real definite desire or purpose in life*, and it is *their own fault if sooner or later they do not attain to their desired goal, no matter what obstacles seem to be in the way. It is only the weak, only those who would never have achieved aught whom we hear bemoaning lost opportunities and wasted lives, telling us what great men and women they might have been had circumstances been more fortunate, while it is really the power of conquering and surmounting apparently adverse circumstances that forms the crowning glory and the main strength of genius.*« (*Peter Parley's Annual*, date unknown:[7] 82-83, emphasis mine)

Although the message is essentially the same as that proclaimed in *All the Year Round*, in this passage it is even more explicit. Right at the beginning, Schliemann's life is introduced as »edifying and instructive« (ibid.) and thus as an example worth emulating. Again Schliemann's career is quoted as a prototype of the realisation of the principle ›labor omnia vincit‹, which is then directly transferred to the reader, the »boy or girl« (ibid.) mentioned in the third sentence. What Schliemann's life exemplifies is that both boys and girls,[8] once they have a »real definite desire or purpose in life« (ibid.) – and the underlying tone here suggests that they had better have one –, are responsible for the achievement of »their desired goal« (ibid.), since if they fail, there is no way to excuse this.

The opposite of what Schliemann represented finds expression in the drunken miller Schliemann claims to have met while still a penniless grocer's shop boy:

---

7 Although the date of publication is unknown, there is evidence in the text that it must have been during Schliemann's lifetime (cf. *Peter Parley's Annual*, 83) and shortly after his Trojan exhibition at the South Kensington Museum (from December 1877 to January 1881) (cf. ibid.: 82).

8 It is noteworthy that both boys and *girls* are addressed here, since what is further delineated by the author as worth aspiring to are qualities that at the time were generally identified as male (i.e., »great men and women« (*Peter Parley's Annual*, date unknown: 82), »the power of conquering and surmounting« (ibid.: 83), and »crowning glory and the main strength of genius« (ibid.).

»But one day there happened an event Dr. Schliemann will remember to the last day of his life. One evening when he was keeping the shop a tipsy miller's boy tumbled in asking for something to drink. Now, this tipsy boy was the son of a clergyman, who had given him a good education and had sent him to college, whence he had been expelled for bad conduct. This so angered his father that he cast him off, saying that he would no longer provide maintenance for the worthless son. Thus cut off from all resources, the youth had embraced the trade of a miller, and, having no more intellectual occupations, he had taken to drinking as a recreation. Now, this vice had not caused him to forget all the Homer he had learnt at college, and after young Schliemann had given him the drink he had demanded, he began, in his half-drunken state, to spout some lines from the immortal bard.« (Ibid.: 85)

What is interesting here are the parallels between the miller's boy's background and Schliemann's own. Although Schliemann had not been sent to college, his father was a clergyman too, and Schliemann had spent some time at a grammar school before he had to start to work as a grocer's shop boy to earn his living. Unlike the miller's boy, however, Schliemann had never been given the chance of a good education, while the miller's boy had neglected it. Nonetheless, as Schliemann's biography demonstrates, Schliemann succeeded in spite of these circumstances, whereas the miller's boy became a desperate drunkard. The miller's boy's life and that of Schliemann thus can be seen as chiasmatically intertwined, in that the former turns, due to »bad conduct« (ibid.), from a privileged child with the best chances for his future into an alcoholic miller's boy and the latter, starting as a penniless grocer's shop boy, finally, by means of self-education, achieves prosperity and worldwide fame. Thus, the miller's boy is used as a foil for Schliemann, consolidating Schliemann's position as a role model who had nothing and achieved everything, as opposed to the miller's boy, who had had everything and lost it. The relevance of the miller's boy anecdote in regard to Schliemann as a role model is underlined by its popular reception in contemporary newspapers and magazines, where it is mentioned in all cases in which Schliemann's biography is presented in more detail.[9]

9 »Dr. Schliemann's Early Career«. *All the Year Round*, June 1874: 200; »Dr. Schliemann«. *Liverpool Mercury* June 8, 1877; »Dr. Schliemann«. *The Graphic*. Jan 3, 1891; »The Late Dr. Schliemann«. *The Times*. Dec 29, 1890: 6.

As opposed to the example of the miller's boy, Schliemann's life story shows that real strength of character lies in those who have gained »the power of conquering and surmounting apparently adverse circumstances« (ibid.) and succeeded against all odds. This is further emphasised by the following passage:

»A wise man[10] once said: ›The longer I live the more certain I am that the great difference between men – the feeble and the powerful, the great and the insignificant – is *energy and invincible determination* [sic]; a purpose once fixed, and then death or victory. That quality will do anything that can be done in this world, and no talents, no circumstances, no opportunities, will make a two-legged creature a man without it‹. Be very sure the wise man is right, and *as you go on in life your own personal experience will confirm and prove his remark.*« (Ibid.: 83, emphasis mine)

Here the importance of a definite goal in life and the determination to attain it is raised above anything else, while »talent« (ibid.), »circumstances« (ibid.), and »opportunities« (ibid.) alone do not suffice to »make a two-legged creature a man« (ibid.). This maxim is directly transferred to the readers by reference to their respective lives and future experiences. The very idea that by means of hard work and perseverance everyone is able to succeed in life, disregarding both their inherent capacities and their surroundings, is paradigmatic for the model of the self-made man. Proof of this is the careers of men like Schliemann, who despite their poor background achieved their aims in life, as formulated by *The Times*:

»I think that *we see in Dr. Schliemann a spectacle*, not perhaps so rare in his own country as it is among us, of the most pure, simple-minded, and ardent devotion to the cause of literature and knowledge in one of its most interesting departments. Dr. Schliemann, with immense labour, as he has told us in that very striking work of his autobiography, qualified himself and provided the means which others would have spent perhaps in ignoble waste; and having in one arduous course of labour thus furnished himself with what was necessary for the purpose of his pursuit, he proceeded to engage himself and to give his life to another course of exertion, perhaps still more arduous, and the result certainly has been that he has given us the most

10 Sir Thomas Fowell Buxton (1786-1845), MP.

splendid example of distinguished sacrifice.« (*The Times*, Jun 26, 1875: 9, emphasis mine)

Again, in this passage Schliemann's »immense labour« (ibid.), his »most pure, simple-minded and ardent devotion to the cause of literature and knowledge« (ibid.), and his prudential handling of his own resources, »which others would have spent perhaps in ignoble waste« (ibid.), are presented as exemplary. Furthermore, Schliemann is referred to as »a spectacle, not perhaps so rare in his own country as it is among us« (ibid.), which makes his achievements even more outstanding and worth emulating from the perspective of contemporary society.

In particular, the quality of perseverance plays a crucial role in this context since it has to be seen as one of the key ideals of Victorian society. This is illustrated by the best-selling conduct book *Self-Help* by Samuel Smiles, first published in 1859 and recently labelled »the Bible of mid-Victorian liberalism« (Cohen/Major 2004: 611). In *Self-Help*, Smiles presents the lives and careers of people who through their extraordinary accomplishments were particularly suitable as role models of the time. While the first edition of 1859 bore the subtitle »with Illustrations of Character and Conduct«, in the second addition of 1866 »Perseverance« was added to the subtitle. This is characteristic insofar as it not only reflects the relevance of perseverance for the book itself, but also for the collective role model presented in the book and eventually conceived by the Victorian reader. The popularity of *Self-Help* during the Victorian Age underlines the importance of this concept at the time: »*Self-Help* was one of the most popular works of nonfiction published in England in the second half of the nineteenth century. It sold 20,000 copies within a year of its appearance and surpassed a quarter of a million by the time of Samuel Smiles's death forty-five years later.« (Sinnema 2002: vii)

What is interesting in this regard is the fact that besides hard work, perseverance also has to be seen as one of the basic requirements of archaeological work. In *Self-Help* this is illustrated by the example of the archaeologist Austen Henry Layard, who became famous through his Mesopotamian excavations in the late 1840s. Characteristically, the chapter presenting Layard and his colleague Henry Rawlinson as role models is titled »Application and Perseverance« (Smiles 1859: x), and the passage describing their achievements is subtitled »Perseverance displayed in the discovery of Nineveh marbles by Rawlinson and Layard« (ibid.) and introduced as follows:

»But there are equally striking illustrations of perseverance to be found in every other branch of science, art, and industry. Perhaps one of the most interesting is that connected with the disentombment of the Nineveh marbles, and the discovery of the long-lost cuneiform or arrowheaded characters, in which the inscriptions on them are written, – a kind of writing which had been lost to the world since the period of the Macedonian conquest of Persia.« (Ibid.: 79)

After a short description of Rawlinson's beginnings of the study of cuneiform writing with the assistance of a »ci-devant clerk of the East Indian House« (ibid.) named Norris, Layard is introduced:

»But to make the learning of these two *self-taught men* of avail, a third laborer was necessary in order to supply them with material for the exercise of their skill. Such a laborer presented himself in the person of Austen Layard, originally an article clerk in the office of a London solicitor. *One would scarcely have expected to find in these three men, a cadet, an India House clerk, and a lawyer's clerk, the discoverers of a forgotten language, and of the buried history of Babylon; and yet it was so.*« (Ibid.: 80, emphasis mine)

Apart from the explicit reference that Rawlinson, Norris, and Layard were »self-taught« (ibid.) men, the fact that all three started out as cadets or clerks to then become famous through the discovery of a »forgotten language and the buried history of Babylon« (ibid.) underlines their achievements as self-made men. This is further emphasised by the description of Layard's character and work:

»[A]fter the lapse of many years, with comparatively slender means at his command, but aided by *intense labor and perseverance, resolute will and purpose, and almost sublime patience, borne up throughout by his passionate enthusiasm for discovery and research*, he succeeded in laying bare and digging up an amount of historical treasures, the like of which has probably never before been collected by the industry of any man. […] And the story of the disentombment of these remarkable works, as told by Mr. Layard himself in his ›Monuments of Nineveh‹, will always be regarded as one of the *most charming and unaffected records which we possess of individual enterprise, industry, and energy.*« (Ibid.: 80-81, emphasis mine)

The qualities mentioned here in regard to Layard are more or less identical to those ascribed to Schliemann in contemporary newspaper and magazine articles. Furthermore, the characterisation of Layard and his work is also reminiscent of reports on Belzoni's archaeological enterprises in Egypt at the beginning of the nineteenth century, which repeatedly accentuate Belzoni's zeal, perseverance, and hard labour.

In addition to these qualities, Schliemann's enthusiasm is generally mentioned as one of his most outstanding character traits and as worth emulating:[11]

»During the fine weather of that and the following year [1873] Dr. Schliemann laboured incessantly with the zeal and perseverance of an enthusiast, and was rewarded for his expenditure of time and money – for the work was carried on entirely at his own cost – by the unique and inestimable collection which by his kindness is now offered to the inspection of the British public.« (*Manchester Times*, Dec 29, 1877)

Here the qualities »zeal« (ibid.) and »perseverance« (ibid.) are explicitly ascribed to the character of an enthusiast[12] who is ready to invest all his personal and pecuniary means to achieve his goal. The fruits gained from endurance and hard work are represented by the »unique and inestimable collection« (ibid.) Schliemann presented to the British public in London. Schliemann's enthusiasm is also pointed out by Gladstone in a eulogy on Schliemann given before the Society of Antiquaries:

»›I must again express my hearty congratulations to Dr. Schliemann on the result of his labours, and my personal gratitude for all that he has done for the enlargement of our knowledge, and for his exhibition, in an age which has perhaps somewhat de-

11 Characteristically, enthusiast/enthusiasm are the most frequently used words in relation to Schliemann and his excavations in the various reports on him in periodicals and newspapers of the time (cf. also Runnels 2002: 17).

12 Cf. also Wieland Schmied's portrayal of Schliemann as an enthusiast in *Troja und Homer: Porträt eines Enthusiasten* (1990).

teriorated in that respect, of the *most noble and high-minded enthusiasm* (Loud cheers).‹« (*Daily News*, Mar 23, 1877, emphasis mine)

Again, Schliemann's »most noble and high-minded enthusiasm« (ibid.) is explicitly referred to as an exemplary quality that Gladstone finds wanting in society at the time. As emphasised by the passionate reaction of his audience, their »loud cheers« (ibid.), reported as a reaction to his exclamation, Gladstone hits the mark with his statement. The appreciation of enthusiasm as such is supported by the various comments on Schliemann's character that celebrate this quality.

As the following comment in the *Freeman's Journal* implies, enthusiasm was also always to some extent associated with dilettantism: »An enthusiast in archæology, Dr. Heinrich Schliemann, has been for several years grubbing amongst the rubbish which he believes the dust of the *reliquias Danaum* – the ruins of sacred Troy.« (*Freeman's Journal*, Aug 25, 1873) The very fact that Schliemann is said to be »grubbing amongst the rubbish« (ibid.) where he »believes« (ibid.) to find the remains of ancient Troy suggests a certain amateur archaeology reminiscent of the tradition of dilettantism in Britain. Unlike other countries, Britain was much more liberal and open to amateurs in general and in the fields of antiquarianism and archaeology in particular (cf. Chapter 3.1). Consequently, public opinion about amateurs was positive, especially as the amateur was closely linked to the self-made man. This also has to be seen as one of the major reasons why Schliemann enjoyed such wide popularity in Britain from the beginning of his career as a self-taught archaeologist,[13] while in Germany he had to struggle hard to obtain acceptance, as a remark by Gladstone reproduced in the *Illustrated London News* reveals:

»He had seen comments upon some of these discoveries of Dr. Schliemann which had filled him with pain, because they had not been conceived in that spirit of generosity and brotherhood which ought to unite whatever differences of opinion might arise in this inquiry. He was only sorry to say that even in Germany, among that

13 This is supported by the following statement by the *Quarterly Review*: »The world was startled out of its apathy to archæological things, and England especially, led by Mr. Gladstone, was disposed more than the half.« (*Quarterly Review*, Jul, 1898: 91)

great and learned fraternity, they were not united by that true brotherhood and spirit of generosity in this matter.« (*Illustrated London News*, Mar 31, 1877: 306)

That the situation was completely different in Britain not only in regard to Schliemann but also to dilettantism in general is emphasised by the following comment: »We would take, as English men of science generally take, a more generous view of his career and achievements than has been the case in Germany, which has hitherto rather played the part of step-mother than of mother, treating him as prophets are treated in their own country.« (*Quarterly Review*, Jan 1878: 91) In spite of this »more generous view of Schliemann's career« (ibid.) of »English men of science« (ibid.) as opposed to that of their German colleagues, the emphasis on archaeological *excavation* as revolutionised by Schliemann also played an important role in Britain, as stated by the *Quaterly Review*:

»Add to this the romance of a man with little education, but with boundless enthusiasm and a deep purse, *startling the learned world, and in a moment throwing its machinery out of gear*. From that time the phrase became familiar that ›the spade is the final arbiter‹. But what the spade does to-day it may undo to-morrow.« (*Quarterly Review*, Jul 1897: 80, emphasis mine)

This passage points out the relevance of the spade in connection to the impact of Schliemann's work on the »learned world« (ibid.). The fact that since Schliemann the spade is said to have become »the final arbiter« (ibid.)[14] – having substituted the hitherto accepted predominance of intellectual and philological studies – is significant in regard to the contemporary background, because it corresponds to the emphasis placed on materiality at the time, as formulated by Gehrke:

»This was not at all merely an affair for specialists in the ivory tower of science but a thoroughly basic question that met with widespread and vital interest of whether

14 Schliemann is generally seen as the founder of spadework in archaeology (cf. Gehrke 2006: 222), as underlined by the *Fortnightly Review*: »The department which Dr. Schliemann has really made his own is not ›archaeological science‹ of any period. It is practical exploration by the spade.« (*Fortnightly Review*, Apr, 1884: 434)

that which was described in the *Iliad* held water. Everything just had to be concretely tangible because it was original! This postulate was reinforced considerably by the historical realism of the 19th century, the penchant of historicism for that which can be documented, its *goût archivique*. This was precisely the context in which that which was concretely tangible could lead to historical reality, thus scientifically demonstrating the authenticity of the author and his tale, a piece of virtually criminological evidence made available by the presence of the findings, which one could extend even more with the help of the spade. The search for Troy for the purposes of such visualisation and confirmation was conducted with passion. It was a topic in which both science and the public showed a lively interest. The virulence of this search is particularly obvious in Heinrich Schliemann, but he was by no means alone. He was only a particular representative of this attitude, as evidenced not least in the widespread interest his discoveries sparked in the public.« (Gehrke 2006: 222, translation mine)

This is further underlined by the following remark by William Simpson, a ›special artist‹ who worked for the *Illustrated London News*, amongst other publications:

»At the same time explorations with the spade are being carried on all over the old world, and amongst these explorers Dr. Schliemann has entitled himself – whatever we may think of his opinions – to be called ›King of Men.‹ The influence of all these various labours is being felt, and archæology is admitted to be advancing in a manner which it has never done before, and we need have little doubt that the more accurate knowledge which we are acquiring will, in all probability, soon clear up many of our doubts as to the old tales and histories which have come down to us.« (*Fraser's Magazine*, Jul 1877: 2)

In Simpson's comment, Schliemann becomes virtually the ›King of the Spade‹, with the power to »clear up many of [society's] doubts as to the old tales and histories« (ibid.). By means of his spade, Schliemann is able to provide material proof for the hitherto indefinite and vague, which, considering the Victorian obsession with materiality and empiricism, must have been fascinating for many.

Furthermore, the fact that Schliemann was publicly received as an archaeological enthusiast is interesting since it links the concepts of the enthusiast as a person prone to intense emotions to that of the archaeologist, commonly regarded as concerned with materiality and rational investigation: Schliemann's »plastic imagination and his fantastic belief in the reality of the battles of Troy described by Homer were more marvelled at than judged in their driving force for the overall image of his personality« (Meyer 1955: 134, translation mine).[15] This is particularly significant since archaeology can already be seen as a field of work combining fact and fiction, reality and fantasy, as suggested in the following statement:[16]

»Imagination is a very important qualification for an archæologist to possess: without this he becomes only a chronicler of facts – a mere writer of catalogues – and must be wholly without the power of seeing the value of the discoveries he may have made. The old theory that a student of antiquities must be a ›Dryasdust‹ ought to be now exploded. There is no department of thought presenting so many influences which are calculated to excite the imagination. The constant contact with relics of former times is an incentive to this faculty, picturing to the mind the former conditions of things and of men. The use and purpose have often to be realised of shreds and patches which have been brought to light, and this is the power which accomplishes it. But, in proportion to the strength of this power, counterpoise of judgement is necessary, otherwise the imagination gets loose and runs riot. Dr. Schliemann is, undoubtedly, an able man; but he must be credited with a vast amount of this sort of unbalanced imagination in order to explain the creations

15 This is illustrated by Schliemann's autobiographies, which »survive written in various languages and extant in various and often contradictory versions, dating form 1852 until 1891. Most often cited is *Heinrich Schliemann. Selbstbiographie bis zu seinem Tode vervollständigt*. This is a version of the autobiography prefixed to *Ilios* (1880-81) and brought up to date by Dr. Alfred Brückner under the guidance of Sophia Schliemann, whose preface to the first edition is dated 23 September 1891« (Calder 1986: 20).

16 The blurring between fact and fiction is also pointed out by Linda Dowling in regard to Bulwer-Lytton and Becker: »It was the poignant ordinariness of broken vessels, utensils, and foundations excavated at such sites as Pompeii and Vulci that made it possible to reconstruct with an almost fictional degree of vividness the circumstances of daily life in classical times. And indeed it was upon such richly suggestive rubble that Edward Bulwer-Lytton erected his novel *The Last Days of Pompeii* (1834) and Wilhelm Adolph Becker raised his quasi-fictional studies of Greece [...] and Rome.« (Dowling 1988: 222)

which he has produced out of the exploration of Hissarlik.« (Fraser's Magazine, Jul 1877: 5-6)

Here William Simpson emphasises the relevance of imagination for archaeological work. At the same time, however, he points out the necessity of a »counterpoise of judgement« (ibid.), since otherwise there is the risk of the »imagination get[ting] loose and run[ning] riot« (ibid.). If we consider in this context the assumption implied above that in the figure of the archaeological enthusiast both emotionality *and* rationality are combined, this passage becomes particularly interesting. As Simpson's statement suggests, although archaeologists are primarily concerned with concrete material remains, they need *imagination* in order to become more than »chronicler[s] of facts […] mere writer[s] of catalogues« (ibid.). This means that contrary to the commonly held belief, archaeological work is not as matter-of-fact as was often assumed. It is this merger of emotionality, imagination, and rationality, in the form of the necessary »counterpoise of judgement« (ibid.) as represented by Schliemann as the embodiment of the enthusiast archaeologist, which can be seen as another reason for Schliemann's wide popularity in nineteenth-century Britain. At a time when emotionality and imagination were considered inferior to rationality and common sense, and thus to qualities that were supposed to be suppressed and controlled in order to secure a stable personality, Schliemann's example of an archaeological enthusiast displaying both was a source of seemingly endless fascination.

What is also interesting with regard to Schliemann's reception as a role model in Britain is the question of why there was a popular need for role models like Schliemann at all if the Victorians were as confident and self-assured as their value system seemed to suggest. The answer to this might be found in the fact that in spite of its firm belief in a rigid moral system, Victorian society was in need of someone to serve as its embodiment. In this context, men like Schliemann, Belzoni, and Layard, who through the coloured and often exaggerated description of their achievements become larger-than-life figures, are extremely important. The combination of this larger-than-life status with an earlier life spent in humble circumstances made men like Schliemann particularly suitable as role models, as a wide public could identify with them, even if, at the same time, it was most unlikely to ever attain the same position.

5.2 Sophia Schliemann: ›Angel Outside the House‹

Madame Schliemann, fulfilling a promise made to her husband, is continuing the excavations at Troy which made his name famous. She is personally superintending much of the work.
(*Woman's Herald*, Oct 26, 1893: 575)

In addition to Schliemann's own status as a self-made archaeologist, which brought him fame in Britain, Schliemann's young Greek wife played a decisive role in his popularisation. As much as Schliemann was a self-made man, the image of Sophia Schliemann, skilfully popularised to support Schliemann's own public success in Britain, can be considered to have been pretty much self-made by him, as pointed out by Traill: »To a considerable extent, this Sophia, whom we all admire, is a creation of Heinrich's vivid imagination and his mastery of what is now called ›hype‹.« (Traill 1993a: 235)[1] In a way similar to that in which Schliemann's image as a self-made man made him suitable as a contemporary Victorian role model, the public image of Sophia also seemed to satisfy – even though less prominently – certain underlying contemporary longings. In the following, I show how Schliemann used the archaeological space to create and promote Sophia as the compromised embodiment of an alternative type of woman to the trad-

1 For a discussion of Sophia's role in Heinrich's excavations, cf. Traill's article »The Archaeological Career of Sophia Schliemann« (1993a): »In summary, Sophia made significant contributions to the excavations at Troy and Mycenae, though these were less dramatic and less extensive than Schliemann and his biographers indicate. [...] I would suggest that she was much more of a wife and mother and much less of an archaeologist than is generally believed. Certainly, if we compare her to her contemporaries Jane Harrison and Amelia Edwards, her knowledge of and interest in the material remains of the ancient world are conspicuously lacking.« (Traill 1993a: 243)

itional Victorian ›angel in the house‹ on the one hand and the New Woman on the other.

Schliemann's first marriage, to the Russian woman Ekaterina Lyschin, had been troublesome right from the start, and Schliemann claimed that he did not find any support in Ekaterina for his plans and wishes. When she refused to immigrate to France with him, Schliemann effected his divorce from her in Indianapolis in 1869, after having fraudulently acquired American citizenship earlier that year.[2] Only three months later, Schliemann married the Greek woman Sophia Engastroménos, who was thirty years his junior. As Mannsberger points out, Schliemann had made sure prior to his marriage that his future wife was to be a young, beautiful, and, most importantly of all, Greek woman with a passion for and knowledge of ancient Greek language, literature, and archaeology (cf. Mannsperger 1992: 67), which once more underlines Schliemann's determination to form his life according to his ideas. Sophia did not fail Schliemann in that respect: From then on she was to be his companion and was presented by Schliemann in the public accordingly, while his first wife, with whom Schliemann had three children, was barely mentioned.[3] Due to Schliemann's successful propaganda, most people did not even know of Ekaterina, and even if they did, it was nothing positive, whereas Sophia was talked of in glowing terms:[4]

»The hardships of the excavations were again, as in Troja, shared by Schliemann's wife, Sophia, a Greek lady whom he had married seven years ago. In consequence of these hardships she was just then confined to her bed by a slight attack of fever, and our acquaintance with the lady remained on that occasion from ear to ear, as it were, for the thin partition wall prevented her, even less than her illness, from taking part in our conversation with her melodious voice. Later on, at Athens, I became better acquainted with this woman of rare talents and goodness, of stately appearance, the most winning manners, and of an apparent quiet disposition. Mdme Schliemann may be called her husband a good genius. If it were true what is said at Athens that

2 Cf. Traill (1982, 1995).

3 In »›Selbstinszenierung‹ Heinrich Schliemanns in der Darstellung der Mitarbeit seiner Frau Sophia« (1992), Mannsperger underscores Schliemann's effort to glorify and popularise Sophia.

4 Schliemann never mentions his first wife in his autobiographical notes.

Schliemann chose the woman for a wife who knew most verses of Homer by heart, a rare chance has made his choice a happy one in other dire tons as well. His divorce from his Russian wife (who lives with two children at St. Petersburg)[5] is also attributed to that lady's lack of appreciation of his favourite poet.« (*Pall Mall Gazette*, May 16, 1891)

In this retrospective comment on Schliemann's wives, bearing the biblical title »›Passing the Love of Women‹«, Sophia is presented as Schliemann's congenial companion, who shared his hardship and supported him in every possible way. She is seen as a woman of »rare talents« (ibid.) and »good genius« (ibid.) to her husband, who allegedly »knew most verses of Homer by heart« (ibid.). This rather unusual knowledge of Homer is implied to have been lacking in Schliemann's first wife, as the author speaks of »the lady's lack of appreciation of his favourite poet« (ibid.). The very fact, however, that Sophia's ability to recite Homer by heart is commented on as »a *rare chance* [that] has made his choice a happy one« (ibid., emphasis mine) reveals the extraordinariness of Sophia as a person, and more precisely as a *woman*, as emphasised by *All the Year Round*: »The fascinations of Troy, however, would not let him rest; *and the enthusiasm of his Athenian wife, a lady who can repeat by heart both Iliad and Odyssey,* hindered the old love from sleeping.« (*All the Year Round*, Jun 1874: 204, emphasis mine) Interestingly, Sophia's ability to repeat the *Iliad* and *Odyssey* is mentioned here as one of the driving forces with regard to Schliemann's fascination with and love for ancient Troy.

The knowledge of Homer and most of the other characteristics mentioned in relation to Sophia – her »rare talents« (*Pall Mall Gazette*, May 16, 1891), her »stately appearance« (ibid.), her »most winning manners« (ibid.), and last but not least the reference to her as her »husband['s] good genius« (ibid.) – are particularly interesting with regard to Victorian gender roles. While the fact that Sophia supported her husband was certainly in accordance with contemporary ideas of femininity in that it corresponded to the concept of the wife and mother as responsible for the well-being of husband and children, the way in which Sophia supported her husband can certainly be seen as progressive for the time. Although she was for her husband a good and supportive companion as prescribed by the Victorian

5 Schliemann's daughter from his first marriage died as a child.

concept of conjugal life, the very fact that Sophia joined her husband in his outdoor expeditions and excavationsmade her the opposite of the idealised ›angel of the house‹-type woman, as formulated by Traill:[6] »In recent years we have come to admire a woman who seemed easily to transcend the limitations imposed on her sex by the era and society in which she lived.« (Traill 1993a: 235) This is emphasised by Amelia B. Edwards's account of her visit to the Schliemanns, published in the *Graphic*:

»It is this faith – not merely in Helen and her *parure*, but in Homer, and in himself – that has carried Dr. Schliemann on from Troy to Argos, and from Argos to Mycenæ. Believing with his whole heart and soul in the scenes and personages of the Iliad, fortified by an indomitable will, and bringing to bear upon these qualities an unusual degree of steady, business-like intelligence, Dr. Schliemann unites in a singular degree the characteristics of the enthusiast and the practical man. […] The portrait of Madame Schliemann is an excellent likeness. She is a tall, fine woman, pale of complexion, with dark eyes and hair, and a stately bearing. I think I understood Dr. Schliemann to say that she was Greek. We had not the pleasure of being introduced to this lady; but we saw her pacing to and fro in the garden, dressed in a long white morning robe that swept the path like a peplum as she walked. None who have read Dr. Schliemann's late account in *The Times* of the painful and laborious way in which Madame Schliemann with her own fair hands assisted him, the other day, in emptying the graves at Mycenæ, can, I think, look upon this portrait without interest and admiration.« (*Graphic*, Jan 20, 1877)

First of all, as in the above-quoted passage, the famous portrait (cf. fig. 3)[7] here referred to by Edwards, which shows Sophia wearing the jewellery of Priam's Treasure, is said to show a woman of »stately bearing« (ibid.). This »stately bearing« (ibid.), suggesting strength and power in character, is later

6 Margarita Díaz-Andreu García and Marie Louise Stig Sørensen (1998) point out that the women doing archaeological work often got in contact with it through their husbands, as was the case with Sarah Belzoni, who contributed to her husband's book *Narrative of the Operations and recent Discoveries within the Pyramids, Temples, Tombs, and Excavations, in Egypt and Nubia* (1820) (cf. Díaz-Andreu García and Sørensen 1998: 4).

7 »[Sophia Schliemann's] portrait later became an icon in the ornamentation of the Trojan queen.« (Cobet 2007: 79, translation mine)

confirmed and extended to her physical capacities by the reference to the »painful and laborious way in which Madame Schliemann with her own fair hands assisted him [...] in emptying the graves at Mycenæ« (ibid.). Both Sophia's strength of character, visible in her portrait, and the physical hardship she endured to support her husband's aim, make her a rather *un*-typical Victorian woman. This impression is further suggested earlier in the same article, where Edwards quotes Schliemann's praise for his wife:

»He then spoke of Madam Schliemann, whose passion for archæological discovery is, apparently, as great as his own. ›Madame Schliemann has a perfect genius for excavating.‹ He said. ›She excavated the tumulus of Achilles on the plain of Troy. That was her own work entirely. She has shared all my fatigues and all my hardships. Nothing daunts her. But for her, I should never have succeeded in preserving the treasure of Priam from the rapacity of the Turkish officials on guard at Hissarlik during the excavations. [...] We could never have done it, but for the help and presence of mind of Madame Schliemann.‹« (Ibid.)

Fig. 3: »Dr. and Mrs. Schliemann«

Leisure Hour, July 7, 1877: 425. Copy owned by the author.

Here Schliemann not only attests to Sophia's »perfect genius for excavating« (ibid.) but also emphasises her independence by referring to excavations that were conducted by her alone. At the same time, Schliemann refers to Sophia's ability to beguile and distract the Turkish authorities by serving them wine while he excavates and hides the Treasure of Priam:

»When I struck the treasure and saw the first glitter of the silver under the rubbish, it was within a few feet of the chair on which the Turkish officer was sitting. I covered it again instantly. I told them all that it was my birthday, and that we would do no more work that afternoon. I invited every one into my hut to drink my health. [...] Our stores were nearly exhausted. But we brought out all the wine and liqueurs we had left, and Madame Schliemann kept filling their glasses, and managed to keep them indoors for several hours, during which time my nephew and I dug out and hid the treasure.« (Ibid.)

In this passage Sophia is presented in the light of her housewifely and ›female‹[8] qualities in that she lures and bemuses the men, although not with her ›female‹ charm, but with alcohol. Her ›female‹ attributes thus indirectly become a means of deception and fraud and make her an accomplice in her husband's crime against the Turkish authorities. The fact that Sophia intentionally distracts the men to enable her husband to commit a theft underlines her readiness to be on her husband's side, a trait emphasised by Traill when he states that »[w]e see her gamely putting up with the inconvenient realities of living at the site of an excavation – uncomfortable lodgings, often shared with exotic and undesirable fauna, blazing sun, freezing cold, rain and mud at Mycenae, and the relentless wins and dust of Troy« (Traill 1993a: 235). At the same time, the very means she uses in order to achieve her aim in this case emphasise that she is not only capable of hard work but also of (›womanly‹) slickness.

However, in addition to these rather negative ›womanly‹ attributes, Sophia is also often portrayed in her role as a mother and housewife looking after Schliemann's self-appointed ›Palace of Ilium‹,[9] as suggested in the following comment: »In his palace at Athens, assisted by the charms and

8 I consider ›female‹ and ›male‹ in accordance with Butler as culturally and thus arbitrarily constructed categories.

9 Also referred to as ›Ilíou Mélathron‹.

linguistic talents of Madame Schliemann and her delightful children, he saw many distinguished strangers.« (*Athenæum*, Jan 3, 1891: 27) The combination of Sophia's ›manly‹ intelligence, her capacity for hard work, and her »stately bearing« (ibid.) on the one hand and her ›womanly‹ beauty, charm, and slickness on the other undermine the traditional female role. With regard to the Victorian ideal of a wife and mother of two children, Sophia *does* and *does not* comply with coeval concepts of femininity and masculinity. In spite of this, or rather *because* of this, Sophia appears to have been extremely popular in Britain, as the various references in the contemporary news media underline.

Sophia is repeatedly referred to as Schliemann's equal, as implied by Gladstone's comment on their recent excavations at a meeting of the Society of Antiquaries in honour of Mrs and Mr Schliemann:

»It was probable that it would be reserved to Dr. Schliemann – *such was his energy, and such was his large fund of buoyance and strength which seemed to abide both in him and Mrs. Schliemann* – to traverse the scenes so as to complete and explain his own discovery (Cheers.).« (*The Times*, Mar 23, 1877: 10, emphasis mine)

Here, Schliemann's personal qualities and capacities, his strength and energy for continuous quest, enterprise, and discovery, are simultaneously ascribed to Sophia.

This near equality of Sophia and Schliemann as perceived by the British public can also be found in a portrait (cf. fig. 3) in the *Leisure Hour* of July 1877 subtitled »Dr. and Mrs. Schliemann« (*Leisure Hour*, Jul 7, 1877: 425). The portrait shows Heinrich Schliemann in contemporary dress on the left-hand side of the picture and Sophia Schliemann wearing the jewellery of Priam's Treasure on the right. The portrait of Sophia most likely is a mirror-inverted copy of the famous photo showing her with the alleged attire of Helen, while Schliemann's portrait also looks like a copy of a contemporary photograph. What is interesting with regard to this portrait, showing Heinrich and Sophia Schliemann side by side, is the question of why Sophia is portrayed with the ancient ornaments, while Schliemann wears contemporary garments. One possible answer to this question could simply be that there was no other picture of Sophia Schliemann available that could be used as a model. However, the question then would be *why* there was no other photograph of her available or publicly known other than the one with the ornaments. The answer to that probably lies in Schliemann's marketing strategies, with which he deliberately used Sophia to promote his work and books. Concerning the picture of Heinrich and Sophia Schliemann published in the *Leisure Hour*, much of Sophia's equality to Schliemann is created by the fact that she is shown wearing Helen's attire, whereas Schliemann does not have to wear the mask of Agamemnon to be identified as a great man. This emphasises that although Sophia's achievements were publicly admired, she was still very much used as a projection screen for Schliemann's success, as is represented by her displaying the ancient jewellery. If Schliemann presented himself in the public as a self-made man, he promoted Sophia according to his idea of a perfect wife and thus objectified her. In this regard Sophia, in contrast to him, becomes the ›other‹-made woman, who though seemingly equal to and independent of her husband is actually his creation. That Sophia is defined through her husband is also illustrated by a comment on Sophia Schliemann in the *Edinburgh Review*:

»We should not here omit to mention that in the course of Dr. Schliemann's operations at Mycenæ, one of the five subterraneous chambers, called Treasuries by Pausanias, was excavated by Madame Schliemann, who, here as at Hissarlik, proved herself the intelligent and devoted partner in her husband's toils.« (*Edinburgh Review*, Jan 1878: 254-255)

Although Sophia is here presented as the excavator of one of five subterraneous chambers at Mycenae, she is then referred back to her husband, in whose toils she proved herself an »intelligent and devoted partner« (ibid.).

However, in line with Schliemann's attempts to promote his own work through Sophia in the public, there are also various instances in which Schliemann repeatedly points out Sophia's sole achievements in regard to their excavations. That the British readily accepted Sophia's progressive position is supported by numerous comments on her achievements in contemporary newspaper and magazine articles and is underlined by the following introduction of her and the audience's reaction to her at a special meeting of the British Archaeological Association in April 1877: »Professor Donaldson said it was only right they should present their sympathies and congratulations to Mrs. Schliemann for the part she has taken in the discoveries (Hear, hear.).« (*The Times*, Apr 12, 1877: 8) Apart from the fact that Sophia's accomplishments are honoured here, the affirmative reaction of the audience suggests a collective agreement in regard to what is being said about Sophia. Her quality and power to act seemingly independently from her husband and her ability to superintend others are also suggested by the following passage from the *Illustrated London News*:

»Madame Schliemann's exploration of one of the so-called Treasuries at Mycenæ, which she superintended while her husband carried on the work in the Acropolis, has also revealed to us the details of one of these peculiar monuments, which are so important, not only as bearing on the old Pelasgic architecture of Greece, but also as bearing on the questions of art and chronology in reference to the objects found at Mycenæ.« (*Illustrated London News*, Feb 23, 1878: 178)

Here the excavation of one of the Treasuries at Mycenae is explicitly attributed to Sophia in that it is referred to as »Madame Schliemann's exploration«. One year earlier, the *Illustrated London News* even refers to Sophia's excavation as »one of the most difficult« of all those she and her husband had made:

»The examination of this Treasury, under Mrs. Schliemann, had been one of the most difficult they had ever made, partly from the nature of the terrain and partly from the obstructiveness of the delegate of the Greek Government, under whose Argus eye all the excavation were conducted.« (*Illustrated London News*, Mar 31, 1877: 306)

Through her archaeological work, Sophia, equal to her husband, has the power to *re*-create the past by bringing to light hitherto hidden relics:

»That the dome of this monument has been long broken in, it was well enough known; but, thanks now to Mrs. Schliemann, the whole of it has been excavated and exposed to view. While her husband was busy within the walls of the Acropolis, she undertook the exploration of this relic of the past and the approach to it, which was formerly completely concealed, is now laid bare.« (*Illustrated London News*, Apr 7, 1877: 330)

Sophia is thus not only allowed the power of *re*-creation, but also the possibility to examine and interpret what she has brought to light. In how far her interpretation carried authority with regard to what her husband presented to the public is hard to say. Regardless of that, in relation to women's traditional role in Victorian society Sophia's position as a female archaeologist at Schliemann's side has to be considered as quite progressive for the time.[10]

First of all, Sophia can be seen in terms of a supportive wife at the side of her husband, as pointed out in a comment on her and her husband in the *Leisure Hour*:

»These tombs [at Mycenae] were the especial objects of Dr. Schliemann's researches, and it is his belief that he has really discovered them. Three years ago this indefatigable explorer, accompanied by his energetic wife, a true ›helpmeet for him‹, commenced his work in the Acropolis of Mycenæ.« (*Leisure Hour*, Jul 1877: 426)

Although in this passage Sophia is referred to as »energetic« (ibid.), in the first place she is defined through Schliemann, as the fact that she is presented as »*his* energetic wife« (ibid.) suggests. This impression seems to be affirmed when she is said to be »a true ›helpmeet for him‹« (ibid.), which emphasises her capacity to support her husband as a wife. At the same time,

10 This puts Sophia Schliemann in the tradition of other female archaeologists and explorers such as Amelia Edwards and others. Cf. Magarita Díaz-Andreu and Marie Louise Stig Sørensen (1998) *Excavating Women: A History of Women in European Archaeology* and Amanda Adams (2010) *Ladies of the Field: Early Women Archaeologists and Their Search for Adventure*. As pointed out by Champion, »[a]mong those who worked in Egypt and the Near East were some of the pioneer women archaeologists, some starting out as travellers or tourists and turning to archaeology, others going out with the aim of research and excavation already in mind. One such was Margaret Benson (1865-1916), daughter of an Archbishop of Canterbury and sister of the novelist E. F. Benson, who directed the excavations of the Temple of Mut at Thebes with her colleague Janet Gourlay in 1895-8« (Champion 1998: 178). Cf. Jane Robinson (1990) *Wayward Women: A Guide to Women Travellers*. Other prominent examples are the explorer and natural historian Isabella Bird (1831-1904) and the traveller and writer Harriet Martineau (1802-1876). Cf. also Jana Esther Fries, Ulrike Rambuschek, and Gisela Schulte-Dornberg (2007) *Science oder Fiction? Geschlechterrollen in archäologischen Lebensbildern*.

it also implies her dependency on him, since, as has been mentioned with regard to Sophia and Schliemann's portrait in the *Leisure Hour*, Sophia is part of Schliemann's creation (cf. Cobet 2007: 69). This point is underlined by Cobet as follows: »On the day before the wedding in Athens on September 24, he wrote to the bride: ›If you marry me, it will be because we want to excavate together, to share our enthusiasm for Homer.‹ Schliemann did a lot to get Sophie to take on ›the position of the desired creature.‹« (Cobet 2007: 69, translation mine)

In spite of the fact that Sophia's presentation to the public was determined and promoted by her husband, her strength of character, her intelligence, and her capacity for hard work are remarkable when set against the contemporary cultural background. What is interesting in this regard is the question as to why a rather *un*typical woman like Sophia would be accepted and even celebrated by the British public, as illustrated by the various comments in magazines and newspapers at the time. The answer to this question is complex. An important aspect is the fact that Sophia primarily moved within the archaeological discourse, which as a counter-discourse to contemporary Victorian discourses not only dissolved class boundaries but also conventional gender conceptions. This is emphasised by Champion when she points out that due to the strong amateur tradition of archaeology in Britain, it was easier for women to take part in it: »Though sometimes sponsored by aristocratic enthusiasts, [the early archaeologists] were themselves frequently not of the wealthy aristocracy or upper middle class as has often been assumed. [...] Women's participation was therefore not ruled out in such contexts; and women had been travelling abroad on their own for a century or more.« (Champion 1998: 176)

In addition, Sophia's Greek descent was crucial with regard to her popularity. First of all, Sophia was a foreigner and essentially belonged to the ›other‹. However, because she was Greek and thus could be connected to antiquity, she became part of what for people was *the most familiar strangeness,* that is, (ancient) Greece, which means that she represented both the ›other‹ and the self. This is crucial for the role Sophia played in regard to British society, as on the one hand her *otherness* allowed the projection of wishes and longings excluded from the Victorian civil discourse and on the other hand her *sameness* made an identification with her possible. This is further enforced by the fact that Sophia combined both the ideal Victorian female who with her ›womanly‹ qualities stood at the side of her husband *and* the rather *extra*ordinary ›manly‹ female who as an equal to her husband excavated ancient monuments: »The hardships of the excavations were again, as in Troja, shared by Schliemann's wife, Sophia, a Greek lady, whom he had married seven years ago.« (*The Pall Mall Gazette*, May 16, 1891)

The combination of both conventional ›womanly‹ and ›manly‹ characteristics, which was favoured by the relative absence of dominant discourses in the archaeological space, made Sophia particularly interesting to British society, since she embodied the opposite of the ›angel of the house‹ without violating the basic moral values of this ideal, as emphasised by Boorstin:

> »Sophia, of course, added the touch of living romance not commonly found at prehistoric diggings. Heinrich became the royal family of archaeology. The young Greek beauty was a welcome variant on the stereotype of fragile Victorian femininity. [...] In London the Royal Archaeological Institute held a special meeting on June 8, 1877, to honor Heinrich and Sophia. The spotlight was on the glowing Sophia, doubly escorted into the hall on the arm of Lord Talbot, the president, and of William E. Gladstone, who had requested that privilege.« (Boorstin 1983: 595)

In this way, the public image of Sophia can be interpreted as that of an ›other‹-made or rather a *man*-made New Woman, representing a new type of woman as a moderate form of what was to evolve into the New Woman later in the century, which once more shows that gender identities become negotiable in the archaeological realm. This suggests that even though people, and in particular men, longed for strong and intelligent women, they hoped for a more moderate type than the *female*-created New Woman. Women like Sophia Schliemann – equal to her husband but still dependent on him – to many (men) seemed to be an alternative to the inferior Victorian ideal of femininity that did not threaten their own manliness.

That the rise of archaeology as an emergent discourse characterised by indefiniteness and often a spatial distance to the home country can be seen as encouraging the acceptance of an alternative to the traditional female role is underlined by other examples of women taking over roles traditionally reserved for men.[11] Very often, however, those women were perceived as a part of their husbands and his (their) achievements. In Sophia Schliemann's case this is further underlined by the fact that nearly all that the public learned of her was through her husband's accounts, as illustrated by the following example of *The Times*:

»New, on the other hand, was [Dr. Schliemann's] account of Mrs. Schliemann's share in the excavation. He said […] ›Mrs. Schliemann always superintended 30 workmen separately, and treated them with mingled kindness and severity. She always had excellent workmen. But she found it very difficult indeed to make them desist from their thievish propensities and to make honest men out of them, and in this respect I have only partly succeeded […]. But, knowing the character of the Greek workmen, Mrs Schliemann knew how to treat her men, and always had most obedient and hardworking men. […] The excavation of the Treasury was exceedingly difficult, because by the sacrilegious attempts made in 1820 by Veli Pasha, the son of notorious Ali Pasha, to force an entrance through the top, the whole upper part of the dome had fallen in, and Mrs. Schliemann was fearfully embarrassed with the removal of the many hundred, of huge blocks with which the interior of the building was filled. The removal of the stone was exceedingly heavy work, particularly in the rainy season, and as all the blocks had to be removed by the long approach, this work much impeded the horse carts; but by perseverance she overcame all difficulties.‹« (*The Times*, Jun 16, 1877: 9)

Even though Schliemann here praises his wife's hard work, her ability to superintend the workmen, and her perseverance, it is *his* account of the excavation, while Sophia's voice is completely absent from the text. Furthermore, this passage also implies that the workmen Sophia superintended were inferior to other men, first because, though hardworking, they were devious, and second, because Sophia is said to have superintended them with »kindness and severity« (ibid.). Significantly, the choice of words used to describe Sophia's surveillance is reminiscent of pedagogy, which not only moves Sophia into the position of a mother-figure superintending

11 Amongst others, there were female travellers and explorers as well as women accompanying their husbands on their »colonial service and missionary activities« (Díaz-Andreu García and Sørensen 1998: 4).

the excavations, but also turns the workmen into children.[12] This illustrates that even though Schliemann praises Sophia for her ability to supervise the workmen, at the same time the way in which he describes her surveillance betrays a certain inferiority of the workmen as opposed to the honest Western European men of Germany or Britain. Consequently, since the excavations being undertaken at the ancient sites were considered crucial and enriching for the Western world, a woman like Sophia Schliemann in the role of an overseer was acceptable and also admired by the British public, which would not have been the case in a different context, as emphasised by the *Athenæum*.

»The reception accorded to Dr. and Madame Schliemann in this country, testifies to the value attached to their discoveries, and I believe that a more careful examination of Dr. Schliemann's statements – thus far only oral – will go to prove ultimately the substantial accuracy of his views [...]. We trust, however, that the work will be continued and completed with equal success, and with results as grand as those for which Greece and the rest of the civilized world is indebted to Dr. Schliemann and his lady, my talented country woman.« (*Athenæum*, Jun 23, 1877: 807)

The example of Sophia Schliemann thus shows how her husband used the archaeological space to create and promote an alternative female identity characterised by the merging of the angel-in-the-house and the New-Woman type. It was this man-made New Woman, strong but loyal to and dependent on her husband represented by Sophia Schliemann that made her popular with British society.

12 The passage also has a strong Orientalist connotation if we consider that in this context the workmen are Greek.

5.3 SEARCH FOR ORIGIN – EXCAVATING THE SELF

> If any other justification were necessary, they might even claim to be in a sense themselves heirs to the ancient civilization of […] [Greece], so completely has it impressed itself on the minds and habits of thought in England and France.
> (*QUARTERLY REVIEW*, APR 1885: 298)

So far I have discussed Schliemann's popularity as a Victorian role model with an emphasis on his status as a self-made man in order to underline the public interest in this status in combination with archaeology. Furthermore, I have underlined how the public image of Sophia Schliemann propagated by her husband further promoted Schliemann's own popular success. On this basis, the present chapter demonstrates how the popularity of Schliemann as a self-made archaeologist can be read as indicative of the relevance of the archaeological discourse for the formation of identity. The status as a self-made man can thus in Schliemann's case be seen as transgressing the conventional notion of the term in that it can also be applied to the self-construction of Schliemann's very identity. Since Schliemann's identity was essentially *self-made*, it serves as an example of a successfully constructed identity as figured prominently in his (auto)biography. In the following, I will show how Schliemann's search for the Homeric Troy can be read as a search not only for his own identity but also for a collective identity and history characterised by both fact and fiction. Schliemann's (auto)biography and its popular reception in Britain are thus specifically illuminating in relation to the connection between individual/collective identity and contemporary discourses. Accordingly, I argue that public interest in Schliemann's search for Troy as related to his individual search for identity was particularly strong because it satisfied the collective wish for confirmation and stabilisation of identity at the time and thus served as a prosthetic memory for the masses.

The Creation of Identity

Heinrich Schliemann's life story has been repeatedly referred to as that of a modern Ulysses travelling the world in search of a home and a purpose in life, which can essentially be summed up as the search for identity within the circumstances of his time (cf. Scheuer 1990: 348):

»Therefore, the path in autobiographical literature leads from the exploration of an ›inner world‹ to the description of world experience; it is also possible to trace a similar path in the history of the ›Bildungsroman‹. The individual is portrayed in a struggle with the world or society. Whereas earlier, in the 18th century, the genre tapped into inner worlds, in which the individual discovered himself through his psyche, we now see the subject in a struggle with the necessities of ›conditions‹.« (Scheuer 1990: 348, translation mine)

Consequently, what Schliemann finally found at Hisarlık seemed to signal the end of his search for his vocation, thus apparently standing as a symbol for ultimate self-fulfilment. After years of wandering and hard work he was eventually able to follow his predestined path, taking up his archaeological enterprise, as summed up in his own account of his excavation:

»When, in the year 1832, at Kalkhorst, a village in Mecklenburg-Schwerin, at the age of ten, I presented my father, as a Christmas gift, with a badly written Latin essay upon the principal events of the Trojan war and the adventures of Ulysses and Agamemnon, little did I think that, six-and-thirty years later, I should offer the public a work on the same subject, after having had the good fortune to see with my own eyes the scene of that war, and the country of the heroes whose names have been immortalized by Homer.« (Schliemann 1976a [1875]: 3)

How much Schliemann's attempt to give meaning to his life was essentially connected to this search for identity is emphasised by various critics,[1] such as Justus Cobet, who point out significant parallels between Schliemann's career and that of Henrik Ibsen's fictional character Peer Gynt of the eponymous play:

---

1 William M. Calder III (1972), Wolfgang Schindler (1976), Jørgen Mejer (1990), and Justus Cobet (2007). Cf. also Michael Siebler (1990: 100).

»Schliemann's letters showed signs of a long phase of reorientation and the search for a new aim in life as early as during the Crimean War. The hero in Henrik Ibsen's drama *Peer Gynt*, a self-centred dreamer, offers us a key to understanding Schliemann's path from business to science. [...] Ibsen's drama was written in 1867, the year before Schliemann went to Troy. The two could not have known anything of one another; all the more valuable should this source be seen for the question of Schliemann's motives as the sign of a particular time in history. To be precise, this ›recent time‹ is the fifties and sixties of the 19th century, which saw the breakthrough of industrialisation and the meteoric expansion of world trade with its massive speculative profits and rapid changes in societal points of reference.« (Cobet 2007: 45, translation mine)

What Cobet and others identified in both Schliemann and Peer Gynt as the reaction to the rapid changes caused by growing industrialisation, modernisation, and an expanding international trade all over Europe was emblematic for British society in general. This explains Schliemann's popularity in Britain and makes his life narrative paradigmatic for the time, as it shows how the individual attempt to find and create identity through origin and continuity at a time characterised by alienation and change mirrored a collective need to retrieve and reaffirm the identity of an entire society:

»Schliemann's life is marked by detours and clean breaks. It is therefore not surprising that he repeatedly felt the urge to give his life literary form in letters, journals, and autobiographies. The medium of autobiography in particular allows one to achieve at least a literary reconciliation between the deeply felt disparity between origin and present times, between self-image and outside judgment. The autobiography can thus be defined as an attempt at self-stylisation and a quest for identity.« (Goldmann 1992: 37, translation mine)

Schliemann's life as such can be characterised in terms of a ceaseless search; at first the search for economic success as a merchant, then the search for his life's profession as an archaeologist, and last but not least the search for the Homeric sites, which can all be seen as part of his larger search for identity. How closely Schliemann's personal search for identity is interwoven with his archaeological excavations is illustrated by the comment of the British archaeologist John Murray on Schliemann's excavations at Hisarlık: »He has still to learn the exactness in the records of his excavations. That would have curbed his imagination, and probably have left us with fewer than seven ›cities‹ one above the other at Hissarlik. If he could but forget the Iliad! But then he would no longer be Dr. Schliemann.« (*Academy*, Dec 11, 1880: 430) Murray's statement is instructive in that it not only reflects prominent traits in Schliemann's character, but also emphasises how these determined his archaeological enterprises. The »exactness« (ibid.) Murray finds wanting in »the records of [Schliemann's] excavations« (ibid.) is as much absent from Schliemann's records of work as it is from the records of his own life.[2] Most of Schliemann's life-story can be read as a record lacking any clear detail which could have benefitted from some curbing of his imagination. At the same time, the imagination Murray alludes to, though negatively connotated here, also refers to the creative and fantastic potential inherent in Schliemann's character and in archaeology in general. Even if Schliemann had been exact in the recording of his archaeological discoveries, for Schliemann, as an archaeologist, there also remained both the necessity for, and the freedom of, interpretation and narration. As has been pointed out above, archaeological work always has to be seen as dependent on a series of hypotheses and thus invariably becomes coloured by imagination to some extent. As part of the past, and as described by Halbwachs,[3] it is always dependent on the interpreting subject's position in the present. In the process of archaeological excavation the past is thus *re*-created by the archaeologist in two respects. First, the archaeologist re-creates the material remains by bringing them to light and, second, he or she re-creates their alleged meaning by interpreting what has been excavated. The archaeologist thus both literally and figuratively *re*-members (cf. Erll 2005: 7) the past by re-arranging relics of the past into a coherent image in the present, thereby conserving a specific memory of that past. That Schliemann's contemporaries shared this concept of the archaeologist as an interpreter of the past is emphasised by the following comment in the *Athenæum*: »The position of Dr. Schliemann, as an interpreter of the remains of antiquity which his sagacity and enthusiasm have brought to light, may well engage our sympathy.« (*Athenæum*, Aug 21, 1880: 245)

2 This is reflected in Schliemann's various autobiographical accounts (cf. also Chapter 5.5).

3 Cf. also Aleida and Jan Assmann and Hans-Joachim Gehrke (cf. Chapter 1).

Confronted with relics of a long lost past, the archaeologist not only becomes an interpreter of antiquities but also a storyteller, a creator of history, imbuing antiquities with meaning by naming, categorising, and finally placing them in a wider context. With the spade as an instrument and stimulus, archaeologists are able to give the past a voice and have its story told through them to the world, as the following passage from *The Times* makes explicit: »Where the antiquarian's spade has not been at work, the smooth land tells no tale.« (*The Times*, Jun 5, 1876: 12) Even though here it is the »land which tells no tale« (ibid.), unless worked on with the »antiquarian's spade« (ibid.), the land itself only offers an accumulation of facts in the form of the material remains of the past. In order to make sense, however, these remains have to be arranged, connected, and interwoven into a story by the archaeologist as the interpreter and narrator. The creation of meaning through interpretation and storytelling presupposes the naming and classification of the objects found, which then finally leads to the establishment of both collective and individual identity. In other words, the interpretation of antiquities through the archaeologist is a process which provides both identity, by naming and classifying the relics, and continuity, by connecting the identified objects through the act of storytelling.

Murray's comment in the *Academy* implies a further aspect crucial to Schliemann's character in relation to his archaeological ventures. As stated above, to dig for the past in the first place always involves a search for the identity of departed peoples and cultures. The process of interpreting archaeological finds thus becomes a means of creating both meaning and history of and for an ancient civilisation. At the same time, the process of interpreting the past affects not only the past but also the present in that the present is always seen in relation to the past while being at the same time defined against it. Consequently, not only is the interpretation of the past dependent on the present but also vice versa. This underscores how the excavation of the past always also involves the search for identity in the present, as emphasised by Murray's statement in regard to Schliemann: »If he could but forget the *Iliad*! *But then he would no longer be Dr. Schliemann.*« (*Academy*, Dec 11, 1880: 430, emphasis mine) For Schliemann the *Iliad* is equated with historical facts, as a consequence of which the Trojan War is conceived of as a past reality that can be proven by means of archaeological research. That this search, however, goes beyond the longing to excavate a lost ancient city is made explicit by the fact that if Schliemann were to forget the *Iliad*, that is, the belief in the historicity of the Homeric poems, »he would no longer be Dr. Schliemann« (ibid.), meaning that he would lose his identity as Heinrich Schliemann. Underlined here is the close interrelation of Schliemann's search for the origin and identity of the Homeric sites *and* the search for his *own* identity. To believe in the *Iliad* was for Schliemann *to be*, while losing faith in it was *not to be*.

In what way Schliemann himself was anxious to present his life-story as a continuum and personal fulfilment is echoed by the various comments in contemporary newspapers and magazines:

»On Dr. Schliemann's return from this voyage he established himself at Paris, with the intention of spending the rest of his days in this city, and occupying himself principally with the study of archæology. He remained there some years, *and then came over him the desire to realise the dream of his life, and to visit at leisure the theatres of those events that had interested his childhood, the fatherland of the heroes whose adventures had charmed and solaced the hard days of his boyhood and youth*. In the summer of 1868 he therefore once more set out to the east, and visited successively *the spots where the poetical memories of antiquity are still living realities*.« (*All the Year Round*, Jun 13, 1874: 203, emphasis mine)

As illustrated by this report on Schliemann's career, the origin of his decision to search for the Homeric Troy is allegedly located in his early childhood. His desire to excavate Troy at some point in his life is presented as the »dream of his life« (ibid.), which connects his childhood fascination with »the fatherland of the heroes whose adventures had charmed and solaced the day of his boyhood and youth« (ibid.) to his later career as an archaeologist. The dream of excavating Homer's Troy becomes in retrospect the narrative thread that runs through Schliemann's life, forming the connecting link between past, present, and future. The dream endows Schliemann's development not only with a distinct origin but also with a *telos* and thus becomes a means of creating continuity and meaning. This is illustrated in Schliemann's ›Autobiographical Notice‹, included in *Troy and Its Remains*:

»At last I was able to *realize the dream of my whole life*, and to visit at my leisure the scene of those events which had such an intense interest for me, and the country of the heroes whose adventures had delighted and comforted *my childhood*. I started, therefore, last summer, and visited in succession *the places which still possess such living poetic memorials of antiquities.*« (Schliemann 1976a [1875]: 8, emphasis mine)

Here the continuity of Schliemann's personal life is emphasised first through the fact that by excavating Troy he »was able to realize the dream of [his] whole life« (ibid.) and second by his statement that he had been fascinated by the Homeric adventures ever since his »childhood« (ibid.). In a further step, Schliemann then links the continuity of his personal life with a collective continuity by turning the Homeric sites into sites of memory. By referring to those sites as »the places which still possess such living poetic memorials of antiquities« (ibid.), Schliemann not only connects the past with the present but also links individual to collective memory. Characteristically, this idea of continuity is repeatedly referred to and thus enforced by Schliemann himself, as reflected in an article in *Belgravia*: »A child's first impressions remain for life, and Henry Schliemann may fairly claim that the pickaxe and shovel were both forged and sharpened in the little village of Ankershagen in Mecklenburg, where he was born in 1822, and where he passed his early youth.« (*Belgravia*, Jul 1877: 76)[4] Again the information provided here is but an echo of what Schliemann himself skilfully spread through the autobiographical prefaces to his books[5] and his various letters, articles, and speeches, such as the following:

---

4 Cf. also Schliemann in *The Times*: »I feel bold to say that the pickaxe and shovel used in the excavations of Troy and Mycenæ were both forged and sharpened in the poor little village in which I was born and educated.« (*The Times*, Jun 4, 1877: 5)

5 This is also prominent in Schliemann's introduction »Autobiography of the Author, and Narrative of his work at Troy« to his excavational account *Ilios. The City and Country of the Trojans* (1880), as suggested by A. H. Sayce's comment on the book: »I had been busy during part of the summer revising the proof of the English edition of Schliemann's *Ilios* which was now ready for the press. Schliemann had prefixed to it an autobiography, full of interesting details relating to his earlier life. This was published as an introduction to his book, but, unfortunately, in a very emasculated form. Mr. John Murray objected to it, partly on the ground that it was unsuitable to a learned work on archaeological discovery, but more especially because it might tend to diminish the respect of

»If I begin this book with my autobiography, it is not from any feeling of vanity, but from a desire to show how the work of my later life has been the natural consequence of the impressions I received in my earliest childhood; and that, so to say, the pickaxe and spade for the excavation of Troy and the royal tombs of Mycenæ were both forged and sharpened in the little German village in which I passed eight years of my earliest childhood.« (Schliemann 1976c [1878]: 1)[6]

Schliemann here depicts his later career as an archaeologist as the »natural consequence of his [early] impressions« (ibid.), which suggests the idea of a causal chain in terms of a rational controllability and planning of one's personal life continuum and again points towards the concept of the self-made man, who with every action from early on has the power to control and shape his own life: »These Ankershagen tales, which first appeared in 1875 and were later developed and incorporated into the autobiographical preface to *Ilios*, created the image of an individual obsessed since childhood with the idea of excavating Troy. So powerful was the image that it lasted virtually unchallenged for a century.« (Traill 1995: 138) In much the same way as the childhood dream to one day become the excavator of Troy referred to in *All the Year Round* is functionalised to create a continuity in retrospect, in the passage from *Belgravia* the very fact that Schliemann claims that the pickaxe and shovel used in his Trojan excavations were »both forged and sharpened« (Schliemann 1976c [1880]: 1) in the village where he was born and grew up is used to create a feigned continuity by linking the past to the present.

---

its readers for the author. Schliemann bowed to the decision of his publisher. I have always regretted this, and still more than loyalty to Schliemann obliged me to return the incriminated proofs to him. There was nothing worse in them than the story of a very human youth, along with one or two statements of a wholly innocent and colourless character, such as the fact that owing to his first acquaintance with business having been at Amsterdam he always instinctively used the Dutch names of the numerals when counting mentally.« (Sayce 1923: 181)

6 Schliemann's »Autobiography of the Author«, published as an introduction to his book *Ilios*, was posthumously completed by Dr Alfred Brückner and published by Sophia Schliemann under the title *Heinrich Schliemann. Selbstbiographie bis zu seinem Tode vervollständigt* (1892).

Returning to the passage from *All the Year Round*, it is also interesting to note that Schliemann's life-story is juxtaposed with the history of the Troad, where Schliemann »visited successively *the spots where the poetical memories of antiquity are still living realities*« (*All the Year Round*, Jun 13, 1874: 203, emphasis mine). Here, in analogy to Schliemann's childhood dream, which shapes his life continuum, the »poetical memories of antiquity« (ibid.) prevail into the present as »living realities« (ibid.) at the spots visited by Schliemann and thus create continuity between ancient and modern times. In much the same way as Schliemann's childhood dream finally materialised and thus proved to be *real* through his successful excavations at Hisarlık, seemingly imaginary »poetical memories« (ibid.) finally turned out to be concrete and »living realities« (ibid.) at the archaeological sites he visited. The juxtaposition of the »dream of his life« (ibid.) with the »poetical memories of antiquity« (ibid.) is illuminating as in both cases the materialisation of these abstract concepts of dream and memories occurs at the archaeological sites. Considering the fact that both Schliemann's childhood dream and the »poetical memories« (ibid.) of the past are dependent on their realisation in order to function as a link between the past and the present and form a continuum between the former and the latter, the archaeological sites become sites of memory where materialisation, realisation, and confirmation take place. With regard to Schliemann's life-story, this is crucial as it is only through the *realisation* of his childhood dream at Hisarlık that this dream actually becomes a means of creating a life continuum. The same is true of the »poetical memories of antiquity« (ibid.), which are connected to the present because the archaeological sites function as their »living realities« (ibid.). In other words, both for Schliemann's life and the »poetical memories« (ibid.) of the past, the archaeological sites become a place where meaning and identity are created. This is emphasised by what Donald Kuspit notes in relation to the metaphorical analogy of archaeology and psychoanalysis:

»But just as the archaeologist builds up the walls of a building from the foundations that have remained standing, determines the number and position of the columns from depressions in the floor, and reconstructs the mural decorations and paintings from the remains found in the debris, so does the analyst proceed when he draws his inferences from fragments of memories, for the associations and from the behaviour of the subject of the analysis. Both of them have an undisputed right to reconstruct by means of supplementing and combining the surviving remains.« (Kuspit 1989: 138)

Just as the archaeologist interprets and recreates a past reality by »supplementing and combining the surviving remains« (ibid.), Schliemann uses both ›factual‹ and ›fictional‹ fragments of his life, insisting on their truth, to create meaning and identity and interweave them into a life-story. The fact that Schliemann uses actual places such as Ankershagen and the alleged site of Troy as sites of memory in combination with objects such as the excavation tools he used in Hisarlık, which he claimed had been figuratively forged in his childhood village to hold his story together, emphasises the importance of *materiality*.

The relevance of materiality is further underlined by the potential attributed to the archaeological sites, as emphasised by the report of a visitor to Mycenae in *Jackson's Oxford Journal*:

»The poets, and after them the schoolmen, have made such a highly mental mythological mystery out of the old times, that my feeling experienced a sudden collapse on confronting these ancient works and location of fable; not for want of material, which is abundant, but because they speak not story of fable, but of unswerving facts.« (*Jackson's Oxford Journal,* Mar 31, 1877)

Significantly, here the archaeological site once more becomes a place where the past turns out to be very concrete, and what had hitherto only been theoretical reasoning suddenly becomes a reality, that is, the »unswerving facts« (ibid.) as perceived by the traveller. This would have been a crucial experience for Victorian contemporaries. Furthermore, the passage is important as it emphasises the relevance of first-hand experience with regard to ancient sites. If we consider the archaeological site as a place where meaning and identity are created both in a literal and in a figurative sense, the immediate experience becomes an essential part in the process of self-identification in that a *realisation*, a *materialisation* with regard to the past and the present self, takes place. In other words, at a place where meaning and identity are created through Schliemann *for* him, and indirectly also for his British audience, first-hand experience – or *reported* first-hand experience – becomes crucial, since it encourages identification.

The relevance of first-hand experience is also implied by the popularity of Schliemann's reports on his excavations in the form of letters, articles, and books continuously published in Britain:

»Only after his conversion to archaeology, which appears to have occurred during his 1868 journey in Greece, did Schliemann begin his metamorphosis into the professional archaeologist. This transformation can be seen clearly in the sequence of publications which began in 1874 with *Trojanische Alterthümer* and its English version *Troy and its Remains*. These books are poorly illustrated, haphazardly organized, and improperly documented by today's standards. They are the works of a rank amateur. Schliemann's enthusiasm, wild fantasy, and patent self-aggrandizement leap off of every page. [...] Only four years later, however, Schliemann's publication of his research at Mycenae, although still heavily colored by his amateurish enthusiasm, is more professional, thoroughly illustrated, well documented and carefully organized.« (Runnels 2002: 17)

Although Runnels notes that Schliemann's writing became more professional over the years, it continued to be characterised by an »amateurish enthusiasm« (ibid.) capable of sustaining a close contact to the common reader. To read Schliemann's vibrant reports and see detailed illustrations of his archaeological finds and sites comes close to actually visiting his excavations and gives the reader the feeling of taking part in the search for ancient Troy, which eventually becomes the substitute for the search for identity in their own lives:

»Schliemann's scholarly books often read like memoirs. Sometimes they are little more than archaeological reports in the form of edited diaries, filled with anecdotes and personal observations, praise of his wife, Sophia, condemnation of Turkish perfidy, approval of a friend, and excoriation of someone, like Frank Calvert, who dared to disagree. All this makes the twelve books delightful reading and important biographical source material. To sum up, roughly 85 percent of the materials for a Schliemann biography are written by Schliemann himself.« (Calder 1986: 20)

Schliemann's capacity to captivate the reader with vivid accounts of his excavations can, for instance, be observed in the description of the discovery of the alleged Priam's Treasure:

»In the new large excavation on the north-west side, which is connected with the one I have just been describing, I have convinced myself that the splendid wall of large hewn stone, which I uncovered in April 1870, belongs to a tower, the lower projecting part of which must have been built during the first period of the Greek colony, whereas its upper position seems to belong to the time of Lysimachus (see plate XIII.). To this tower also belongs the wall that I mentioned in my last report [...]. In excavating this wall further and directly by the side of the palace of King Priam, I came upon a large copper article of the most remarkable form, which attracted my attention all the more as I thought I saw gold behind it. [...] In order to withdraw the Treasure from the greed of my workmen, and to save it for archæology, I had to be most expeditious [...]. While my men were eating and resting, I cut out the Treasure with a large knife, which it was impossible to do without the very greatest exertion and the most fearful risk of my life, for the great fortification-wall, beneath which I had to dig, threatened every moment to fall down on me. But the sight of so many objects, every one of which is of inestimable value to archæology, made me foolhardy, and I never thought of any danger. It would, however, have been impossible for me to have removed the Treasure without the help of my dear wife, who stood by me ready to pack the things which I cut out in her shawl and to carry them away.« (Schliemann 1976a [1875]: 323-324)[7]

This passage illustrates Schliemann's gift for well-written and detailed first-hand description of his archaeological excavations. Apart from the spectacular discovery presented here, he constructs a very clear distinction between identity and alterity by creating a dichotomy between ›us‹ and ›them‹, ›good‹, and ›bad‹, that is, Schliemann and his wife versus the greedy workmen. By defining his and his wife's Western identity against the Eastern alterity, Schliemann, in an act of Orientalism, strengthens Western identity and at the same time encourages British contemporaries to identify with him.

Schliemann's reports can thus be understood as inviting the reader not only to join him in his venture but also to identify with him and his search for meaning and identity: »With the ›Odyssey‹ in his hand and a magnificent panorama before his eyes, he wondered whether he was the same person that years before had sold herrings and walked penniless in Ham-

7 Note that the description of the discovery of the treasure is nearly identical in choice of words to Schliemann's later description in *Ilios* (1880).

burg.« (*Belgravia*, Jul, 1877: 84) Here the reader is confronted with Schliemann as a man trying to link his present life not only to his own past but also to the landscape and its mythical aura, evoked by the Homeric text in his hand. The fact that Schliemann is said to be wondering »whether he was the same person that years before had sold herrings and walked penniless in Hamburg« (ibid.) is significant as it once again points towards Schliemann's search for continuity in his life. In fact, this is a question that many contemporaries might have asked themselves after the profound sociohistorical changes experienced in the wake of industrialisation. Like Schliemann, they might have tried to connect their present life to the past, trying to make sense of it in order to eventually be able to establish an identity-generating continuity. For Schliemann, however, it is only *after* his successful excavations that he is finally able to create or rather *construct* a life continuum by claiming that he has been dreaming of becoming an archaeologist and the excavator of ancient Troy since his early childhood:

»Schliemann's childhood dream of Troy is a biographical construction of the fifty-four-year-old who has just come out with ›Ilium's Great Tower‹ and ›Priam's Treasure‹, the myth of origin of the archaeologist that was incorporated into the *Autobiography* of 1880, canonized by Sophie Schliemann with the publication of the monograph in 1892.« (Cobet 2007: 22, translation mine)

However, as Schliemann's popular success suggests, for his readers it did not seem to matter whether what he claimed was real; rather, what was important was whether it *felt* real to them because they could identify with it.

This also seemed to be the case with the inferences drawn by Schliemann with regard to his discoveries. People actually wanted to believe in Schliemann's stories because for them they meant the realisation of a myth that offered stability and orientation and thus the retrieval of identity, as pointed out in an article in the *Athenæum*: »We were concerned with the ›Troy of Priam‹ – ›Homeric Troy‹, which is, of course, all the world cares about.« (*Athenæum*, Mar 27, 1875: 431) As suggested by this statement, people were not only waiting to hear that Schliemann had actually discovered the »›Homeric Troy‹« (ibid.) – they were waiting for Schliemann to tell *his*tory, but they also wanted this story to be convincing. The need for someone like Schliemann to interpret his archaeological discoveries is also emphasised in a review of his book *Mycenæ*: »Here we are without assistance from the treasuries, and must do what we can by collating the relics of the tombs of Dr. Schliemann with tradition and history.« (*Athenæum*, Dec 22, 1877: 820) Although Schliemann does nothing other than »collating the relics [...] with tradition and history« (ibid.), the point is that he has no qualms about actually *doing* it:

»I have proved that in a remote antiquity there was in the plains of Troy a large city, destroyed of old by a fearful catastrophe, which had on the hill of Hissarlik only its Acropolis, with its temples and few other large edifices, whilst its lower city extended in an easterly, southerly, and westerly direction, on the site of the later Ilium; and that, consequently, this city answers perfectly to the Homeric description of the site of *sacred Ilios*.« (Schliemann 1976c [1880]: 277)

While others were much more reluctant to connect the relics unearthed by Schliemann to the Homeric legends, Schliemann, in the same way as he believed in the reality of the Homeric stories, convinced others to believe in *his*tories:

»We shall perhaps never know what truth there is in the Homeric legends or in the story of Agamemnon's death: but what we do know, is that there was at Hissarlik a prehistoric city of great opulence, which was destroyed by fire, and that in the tombs at Mycenæ which tradition assigns to the Royal House of Agamemnon were found bodies interred with every accompaniment of the Royal state and magnificence.« (*The Times*, Dec 29, 1890: 6)

In other words, through his adherence to the belief in the reality of Troy, Schliemann answered a public longing for an authoritative statement confirming the reality of the Homeric texts. This is made explicit in a lecture by the archaeologist Charles Waldstein reported in *The Times*:

»[Schliemann] could by no means accept the view that the city of Homer was a phantom. The descriptions of the poet were of too simply a sensuous and realistic character to give probability to such a theory, and, apart from pure and obvious allegory like that of Bunyan, he knew of no poet or group of poets who had given what so clearly bore the aspect of historical narratives without a basis of reality. All the descriptions, all the epithets applied to gods and men, carried with them the conviction that they derived from the concrete and not merely a poet's fancy.« (*The Times*, May 26, 1890: 10)

Taking up the tradition of the ancient Greek historians, Schliemann thus succeeds in giving substance to what by some is considered vague and insubstantial:

»In their eyes Agamemnon was not, as one school of modern critics regard him, *a mere shadow projected on the blank background of an unknown past*, and of which we shall never grasp the *substance*. This *magni nminis umbra* [sic] to the ancients suggested a real personality – a king whose disastrous fate coming so soon after his triumphant return from Troy, served in after ages as the favourite theme of epic and tragic poetry.« (*Edinburgh Review*, Jan 1878: 221-222, emphasis mine)

Schliemann seemingly provided material proof and concrete stories for an indefinite past hitherto only described by myths. Against the background of the nation's subtle feelings of insecurity and cultural uprooting, Schliemann thus offered comfort in this regard by using *his*tories. This is also suggested by the following statement in the *Illustrated London News*, which, although referring to modern theories that doubt the historicity of Homer's texts, also mentions the »realistic view« (*Illustrated London News*, Mar 24, 1877: 282) of Schliemann's theories and the temptation of believing in the Homeric reality:

»The modern theory that the story of the Iliad is nothing more than a myth finds no entrance into the mind of Dr. Schliemann […] and if we adopt the realistic view of Dr. Schliemann as to the Iliad, *it is a pleasing probability to indulge in that the cup mentioned by Homer is, after so many years, again visible and may be touched and felt.*« (Ibid., emphasis mine)

The passage's emphasis on the allure of the possibility of *seeing*, *touching*, and *feeling* a relic from the past which had lain buried for centuries and which might be the very same as that mentioned by Homer again points towards a general longing to find continuity in the form of material remains that can be literally *grasped*.

However, for Schliemann it was not enough to visit the archaeological sites, since even though they are the »living realities« (*All the Year Round*,

Jun 13, 1874: 203) of the past and thus function as material representations of it, their identity remains indefinite at first. The creation of meaning requires further investigation to make possible the naming, classification, contextualisation, and finally the identification of the indefinite ancient monuments. In this sense, digging for Homeric Troy becomes the search for meaning and identity with regard to Schliemann's own life, to Greek history, and also to a collective Western history. Ancient Greek history as presented by Schliemann in the press and his books serves as a prosthetic memory for both Schliemann and contemporary British society. Against this background, Schliemann's excavations in Asia Minor function not only as the individual realisation of a childhood dream and the creation of personal identity and continuity, but also as a means of providing material proof of the role of the Homeric myth as a crucial part of Western culture in general and British collective self-understanding and self-identification in particular. This is suggested by Lord Houghton's introduction to a talk given by Schliemann in front of the British Archæological Association:

»Lord Houghton, in opening the proceedings, said that the subject which would be brought before them that evening referred to ancient history. In a certain sense it was hardly one which an archæological society usually entertained, as it did not concern matters more immediately connected with the history of its own country; but there were exceptions in this respect, and one was as regarded the occupation of this country by the great Roman nation. Although this was the general habit *it need not separate them from the great sources of art which formed the basis of education of the youth of this country, viz., the antiquities of Greece and Rome.* (Cheers.)« (*The Times*, Apr 12, 1877: 8, emphasis mine)

The relevance of ancient Greece for British self-understanding is further emphasised by the following statement by Dr Birch,[8] commenting on a paper read out by Schliemann about his recent excavations and given at the British Archæological Association on April 11, 1877: »With regard to Dr. Schliemann's discoveries, the whole world must be indebted to him for the efforts he had made. (Applause.) *Nowhere were these efforts more appreciated than in this country, where there was to be found a deep love and feeling for Greek art, which had been imbued by classical education*

8 Samuel Birch was the keeper of the Oriental Antiquities in the British Museum.

(Cheers.)« (*The Times*, Apr 12, 1877: 8, emphasis mine) Apart from claiming a particular interest and appreciation in Britain for Schliemann's achievements, Birch thus also stresses his country's »deep love and feeling for Greek art« (ibid.). The words »love« (ibid.) and »feeling« (ibid.) used in this context are instructive as they suggest a certain emotionality beyond the rational surface in connection with the esteem for classical Greece, which reflects a certain ›irrationality‹ of the collective fascination with classical times, climaxing in the passionate participation of many British figures on the Greek side in the Greek War of Independence between 1821 and 1829.

With regard to the popularity Schliemann enjoyed in Britain, the emotional affiliation with ancient Greece is particularly crucial insofar as it helps explain the wide acceptance Schliemann's theories gained in Britain despite the lack of convincing proof behind them. On the basis of its status as the *most familiar strangeness*, Classical Greek culture had become the ›defining culture‹ in Britain, and thus anything associated with that culture was perceived to be precious and desirable, particularly at a time of cultural insecurity.

Homer and the Bible

The cultural relevance of ancient Greece and the Homeric poems becomes particularly clear when viewed against the contemporary sociohistorical background. Schliemann began his excavations in the Troad at a time when British society had passed through major sociohistorical changes, which also involved the demolition of traditional value systems. This was particularly true with regard to the Christian tradition and belief system, which, though at first seemingly confirmed by Layard's excavations in Mesopotamia since the late 1840s,[9] was questioned in principle by Darwin's theory of evolution as well as recent geological insights.[10] Apart from the fact that the biblical assumptions regarding the creation of mankind were challenged by theories put forward by Darwin and others, the author-

9 Layard's excavations of the ancient cities of Niniveh, Nimrud, and Babylon seemingly offered material proof for the stories told in the Bible (cf. Malley 1996a, 1996b, 2004, 2012).

10 Cf. Chapter 2.2.

ity of the Bible as such was increasingly undermined. This also affected the Bible's cultural function as a traditional and thus stabilising entity providing meaning both for the present and the future. As a consequence, people were confronted not only with the instability of their previously unchallenged belief systems but also with the loss of a guideline connecting past, present, and future (cf. Parsons 1988: 204).

It is against this background that we have to analyse Schliemann's reception in Britain, since it is crucial to our understanding of his popularity from the 1870s onwards. At a time when the authority of the Bible as the fundamental written record of Western civilisation was being questioned, people felt the need for alternative forms of stability and continuity. In this context, Homer's poems suggested themselves as a substitute for the biblical text, as Cobet states:

»The rough pattern of the process by which the tradition was eventually established in the form that has come down to us is the same for both stories: Like under a burning glass, the community-building event from the early times is focused on a single great battle under a great leader at a single location. This location is authenticated by the ruins of massive city walls.« (Cobet 1992: 362, translation mine)[11]

This cultural exchangeability was possible due to the basic structures and functions the former and the latter text shared. Like the Bible, the Homeric texts had been part of Western tradition as long as people could remember and told stories of a mythical and heroic past that people could relate to. These two aspects of tradition and a mythical past are crucial for the gen-

11 See also Cobet's comment about the relevance of the biblical and Homeric sites as sites of belonging: »To explain it I will select the reflection on dealing with two ›places of linkage‹ prominent in our tradition: Homeric Troy and the Jericho of the Old Testament. […] Significant historical events as told in the great writings of the Greeks and the Jews for the European tradition seem to take unbreakable shape in the archaeological monuments – evidently a means of escaping the doubts of historical criticism, which has developed into a permanent part of modern historical culture precisely in the exegesis of the books of the Old Testament and in the interpretation of the Homeric epic poems since the 16th and 17th century, but especially since the Enlightenment and the Age of Historicism.« (Cobet 1992: 361, translation mine)

eral esteem and acceptance of both the Bible and Homer as they seem to satisfy the collective need of a society in the process of change. The relevance of the Bible and the Homeric texts for society is emphasised by various comments in contemporary newspapers and periodicals, which equate the importance of the Bible to that of Homer's poems for Western culture. The following statement from the *Leisure Hour* of 1874, for instance, illustrates this:

»No story recorded in verse or prose, mythological or historical, with the exception of the Bible which forms the basis of Christianity, has exercised so widespread and powerful an influence over the human intellect, and so extensively and deeply permeated the ancient and modern literature of the western world, as the story of the siege and capture of Troy.« (*Leisure Hour*, Jul 1874: 439)

Three years later, something very similar was proclaimed by the *Illustrated London News*: »There is no story, outside of the Bible, which has appealed to such a multitude and variety of human sympathies, during such a long period, and in so many different parts of the world, or stages of moral and intellectual culture, as this most widely popular ›tale of Troy divine‹ has done.« (*Illustrated London News*, Mar 24, 1877: 281) Both comments begin by setting Homer's tale of Troy against the Bible by informing the reader that »[n]o story [...] with the exception of the Bible« (*Leisure Hour*, Jul 1874: 439) has been as widely received as Homer's text and that »[t]here is no story, outside of the Bible« (*Illustrated London News*, Mar 24, 1877: 281) which appeals to so many people. This is significant because putting Homer's tale next to the Bible not only underlines the importance of the Homeric text for the »western world« (ibid.) from a contemporary point of view but also makes explicit its relevance as seen through the eyes of nineteenth-century periodicals as representative of nineteenth-century society.

Also striking is the implicit reference in both instances to continuity and tradition. In the *Leisure Hour* the reader is informed that no other story except the Bible has »so extensively and deeply permeated the *ancient* and *modern* literature of the western world« (*Leisure Hour*, Jul 1874: 439, emphasis mine), and the *Illustrated London News* states that apart from the Bible no story »appealed to such a multitude and variety of human sympathies, *during such a long period*, and in so many different parts of the world, *or stages of moral and intellectual culture*« (*Illustrated London News*, Mar 24, 1877: 281, emphasis mine). By mentioning Homer's tale as having influenced both »ancient and modern literature« (ibid.) and emphasising its appeal to human beings »during such a long period« (ibid.) and »stages of moral and intellectual culture« (ibid.), these statements reveal its connective and lasting potential. Like the Bible, Homer's tale of Troy is a text which has prevailed for centuries and thus has the potential to connect the past, the present, and the future.

In addition to this diachronic relevance of the Homeric poem, the statements in the *Illustrated London News* and the *Leisure Hour* also refer to the poem's synchronic importance by referring to it as being »most widely popular« (ibid.) due to its »widespread and powerful [...] influence over the human intellect« (ibid.), »appeal[ing] to [...] a multitude and variety of human sympathies« (ibid.) »in so many different parts of the world« (ibid.). Again, this is fundamental since it establishes a sense of community and collective identity. Thus, taking the diachronic and synchronic relevance of the Homeric poem as emphasised in the *Illustrated London News* and the *Leisure Hour* into account, the *Iliad* can be said to function as a means both of creating continuity by linking the past to the present and of establishing a collective identity by underlining its popular reception.

With regard to British society, Schliemann's excavations in the Troad consequently became important for various reasons. First of all, the fact that Schliemann's work was related to Homer made his activities interesting for a society in which Homer's stories were held in similar standing to those of the Bible. Moreover, Schliemann's enterprise seemed to be suitable for a wider public in Britain, since for many Homer seemed to be the answer to their search for tradition and identity. In addition to this rather abstract fascination with its focus on the literary text, Schliemann's excavations in Hisarlık seemed, similar to Layard's excavations in Mesopotamia with regard to the Bible, to promise *material* proof of the Homeric poems. Thus, as was the case with Layard's excavations, the public once more could hope to be presented with archaeological relics that proved the reality of what had until then been passed on as a historical myth of ancient Greece. This is underlined by the statement by Lord Stanhope, president of the Society of Antiquaries, defending the »proposal on the part of the Society of Antiquaries that examinations should be made at the public coast of the tumuli in the plains of Troy« (*Leeds Mercury*, Mar 27, 1873):

»It is impossible to foresee, even with any approach to accuracy, what may or may not result from any Asiatic excavation. When Mr. Layard commenced these researches, pursued with so much of patient skill and attended, with such triumphant success, did any one anticipate even in the slightest degree the recovery of those gigantic winged bulls which are now among the most striking historical monuments with which the British Museum is adorned?« (Ibid., Mar 29, 1873)

As was the case with Layard's excavations, supporting Schliemann's excavations became a matter of faith. It meant believing in something that was »impossible to [be] foresee[n]« (ibid.) or »anticipate[d]« (ibid.) by anyone, and it also meant believing in a solid and great reality beneath the surface as represented by the »recovery of those gigantic winged bulls« (ibid.) in Mesopotamia. Both of these aspects, *believing* in something and *believing in a great reality beneath the surface* of things, were crucial for the public interest Victorian society took in Schliemann's excavations, since they seemed to offer comfort at a time when many felt the loss of traditional values and belief systems. In this regard, the faith in Schliemann's excavations at Hisarlık assumed a stabilising function for society at a time of instability. To announce the possibility of proving the historicity of what Homer told in his *Iliad* was to promise proof for one of the last myths as yet left unscathed by modern times.

In this context, William Simpson's[12] comment on Schliemann's discoveries in Mycenae is instructive: »In adopting tradition as an authority – and it is the only evidence Dr. Schliemann has been able to give as to Agamemnon's tomb – it is clear that he did not consider the difficulties and even absurdities which such a line of arguments might lead to.« (*Fraser's Magazine*, Dec 1877: 693) Even though this is intended as a criticism of Schliemann's approach, Simpson's comment says as much about the people supporting Schliemann's work as about Schliemann himself. Read against the background of British contemporary society, the very fact that Schliemann »adopt[ed] tradition as an authority« (ibid.) is exactly what made him and his enterprise so fascinating and interesting for a wide public in Britain. As remains to be shown, it was this very faith Schliemann held in the authority of tradition which captivated the masses.

---

12 William Simpson was one of Schliemann's ardent critics (cf. Chapter 5.6).

The general interest in and fascination with Schliemann's excavations is further illustrated by the statement made by the Chancellor of the Exchequer, who, although of the opinion that the exploration in the Troad »has no practical object, but aims at satisfying the curiosity of those who believe that the narrative of Homer was a true history and not the creation of a poet's imagination« (*Leeds Mercury*, Mar 27, 1873), did not reject the excavation *per se*. However, rather than approving of the spending of public money for the excavations, the Chancellor made reference to Layard's work in Mesopotamia, which was financially supported by the *Daily Telegraph*: »The *Daily Telegraph*, with my hearty approbation, is exploring, without any assistance from the public purse, the secrets that lie buried under the mounds of Mesopotamia.« (Ibid.) Again, the fact that the Chancellor referred in connection with Schliemann's excavations at Hisarlık to Layard's exploration in Mesopotamia helps explain the public interest in the Schliemannian search for ancient Troy. By mentioning Layard in the same instance with Schliemann, the Chancellor inevitably put Schliemann in the tradition of a great archaeologist before Schliemann had actually made his popular discoveries. Furthermore, Schliemann's search for the Homeric Troy is in this way linked once more to Layard's discoveries in Mesopotamia, which were brought forward and marketed by Layard as material proof for certain stories told in the Bible. Thus, if Layard had finally succeeded in producing »the secrets that lie buried under the mounds of Mesopotamia« (ibid.), why should Schliemann not accomplish the same at Hisarlık?

Ancient Greece and Modern Britain

It is in this context that the figure of Schliemann's young Greek wife, Sophia, plays a crucial role in her function as a connecting link between modern Western civilisation and ancient Greece. On June 9, 1877, *The Times* gave the following report of a meeting of the Royal Archæological Institute in honour of Sophia:

»It will be remembered that a few weeks ago Dr. Schliemann received, at a special meeting of the Royal Archæological Institute, diplomas of membership for himself and Mrs. Schliemann, on account of their great services to Homeric archæology. The hope was then expressed that the Institute might soon have the pleasure of personally welcoming Mrs. Schliemann among them. Arrangements were accordingly made soon after that lady's arrival in this country for another special meeting of that learned body yesterday, at which she would read a paper on ›The high culture of the Ancient Greece; the Agents who contributed to it; the Reasons of its decay; the Advantages of the Language of Plato; and further of the share the authoress has taken in the Discoveries at Troy and Mycenæ.‹« (*The Times*, Jun 9, 1877: 9)

While the keywords announced in connection to her paper, such as »high culture« (ibid.), »decay« (ibid.), and »the Advantages of the Language of Plato« (ibid.), are illuminating already insofar as they all correspond and could be transferred to coeval British discourses, extracts from the paper given by Sophia in front of the Royal Archæological Institute will help make certain points even clearer:

»›At a time when the rest of the world was still living in barbarism's dark night, my ancestors the ancient Greeks, had in science and arts reached such a pitch of perfection as can never be surpassed by man. [...] Our political institutions, our statesmen, our orators, our philosophers, and our poets have in all later ages been objects of wonder and admiration to the world at large; *they have for thousands of years been the ideals of perfection to those who aspired to culture; in fact, so much so, that even at the present day no one is considered to have a high education unless he be thoroughly acquainted with them.*‹« (Ibid., emphasis mine)

Significantly, Sophia begins her paper by relating herself directly to the ancient Greeks when she speaks of them as »›my ancestors‹« (ibid.). She thus skilfully creates a continuity between ancient and modern times. At the same time, by mentioning the ancient Greeks as her ancestors, she refers back to a popular photograph of herself which shows her wearing pieces of jewellery from the so-called treasure of Priam excavated by Heinrich and her at Hisarlık in 1873. Similarly to what she utters in her speech in front of the Royal Archæological Institute, Sophia creates a subtle connection between the past and the present by wearing the ancient Greek jewellery as a modern Greek woman. The very fact that she, as a woman personifying both ancient and modern Greece, holds her speech in Britain for a British audience furthermore makes an identification with her possible.[13] However, before giving the audience the chance to join her by identifying themselves with the classical Greek tradition, Sophia does not miss the opportunity to praise her ancestors with the hyperbolic statement that the »pitch of perfection [achieved by them] can never be surpassed by man« (ibid.). She emphasises the superiority of her Greek ancestors by contrasting it with »barbarism's dark night« (ibid.) »the rest of the world was still living in« (ibid.). Considering the current debate on ancestry and the origin of mankind stirred up by Darwin and others, the fact that Sophia refers to the world's barbarism at a time of enlightenment in ancient Greece must have deeply impressed her audience for at least two reasons: first because her audience would probably have been anxious to identify with Sophia's ancestors rather than with the barbarians beyond Greek borders, and second because Sophia's comment might just as well have painfully reminded people of the fact that the barbarians mentioned by her were actually *their* ancestors and in character not very distant from Darwin's primitive ape-man.

Before relieving the audience of this unpleasant association, Sophia continues to relate herself back to the ancient Greeks by pointing out their greatness. While at first she retains the categories of ›me‹ and ›them‹, these categories are dissolved when she proceeds by speaking of »[*o*]*ur* political institutions, *our* statesmen, *our* orators, *our* philosophers, and *our* poets« (ibid., emphasis mine), thereby creating a collective identity. This is crucial insofar as in the following sentence Sophia brings out the ancient Greeks' lasting importance for and influence on humanity, which reaches right into modern times, since »even at the present day no one is considered to have a high education unless he be thoroughly acquainted with [her Greek ancestors]« (ibid.). By stressing the enduring influence classical Greece has had on »the world at large« (ibid.) throughout the centuries up to the present, Sophia links her ancestors' greatness to modern times and thus eventually allows her audience to join her in identifying with the supremacy of classical Greece. As with the rest of the Western world, Britain could claim the following to be true with regard to the status classical Greece has had in their culture for centuries: »[The ancient Greeks] have for thousands of years been the ideals of perfection to those who aspired to culture.« (Ibid.)

13 Sophia Schliemann was able to fluently recite Homer in ancient Greek (cf. Chapter 5.2).

Taking all this into consideration, it becomes clear that in her speech Sophia establishes a continuity not only between herself and the ancient Greeks but indirectly also between the latter and Britain by referring to the importance ancient Greece still has for those aspiring to an elite education and culture, which Britain claimed for itself. The link between Britain and Greece that Sophia creates is strengthened in the further course of her speech when she comments on the British role in the Greek War of Independence:

»We Greeks owe to England everlasting gratitude, because without the generous assistance of the great country Greece could never have attained her independence. Only lately, again, England has with generous liberality ceded to us the beautiful Ionian Islands. But it is said that gratitude is a lively anticipation of future favours, and so I venture to hope that England will not desert the cause of Greece in the present eventful crisis. I conclude with an appeal to the English ladies to teach their children the sonorous language of my ancestors, so that they may be enabled to read ›Homer‹, and our other immortal classics in the original.« (Ibid.)

While in the first part of her talk Sophia emphasised the importance of ancient Greece for Western civilisation, including Britain, here she underlines the importance of Britain for Greece through its support against the Turks. This is particularly interesting since it presents the two nations as united in the fight for the same aims and in their adherence to the same cultural ideals, thus creating the feeling of a collective identity. This is enforced by Sophia's plea for Britain's continuous support in the future *in combination with* her »appeal to the English ladies to teach their children the sonorous language of [her] ancestors, so that they may be enabled to read ›Homer‹« (ibid.). By juxtaposing these two aspects, the past and the present, British and Greek identity are skilfully intermingled, and this invites the British audience to legitimately identify with ancient Greek culture, as Shelley had done in his preface to *Hellas: A Lyrical Drama* half a century earlier:

»We are all Greeks. Our laws, our literature, our religion, our arts, have their root in Greece. But for Greece – Rome the instructor, the conqueror, or the metropolis of our ancestors would have spread no illumination with her arms, and we might still have been savages and idolaters.« (Shelley 1821: 286)

Thus, as a closer analysis of Sophia's speech shows, she succeeds in connecting not only ancient and modern Greece but also modern Greece and Britain and eventually also modern Britain and ancient Greece. In this sense the search for the remains of Homer's Troy can also be read in terms of a search for Britain's own roots. As such, it underlines how the negotiation of individual identity (Sophia Schliemann) is essentially connected to collective identity (British identity) and vice versa.

An alleged connection between Britain and ancient Greece was further repeatedly created through references to the legendary Brute of Troy, which functioned as a founding myth for Britain:[14] »There is an allegorical truth in the legend that Brute the Trojan settled at last in an isle of the northern sea; for indeed we stand nearer to the physical qualities and the mental attitude of a Homeric age than to the southern races now.« (*Quarterly Review*, Jul 1898: 109) The identification of a British collective, referred to by »we« (ibid.), with the »Homeric age« (ibid.), in contrast to the modern »southern races« (ibid.), is significant as it once again underlines the connection between Heinrich's search for Homeric Troy and the search by British society for identity and origins symbolised by »Brute the Trojan« (ibid.). The extent to which the popular reception in Britain of Schliemann's excavations was influenced by the fact that they were undertaken on Greek soil is alluded to in an earlier extract from the *Quarterly Review*:

14 Cf. the mythical foundation of Rome by Aeneas, the Trojan: »The Romans want a link back to Troy, but they do not want it to be too direct: in the developed version of the foundation myth Aeneas does not simply found Rome, but founds Lavinium, and then from Lavinium is founded Alba Longa, and from Alba Longa is founded Rome.« (Feeney 2007: 1992) For Britain, the legendary Brutus of Troy, an alleged descendant of Aeneas, assumed the role of a founding myth. The legend of Brutus »was started by Nennius in the ninth century. Geoffrey of Monmouth, a prolific myth-maker of the early twelfth century, elaborated by explaining that Brutus landed at Totnes, overcame the giant Gogmagog, and founded London under the name of Trinovantum or New Troy. The story held credence until Tudor times, with John Leland protesting its authenticity as late as 1544« (»Brutus«, The Kings and Queens of Britain, Oxford Reference Online).

»Had Dr. Schliemann's excavations been carried on in some ruined town of Mexico, or abandoned site in eastern Asia, and there produced these results, the scientific archæologist would have felt perhaps equal interest. [...] *But the interest of the educated public would have been far more languid than it is at the present. Now every one feels that what lies at the root of Greek art lies at the root of all art, and carries the seeds of modern civilisation.*« (*Quarterly Review*, Jan 1878: 62, emphasis mine)

Here the public interest in Schliemann is explicitly accounted for by the matter and site of his archaeological enterprise, which is especially fascinating to the public because »Greek art [...] carries the seeds of modern civilisation« (ibid.).[15]

In addition to Britain's affinity for ancient Greece on the basis of mythical associations, the reference to Brute the Trojan in the *Quarterly Review* implies that there were also attempts to link Northern Europeans in general and the British in particular to the ancient Greeks by means of underlining alleged anthropological similarities. This was frequently attempted through interpretions of the human remains, and human skulls in particular, found at Hisarlık and Mycenae by means of phrenology. This is underlined by the description given in *The Times* of three human skulls which were part of Schliemann's Trojan collection on display at the South Kensington Museum in London: »Here are also three skulls, two found in Priam's City, and the other in that buried beneath it. All three are of *marked Aryan type*. Two of the crania have plainly been fractured in battle.« (*The Times*, Apr 18, 1878: 7, emphasis mine) The very fact that the Trojan's skulls are said to be of »marked Aryan type« (ibid.) is significant, first because it points towards the Victorian obsession with phrenology, and second because it suggests a racial kinship between the Trojans and Northern Europeans. The question of who the ancient races of Troy and Mycenae really were and whether any relations and similarities could be established between them and ancient Northern tribes was of particular interest to both science and the public, a fact underscored by a lively discussion of the topic which took place in the daily press and periodicals for many years:

15 Cf. also the Sherd of Amenartas in Haggard's *She,* which delineates the history of the Vincey family from modern Britain to ancient Greece and Egypt.

»The objects found at Hissarlik are of greater antiquity than those of Mycenæ; they do not bear the stamp of Assyrian art, but they are also not of Phœnician origin: these two points are now thoroughly established. On the other hand, the symbolical figures engraved on many of them have an unmistakable Aryan character [...]. This fact is confirmed by the ethnological characters of the skeletons found by M. Schliemann at Troy, which also belong to the Aryan race.« (*Athenæum*, Mar 22, 1879: 385)[16]

Here again, the »ethnological characters of the skeletons« (ibid.) found at Hisarlık are presented as proof of the skeletons' »unmistakable Aryan character« (ibid.).[17] Apart from these anthropological questions relating to the human remains found at Schliemann's sites, certain symbols allegedly characteristic of the Aryan race were foregrounded to support the assumption of an Aryan population in ancient Troy and Mycenae:

»The peculiar cross, with the arms curved or straight, and bent at right angles, or crossed by lines at the four ends, and often marked with four studs or points, is called in Sanscrit *swastika*, and it is said to represent the machine used by the primitive Aryan tribes to generate fire by friction. [...] Its wide-spread use makes it a connecting link in ethnographic science; and its constant occurrence on the Trojan terracottas, in all the strata, is claimed as a decisive proof that all the successive inhabitants of the hill were Aryan races.« (*Quarterly Review*, Jan 1874: 565)

In addition to linking the use of distinct symbols to ethnographic issues, this passage emphasises the fact that »all the successive inhabitants of the hill were Aryan races« (ibid.), which suggests both racial continuity and stability.

---

16 A similar conclusion on the subject was published in the *Daily News*: »A careful consideration of these relics of ancient humanity leads to the conclusion that the ancient inhabitants of Hissarlik – whether the Trojans of Homer or not – *were undoubtedly of noble race. The brain capacity is large, and the facial angle of the high type of the Aryan.*« (*Daily News*, Dec 26, 1877, emphasis mine)

17 The German physician and anthropologist Dr Rudolf Virchow joined Schliemann in 1879 at Hisarlık to undertake an anthropological investigation at the site.

While the hitherto mentioned aspects are mainly concerned with racial similarities which seemed to confirm a likeness between these ancient people and modern civilised men, at least on an anthropological level there were also other attempts to connect the finds at Troy and Mycenae to early European history. Accordingly, some people were convinced that art and artefacts found at Mycenae were very similar to or even identical with those found in Northern Europe: »When Dr. Schliemann first exhibited before the Society of Antiquaries photographs of the objects he had discovered, the most learned and distinguished members of that Society were at once struck with the analogy between them and the work of what is called the Bronze age in Northern Europe.« (*Quarterly Review*, Jan 1878: 78-79) The archaeologist Hodder Westropp went as far as to state that the human remains found at Mycenae were actually the remains of a Gothic invasion of the region:

»I would not venture to make this request but that some eminent scholars and archæologists to whom I communicated this view have received it with approval, and have urged me to place it before the public. I put it forward merely as a suggestion to be discussed, and shall be happy to withdraw it should any cogent argument be brought forward against it. The view I would venture to suggest is this – that the bodies found at Mycenæ are those of some Northern chiefs who had been killed or had died in Greece, and had been buried at Mycenæ with all their treasures. The reasons for my view are the following: The tombs and objects discovered present most marked features characteristic of the north of Europe.« (*Athenæum*, Aug 17, 1878: 217)

What these different theories have in common, regardless of whether they were based on alleged racial or cultural similarities, is that they all try to connect Schliemann's discoveries to Northern Europe in some way or other. This is what made them influential regarding the public reception of Schliemann's excavations in Britain. Taking into account the assumption delineated above that public interest in Schliemann's excavations represents a collective search for the origin and identity of Victorian society, theories that connected the ancient sites at Hisarlık and Mycenae with the Victorians' own culture and race were likely to appear very attractive to British contemporaries.

5.4 Archaeology and Prosperity

> Gold and other precious metals – weighing altogether about twenty pounds – have also been found, mostly in the more recent excavation.
> (*Athenæum*, Mar 10, 1877)
>
> You will have a full account of the treasures found in the tombs at Mycenæ, as soon as Dr. Schliemann can leave the mouth of the rich mine which he has opened and changed the pickaxe for the pen.
> (*The Times*, Dec 7, 1876: 8)

In his article »Schliemann: Hero or Fraud?« (1998), Donald F. Easton formulates and provides an answer to the following question: »The perfect archaeologist needs a number of attributes. Which did Schliemann have? Money and drive: these are the first essentials, and Schliemann had plenty of both.« (Easton 1998: 338) A recurring topic in connection to archaeology is prosperity and wealth, something which can eventually be traced back to the close amalgamation of consumption and identity attested to by Storey that was particularly prominent in the Victorian Age.[1] In the follow-

1 The importance of consumption in Victorian Britain is emphasised by Thomas Richards in *The Commodity Culture of Victorian England*, where he refers to the Great Exhibition of 1851 as »a monument to consumption« (Richards 1991: 3): »Until the Exhibition the commodity had not for a moment occupied center stage in English public life; during and after the Exhibition the commodity became and remained the still center of the turning earth, the focal point of all gazing and the end point of all pilgrimages [...]. The Great Exhibition of 1851 was the first outburst of the phantasmagoria of commodity culture. It inaugurated a way of seeing things that marked indelibly the cultural and commercial

ing, I will show how Heinrich Schliemann's identification as an archaeologist was closely linked to his material success and exemplifies connections between prosperity and identity formation.

As has been shown, Schliemann's archaeological career is closely related to a search for origin and identity, which Schliemann finally appears to have found in and through his work at the excavation sites. In particular the excavations at Hisarlık were presented by Schliemann and the contemporary press as the fulfilment of his childhood dream, and thus as the realisation and achievement of selfhood. What is striking with regard to both Schliemann's own comments and the many different newspaper and magazine articles written about his excavations is the fact that they frequently refer to the expenses incurred by Schliemann for his archaeological enterprises as well as to Schliemann's earlier career as a merchant. This is made explicit in the following statements made by Schliemann at a Grocers' Company Banquet in London:

»I feel very much pleasure in thinking that I am myself a grocer, and that in praising here the grocer's business, I praise a trade which I have followed up with unremitting zeal for a period of twenty-eight years. [...] But my business has never prevented me from continuing my studies, and when, in April, 1864, I thought I had money enough to retire from commercial pursuits, I found myself also in possession of sufficient theoretical knowledge to devote the remainder of my life to Homeric archæology. The habit I had acquired in my long career as a grocer not to do anything superficially, but to proceed in everything with tact, system, and perseverance, has been of immense advantage to me in my archæological explorations; and I feel bold to say that had I not been a grocer, I could never have succeeded in discovering Troy or the five Royal sepulchres of Mycenæ. I deem it superfluous to say anything to the praise of commerce, because, without commerce there could be no ambition, and without ambition, there would be no science. Thus, without commerce, men would be brutes.« (*Leisure Hour*, Jul 1877: 480)

This passage points out various important aspects that emphasise how commercial success and archaeology were interlinked in Schliemann's life. Schliemann begins his panegyric on the grocer's business by mentioning

life of Victorian England and fashioned a mythology of consumerism that has endured to this day.« (Ibid.: 18) that he himself had been a grocer for twenty-eight years and still *is* one today. That he still presents himself as a grocer despite at this point having been out of the business for more than ten years shows the ways in which for him his former profession as a grocer is connected to his present life as an archaeologist. This is underlined by Schliemann's statement that »had [he] not been a grocer, [he] could never have succeeded in discovering Troy or the five Royal sepulchres of Mycenæ« (ibid.). In particular, Schliemann refers to the fact that as a grocer he learned »not to do anything superficially, but to proceed in everything with tact, system, and perseverance« (ibid.), which he considers to be of »immense advantage« (ibid.) for his work as an archaeologist. By claiming that the skills he acquired as a grocer were extremely helpful for his »archæological exploration« (ibid.), Schliemann skilfully links his earlier to his later career, thereby once more characteristically creating an apparent continuity. In addition to that, Schliemann goes so far as to claim that he would never have succeeded in excavating Troy and the sepulchres at Mycenae had he not been a grocer in his earlier life. This is extremely significant since it provides an explicit link between Schliemann's commercial success and his archaeological success.

If we further consider that Schliemann's search for Troy and Mycenae can also be read as a search for his own self and identity, his commercial success becomes an indirect means of achieving selfhood through archaeology. This is crucial as it also suggests that Schliemann's life seems to illustrate how economic and material success finally leads to self-fulfilment and self-discovery. In this context, money becomes a means of achieving identity, something which suggests that identity is itself a commodity that can be *bought* as well as produced through consumption, as postulated by Storey when he states that »[c]onsumption is perhaps the most visible way in which we stage and perform the drama of self-formation. In this sense, then, consumption is also a form of production« (Storey 2008: 78, cf. also Chapter 1). This is particularly relevant against the background of contemporary Victorian society, in which on the one hand economic success and prosperity were the determining forces, while on the other hand people felt alienated and uprooted in a quickly and drastically changing society (cf. Daly 1994: 26). What Schliemann's life and career seem to symbolise in this context is that financial success can finally lead to self-fulfilment and the (re)establishment of a sense of identity that seems to have been lost. For a materialist society, characterised by the fact that presumably anything could be achieved by hard work and money, while at the same time lacking a fundamental feeling of identity, this connection must have seemed very attractive. If Schliemann had managed to achieve self-fulfilment and establish identity through hard work and financial success, it would stand to reason that the same was possible for a prosperous and progressive society such as Victorian Britain:

»[Schliemann's] example at the present time appears more worthy of note, from the circumstance that he is not a man trained to the profession of literary and academic scholarship; that he has never been a professor of any of the German or other Universities: but that his youth and part of his manhood, in Hamburg and in London, as we understand, *were incessantly occupied with commercial business.*« (*Illustrated London News*, Mar 24, 1877: 281, emphasis mine)

Apart from the fact that Schliemann here is stylised as a self-made man, his incessant occupation with »commercial business« (ibid.) during his earlier career is explicitly referred to. Furthermore, the fact that Schliemann is once more mentioned as an »example« (ibid.) underlines the attempt to make Schliemann's individual success resonate at a collective level.

That progress and prosperity become a means of establishing identity is further emphasised by the last two sentences of Schliemann's statement at the Grocers' Company Banquet. Here Schliemann proclaims that »without commerce there could be no ambition, and without ambition, there would be no science« (*Leisure Hour*, Jul 1877: 480), which leads him to the conclusion that »without commerce, men would be brutes« (ibid.). Considering the assumption that material success becomes a means of *purchasing* identity, as illustrated by Schliemann's career, his claim that commerce is eventually the driving force of science, which helps distinguish men from »brutes« (ibid.), is significant. As outlined in Chapter 2.2, Victorians were equally fascinated by theories of degeneration as they were unnerved by the degenerate, uncivilised, and barbarous and were thus anxious to maintain boundaries between themselves, that is, the civilised and modern, and the ›other‹, be it the uncivilised past *or* the present. In this context, Schliemann's statement can be read as an attempt to affirm identity by contrasting modern, progressive society with an uncivilised alterity.

The very fact that for Schliemann commerce is the basis for a civilised society is telling, since in relation to Schliemann's own life story material success becomes a means of establishing and affirming identity against the ›other‹. In this context, the following statement in *The Times* commenting on Schliemann's Trojan excavations is enlightening: »It is difficult, also for any Oriental to imagine that people can spend large sums of money in digging for mere rubbish.« (*The Times*, Jul 24, 1876: 6) Here the »large sums of money« (ibid.) spent by Schliemann are explicitly referred to while the »Oriental[s]« (ibid.), in an instance of blunt Orientalism, are portrayed as ignorant. By contrasting the huge amount of money spent by Schliemann on his excavations with the alleged ignorance of the Orient, the statement valorises Western culture and its interest in the remains of the past, which for the Oriental is apparently nothing but rubbish. At the same time, it suggests that Schliemann, representing British society, knew what was worthwhile spending money on. What in the eyes of »any Oriental« (ibid.) appears to be mere rubbish is in fact the fragments out of which identity and continuity are eventually recreated.

The connection between money and archaeology, representing in Schliemann's case a path to self-fulfilment and the creation of identity, is further underlined by the numerous references in contemporary newspapers and magazines to the extraordinary amount of money invested by Schliemann in his excavations, such as the following:

»Whatever opinion we may form as to the scientific and historical result of Dr. Schliemann's discoveries, and however we may feel disposed to dissent from some of his conclusions – tinctured as they are with an enthusiasm natural enough under the circumstances – there can be but one opinion as to the gratitude we owe him for the unwearied zeal with which he prosecuted his labours, at a very heavy expense, during a period of nearly two years on the supposed site of Troy; as well as for the candid and complete manner in which he has communicated the results of those labours to the public, and afforded them amplest means of drawing their conclusion form the materials thus placed at their disposal.« (*Edinburgh Review*, Jan, 1878: 508)

Here and in many other articles on the topic,[2] Schliemann's archaeological work is explicitly connected to hard labour and zeal as well as to »a very heavy expense« (ibid.) incurred by him for his excavations. This is also underlined by another statement relating to Schliemann's excavations at Mycenae: »It is not by chance that the statues of Pæonius or the remains of the King of Mycenæ have been found, but by the *careful application of money*, skill, and labour to unroll the faint clue afforded by classical writers.« (*Pall Mall Gazette*, Dec 27, 1877, emphasis mine) As these examples illustrate, Schliemann's archaeological success was often discussed in the public sphere in terms of heavy expense and hard work.

Another interesting aspect that links Schliemann's archaeological career to material success is the alleged significance good fortune, in the sense of serendipity, played in his life. In 1874 the Assyriologist A. H. Sayce commented on Schliemann's discoveries at Hisarlık as follows: »Dr. Schliemann will have to be added to the list of fortunate explorers who have succeeded in finding exactly what they want.« (*Academy*, May 7, 1874: 258) What is articulated here with regard to Schliemann's discoveries is what Cobet identifies as a *leitmotif* in Schliemann's life: »The topos of great misfortune that turned out to be the fortune of his life runs like a leitmotif through Schliemann's biographical texts.« (Cobet 2007: 26, translation mine) Interestingly, however, the occurrence of unexpected fortune in the moment of apparent failure (cf. Cobet 1990: 164) was in Schliemann's case always linked to hard work and material success. It is this that finally leads to happiness through the realisation of Schliemann's lifelong dream that eventually makes possible the establishment of identity:

---

2 The following provide further references to the money invested by Schliemann in his excavations: »It is perhaps generally known that the excavations at Mycenæ are carried on at the sole expense of Dr. Schliemann.« (*The Times*, Dec 7, 1876: 8); »The heavy expenses which Dr. Schliemann has borne in the cause of science need not be particularized.« (*The Times*, Mar 17, 1877: 12); »Mr. Newton then, alluding to the value of Dr. Schliemann's discoveries at Mycenæ, expressed the hope that they would be followed up by the exportation, undertaken by some British millionaire emulating Dr. Schliemann.« (*The Times*, May 5, 1877: 12); »Archæology, however, is indebted to Dr. Schliemann for the zeal with which he expended a large portion of his private fortune.« (*Athenæum*, Nov 14, 1874)

»We may adopt, though with a smile, the words of the enthusiastic friend[3] and interpreter of Dr. Schliemann, that ›the hill of Ilium had been a solitude for 1500 years, till a man and a woman encamped there three years ago‹ – like another Deucalion and Pyrrha – to evoke the forms of heroic life from the buried stones; and the course of life by which Dr. Schliemann was prompted to and prepared for the work, which he executed with the unaided resources supplied by his own life-long industry.« (*Quarterly Review*, Apr 1874: 535)

Schliemann's »life-long industry« (ibid.) provided the »unaided resources« (ibid.) which finally become the key to happiness and self-fulfilment. What is also crucial in this context, as pointed out by Cobet, is the fact that Schliemann himself does not tire of emphasising the significance of unexpected fortune at moments of deep misfortune and despair in his life. The impression Schliemann's good fortune made on British society is reflected by a comment in the Irish *Freeman's Journal*:

»There is a special piece of good fortune in the luck which has given all these beautiful articles of gold and silver to the German, when our own excavators, Mr. Layard and Mr. Smith, can find little else than painted tiles and nasty winged bulls on large stone slabs, in the palace of Nimroud and Sennarcherib. Then the glory of actually finding those precious relics of *Trojæ sub mœnibus altis* this is hard to bear.« (*Freeman's Journal and Daily Commercial Advertiser*, Aug 25, 1873)

Here the exceptional good fortune of Schliemann is emphasised in a contrast of his discoveries with those of Layard and Smith, who are seen to have excavated *only* »painted tiles and nasty winged bulls« (ibid.). Apart from Schliemann's »good fortune« (ibid.) underlined in this comment, the relevance of gold and silver for a popular and successful discovery is also revealed.

As a whole, Schliemann's life stands not only as an example of fortune and happiness achieved through hard work, but also as an instance of

3 The enthusiastic friend referred to in this quotation is the French Orientalist Émile Burnouf: »M. Émile Burnouf, in an article in the ›Revue des Deux Mondes‹, for January 1, 1874, ›Troi d'apères les dernières Fouilles faites en Troade‹, giving an excellent account of Dr. Schliemann's labours.« (*Quarterly Review*, Apr 1874: 535)

someone who eventually achieved his aims in life in spite of seemingly desperate moments and situations. This optimistic view of life, based on the strong belief that progress and prosperity would finally lead to the fortune and happiness illustrated by Schliemann's career corresponds to the core values of Victorian society. Consequently, Schliemann's popularity in Britain at the time mirrors the British obsession with self-fulfilment by means of progress, development, and prosperity.

Interestingly, the public associated archaeology in general and Schliemann's excavations in particular not only with the recreation of the past but also with treasure hunting.[4] The fascination with Schliemann's excavations in terms of a treasure trove is supported by the many references to the »[g]old and other precious metals« (*Athenæum*, Mar 10, 1877) excavated by him at Hisarlık and Mycenae:

»Much curiosity was excited, towards the close of the last summer, by the announcement, which appeared first in the German newspapers, but soon found its way into those of this country also – that a German savant, who was known to have been engaged for a considerable time past in researches on the plain of Troy, had not only determined beyond a doubt the site of that far-famed city, but had brought to light the very place of King Priam himself, and, what was more, had found upon the *site a large portion of the treasure in gold and silver that had once belonged to the Trojan monarch*, and which the Greek invaders, as it appeared, had omitted to carry off.« (*Edinburgh Review*, Apr 1874: 506, emphasis mine)

Even though the so-called Priam's Treasure was considered precious at the time and excited huge public interest, as far as material wealth was concerned Schliemann's discoveries at Mycenae surpassed his earlier excavations. This is underlined retrospectively in a comment on recent excavations at Olympia: »But at Olympia the results appealed most to the limited class of students of ancient art. The gold of Mycenæ, on the contrary, glittered in the eyes of everyone.« (*Quarterly Review*, Jul 1897: 79)[5] The com-

4 Before his archaeological career, Schliemann had also profited from the California Gold Rush when he travelled and lived in the USA.

5 What is interesting in this context is the fact that the first reports of Schliemann's excavations appear in the *Illustrated London News* only after Schlie-

mentary that »the gold of Mycenæ [...] glittered in the eyes of *everyone*« (ibid., emphasis mine) emphasises that the gold discovered by Schliemann at Mycenae and Hisarlık played an important role in the public fascination with his excavations.[6] In particular, the golden death masks that had been found by Schliemann at Mycenae »excited [...] much curiosity« (*Athenæum*, Mar 10, 1877: 327) among the public.[7] The importance of gold for the impact of Schliemann's excavations is further alluded to in the following passage: »[I]t seems as if this exploration of Tiryns, though by no means so fruitful in gold work and other museum curiosities as those of Mycenae and Troy, is, from the antiquarian point of view, really the most valuable of them all.« (*Academy*, Nov 21, 1885: 345) What is interesting here is the implicit differentiation between the excavation sites in terms of what is valuable »from the antiquarian point of view« (ibid.) as opposed to what is suited to the public, that is, »fruitful in gold work and other museum curiosities« (ibid.). Accordingly, since gold had been found at both Mycenae and Troy, these sites were capable of attracting a wide public interest, as confirmed by the many magazine and newspaper reports about Schliemann's excavations at these places.

Against the background that Schliemann's archaeological career was based on his commercial success, the gold finds at Mycenae and Troy are significant, since they seem to suggest that his successful business career helped him achieve even greater wealth through the findings at these sites.

---

mann's discoveries at Mycenae. His excavations in the Troad are then reported in retrospect.

6 »Gold« or »golden« are among the most frequently used words in contemporary newspaper and magazine reports about Schliemann's excavations.

7 While fascinated by the gold of the mask, people were generally appalled by the ugliness of the death masks found at Mycenae: »Besides the famous gold masks, which are ugly enough in all conscience, however mighty the faces they may have covered, there was a well-modelled cow's head in silver with golden horns, cups and bowls of various shape, some much battered belts, breast ornaments, bracelets, studs, buttons, and so on« (*All the Year Round*, June 2, 1877: 331) and »[a]fter reading Dr. Schliemann's glowing description of these masks on the first announcement of their discovery, we confess that it was not without a shudder that we first beheld these hideous libels on the ›human face divine‹« (*Edinburgh Review*, Jan 1878: 236).

How this might have symbolically satisfied a collective Victorian leaning towards consumption is pointed out by Daly: »If Britain had pushed its way to the front of world trade by developing its *productive* power, after mid-century *consumption* assumed a new importance, and by the 1880s consumerism had fundamentally transformed Victorian culture.« (Daly 1994: 26) At the same time, as far as contemporary society was concerned Schliemann's treasure trove seemed to confirm the belief in a prosperous past, as illustrated by a comment in *Jackson's Oxford Journal*:

»Magnificent pectorals, baldrics, girdles, and masks all of solid and pure gold, and all wrought in any infinity of designs and patterns, which covered them as completely as embroidery would, *showed that the age of gold was a reality, and thus places one more barrier in the way of those levelling reasoners, who assert that things that are not probable, according to our experience, never existed, except in the excited imagination of the poet or enthusiast.*« (*Jackson's Oxford Journal*, Mar 10, 1877, emphasis mine)

Here the ancient golden relics not only function as proof of bygone wealth and prosperity, but also as confirmation of the existence of a remote past. Furthermore, these relics prove that people »who assert that things that are not probable, according to our experience, never existed« (ibid.) are wrong in their assumption. Against the background of Schliemann's life, in which commercial success made an archaeological career possible, which eventually became a means of the retrieval of identity, this can be read as a key passage. At the same time, it helps explain the Victorian fascination with Schliemann in that it mirrors the contemporary obsession with prosperity and materialism in combination with an underlying search for identity.

5.5 Dr Henry Schliemann: The Art of Self-Promotion

> It is impossible to separate Dr. Schliemann's work from the personality of the worker.
> (*Quarterly Review*, Jul 1881: 303)

So far the emphasis has been on Schliemann's status as a Victorian role model and his association of ancient Greece with the *most familiar strangeness*, which were both decisive factors for his popularity in Britain. In addition to that, however, Schliemann's popular reception was determined by his self-staging in the British public. In the present chapter, I show how Schliemann used the popularisation of his image in the public eye to consolidate and perpetuate his self-made identity. In this context, Schliemann's self-promotion in Britain can be seen as an *act of identity* characterised by the celebration of identity and the continuous repetition of certain topoi related to his life and career. On the basis of Schliemann's public presentation, the chapter will thus confirm the relevance of performativity for the formation and consolidation of identity.

In 1880 Schliemann moved with his family into the so-called ›Ilíou Mélathron‹ in Athens, an ostentatious neo-Renaissance palace designed for him by his architect friend Ernst Ziller and built from 1878 to 1879. In the following years Schliemann devised a magnificent mausoleum for himself based on the Athenian central cemetery Proto Nekrotafeion Athinon, where his body was indeed entombed after his death in 1891. In accordance with Schliemann's wishes, one side of the mausoleum depicts the battlefield from the *Iliad*, while the other side portrays Heinrich and Sophia Schliemann at an excavation site surveying the workmen. Commenting on the couple's frieze, Mannsperger refers to it as Heinrich's ultimate act of self-staging, climaxing in an apotheosis: »The self-glorification culminated in Schliemann's mausoleum: Already a ›temple‹ in itself, it has a ›Schliemann frieze‹ showing him and his wife in full figure at the archaeological site in place of a frieze of gods – the self-aggrandisement ended in divinisation.« (Mannsperger 1992: 71, translation mine) This is underscored by the illustration of the Trojan War on the other side of the mausoleum, which creates both a connection and a parallel between the Trojan War and Schliemann, the past and present. Implicitly, connotations of the Trojan War such as heroism, magnitude, courage, and power are thus transferred to the Schliemanns and their life. Furthermore, through its status as a (founding) myth for the Western world, the juxtaposition of the Trojan War with the Schliemanns turns Schliemann's life into something fundamental and eternal. This is further enforced by the fact that the frieze shows Schliemann and his wife excavating ancient Greek remains that are most likely meant to be those of Troy. As the excavator of Troy, Schliemann thus stages himself as an agent in the solidification and perpetuation of Greek history in the present by bringing its remains to light and thus reviving the past in the present. At the same time, cultural memory, represented by the legendary Trojan War, and collective memory, represented by Schliemann's life and career, are interlinked and presented as parts of the same thing, underlining once more how Schliemann succeeded in connecting his identity to a collective, heroic Greek past and how much this contributed to both his self- and public image. Schliemann's continuous effort to link his own identity to ancient Greece is furthermore made explicit in the naming of the two children he had with Sophia, who were called Andromache and Agamemnon.

The two pompous buildings commissioned and devised by Schliemann in Athens, the ›Ilíou Mélathron‹ and the mausoleum, and the naming of his children are characteristic of Schliemann insofar as they can be read as representing his lifelong efforts at self-promotion. Schliemann was not only a self-made man, but he also knew how to present himself to the public and so play to the gallery. A comment on Schliemann's life in *All the Year Round* reads: »His life has not yet been incorporated into biographical dictionaries and encyclopædias, and we should be quite ignorant of his early career had not a short memoir prefixed by himself to an earlier and almost unknown work, afforded us materials for the following brief account.« (*All the Year Round*, Jun 1874: 199) As suggested here, while Schliemann's biography was not yet presented in books other than those authored by himself, at this early point of his career the few facts of his life known to the public originated from his own prolific pen.[1] This is significant because it shows how Schliemann manipulated his public life story from the beginning to make it appear as grandiose and glorious as his ›Ilíou Mélathron‹ and mausoleum: »[Schliemann's] motive, after all, to give shape and an overall theme to his life, seems obvious and venial enough, especially when we consider his autobiography in its contemporary context. Romantic enhancement of mundane biographical details seems to have been more acceptable in the nineteenth century than today.« (Traill 1993b: 30) Consequently, Schliemann did not miss any opportunity to refer to his achievements by always elaborating on the same scenes from his childhood and youth in connection with his later career as an archaeologist in his letters, articles, lectures, and prefaces to his excavation accounts:

»The touching story of how Schliemann clung to his childhood resolve throughout his life and made a fortune as a merchant with the express purpose of being in a position to fulfill his dream has warmed the hearts of millions. It *is* a good story. Unfortunately, as a number of scholars have now pointed out, it is quite untrue. After the shock of realizing that we have been duped has worn off, however, we might feel inclined to forgive Schliemann his creative fiction on the grounds that *se non è vero, è molto ben trovato.*« (Ibid: 29)

At the same time, he was very careful to justify any contradictions *a priori* – and there were many[2] –, as underlined in his »Autobiographical Notice« to *Troy and Its Remains*:[3]

---

1 Cf. also Calder's above-quoted comment on the presentation of Schliemann's bibliography in general: »To sum up, roughly 85 percent of the materials for a Schliemann biography are written by Schliemann himself.« (Calder 1986: 22)

2 Articles that focus on Schliemann's credibility are Traill (1983) »Schliemann's ›Discovery‹ of ›Priam's Treasure‹: A Re-examination of the Evidence«, Traill (1986) »Schliemann's Mendacity: A Question of Methodology«, Easton (1984) »Schliemann's Mendacity – a False Trail?«, Easton (1992) »Was Schliemann a Liar?«, Easton (1994) »Priam's Gold: The Full Story«, and Easton (1998) »Heinrich Schliemann: Hero or Fraud«.

3 Arentzen points out that »*Trojanische Alterthümer* [*Troy and Its Remains*], like Layard's *Nineveh*, is not a genuinely scholarly publication. It is a collection of articles that were originally written, on the spot, for the *Augsburger Allgemeine*

»The present book is a sort of Diary of my excavations at Troy, for all the memoirs of which it consists were, as the vividness of the descriptions will prove, written down by me on the spot while proceeding with my works. If my memories now and then contain contradictions, I hope that these may be pardoned when it is considered that I have here revealed a new world for archæology, that the objects which I have brought to light by thousands are of a kind hitherto never or but very rarely found, and that consequently everything appeared strange and mysterious to me.« (Schliemann 1976a [1875]: 12)

Although here Schliemann is commenting on possible contradictions in the classification of his findings, his statement can also be read as a precautionary warning against any contradictions that might arise in his accounts, including the description of his life and his excavations. Skilfully, the self-made man thus created his own *his*tory, which was eagerly consumed by the public:

»As soon as I had learnt to speak, my father related to me the great deeds of the Homeric heroes. I loved these stories; they enchanted me and transported me with the highest enthusiasm. The first impressions which a child receives abide with him during his whole life; and, though it was my lot, at the age of fourteen, to be apprenticed in the warehouse of E. Ludwig Holtz in the small town of Fürstenberg, in Mecklenburg, instead of following the scientific career for which I felt an extraordinary predisposition, I always retained the same love for the famous men of antiquity which I had conceived for them in my first childhood.« (Ibid.: 3)

Contemporary newspapers and magazines supported this process by bringing Schliemann's self-made life story to the market and presenting it using recurring facts and motives introduced by Schliemann himself.[4] The rele-

---

Zeitung. [...] Accordingly, we see ideas leaping forth and later disappearing, without any explanation« (Arentzen 2001: 180).

4 Calder names the following aspects as responsible for Schliemann's lasting popularity: »What is there about the life of Schliemann that makes him so popular. There are three obvious attractions: 1. Heinrich Schliemann is the poor boy who became rich and famous through hard work and good luck. He has almost become a Jungian archetype for this beloved folk figure. 2. Schliemann traveled in the most exotic place of the world of his time. [...] But, with this, he

vance of repeating and performing certain motives of Schliemann's biography to consolidate identity is evident here and reminiscent of the working of Assmanian cultural memory, which is perpetuated by regular repetitions in the form of rites and festivities. At the same time, it emphasises the performative and constructed character of identity as postulated by Butler and Storey.

The most prominent scenes repeatedly referred to in relation to Schliemann's life are (1) the illustration of the burning Troy in Georg Ludwig Jerrer's *Weltgeschichte für Kinder* (1828),[5] (2) the Latin essay Schliemann wrote about the Trojan War and presented to his father, (3) the drunken miller's boy reciting Homer, (4) the innumerable languages learned by Schliemann, (5) his commercial success, (6) the fortune he achieved

---

had a beautiful, faithful, intelligent wife, Sophia, a companion and colleague, a fellow archaeologist. 3. Heinrich Schliemann, the uneducated businessman, who left school at the age of fourteen, proved all the professors wrong by stubborn, simple faith worth more than their books, their advanced degrees, and hard-earned experience.« (Calder 1986: 19)

5 Traill (1993b) had Schliemann's signature in the copy of *Weltgeschichte für Kinder* checked by a handwriting expert, who confirmed that this was not the handwriting of an eight-year-old boy. Schliemann had claimed to have been inspired by Jerrer's *Weltgeschichte*, which he had owned as a child: »Accordingly, until it can be shown that at the age of eight Schliemann had an unusually mature hand, we should assume that the signature is that of the adult Schliemann. It follows therefore that there is no *contemporary* evidence to support Schliemann's claim that at the age of eight he was exposed to Jerrer's account of Troy and to the famous engraving accompanying it.« (Traill 1993b: 40) In his article »Der Mythos von Trojas Untergang in Schliemanns Autobiographie. Zur Archäologie eines Erinnerungsbildes« (1992), Stefan Goldmann focuses on Schliemann's childhood memories. Goldmann also interprets Schliemann's fascination with Oliver Goldsmith's *The Vicar of Wakefield* (1766) and Walter Scott's *Ivanhoe* (1820) in the context of Schliemann's childhood experiences. »The memory and the picture taken from the European treasure trove of images covers and hides a deep rupture and loss in his life's story. It constitutes the attempt to create continuity and literary coherence at precisely the point where there had once actually been discontinuity, a catastrophic break.« (Goldmann 1992: 47, translation mine)

through faith, ambition, and hard work, and (7) his beautiful Greek wife, who knew Homer by heart. These alleged facts and motifs dispersed by Schliemann form the framework of his public life story and thus can be found in the majority of the articles that comment on Schliemann's life at the time, as the following passage from *The Times* serves to illustrate:

»Dr. Schliemann's success, like all great achievements, is a triumph of faith; but although this is its most striking characteristic the fact affords no reason for representing, as some do here and elsewhere, that the discoverer is a mere ignoramus. It is true that Dr. Schliemann is a self-taught man, that he began life as a shop boy, was subsequently a cabin boy, then a clerk, and spent his best days in Russian commerce; but on the other hand, as we learn from his biography, at the age of 10 he presented his father with a Latin essay on the Trojan war, later ›prayed God that by His grace he might have the happiness to learn Greek‹, and in addition to amassing a large fortune, has mastered the English, French, Dutch, Spanish, Italian, Portuguese, Russian, Swedish, Polish, Greek, and Arabic languages. Even as to archæology itself, Dr. Schliemann's studies have not been inconsiderable. For more than ten years he has applied himself to it, especially in its connexion with the Homeric poems, without relaxation, and his knowledge of the Iliad and Odyssey is fairly represented by the popular belief that he knows the whole of them by heart. He was attracted to the lady who is now Mrs. Schliemann by the ability to translate the ›song divine‹, and has since cultivated her powers by refusing in their walks to enter upon other subjects before she had repeated a certain number of lines.« (*The Times*, Mar 17, 1877: 12)

In this passage Schliemann's life is summed up by five of the seven aspects characteristic of biographical notes on Schliemann. Although the arrangement of the different motifs varies in the many articles commenting on Schliemann's life, there are usually at least four of the seven aspects mentioned in relation to his career.

How much Schliemann's public life story was the sum of carefully selected and manipulated facts and incidents is also suggested by the fact that his first wife, Ekaterina Lyschin, is almost never mentioned in articles dealing with his life.[6] That the public image of Schliemann promoted by the

6 In the newspapers and magazines on which this analysis is based, Schliemann's first wife is only mentioned once: »[Schliemann's] divorce from his Russian

contemporary press was dependent on what Schliemann presented to them is emphasised by the fact that it essentially mirrors Schliemann's autobiographical notes:

»The marriage with Katharina is mentioned neither in the biography of 1868 nor in that of 1880, which goes into much more detail with regard to his private life, in the latter case perhaps at the request of the publisher. [...] This biography [that published in 1880] later introduces Sophia seemingly in passing and characterises her as an accomplice in Troy, as though she had always been Schliemann's wife.« (Cobet 2007: 44, translation mine)[7]

In addition to his first marriage, Schliemann's sojourn in America is usually not mentioned in comments on his life, presumably because it did not fit into Schliemann's carefully selected *his*tory, which was supposed to reflect a continuous development and unfolding of his childhood dream, as emphasised by the *Edingburgh Review*: »His name will live as that of one in whom genuine enthusiasm, springing from the memories of childhood, became in manhood the spirit of personal and indefatigable exploration.« (*Edinburgh Review*, Apr 1881: 547) That Schliemann applied the same strategies to his excavation accounts, in that he not only manipulated and created his own life story but also (the) *his*tory of finding Troy, is pointed out by Allen:[8]

»Everybody knew [that Schliemann discovered Troy] because Schliemann himself made absolutely certain that the story of his discovery became part of modern mythology: the heroic archaeologist connecting the quotidian present with a golden, legendary past. In his archaeological publications, he reinvented his own life to create this myth.« (Allen 1999: 1)

---

wife [...] is also attributed to that lady's lack of appreciation of his favourite poet« (*The Pall Mall Gazette*, May 16, 1891), quoted above. Implicitly, his ex-wife's disinterest in Homer is used here as a justification for Schliemann's divorce.

7 The biographies referred to are the biographical notes in Schliemann's *Ithaca* (1868) and *Ilios* (1880).

8 Cf. Traill (1983, 1984a, 1984b, 1995, and 1996); Calder (1972); Easton (1992, 1994, and 1998).

As his fame and the appreciation of his work suggest, Schliemann's version of his life and work were extremely popular both in his lifetime and after his death.

While by now Schliemann's motives for his selected and constructed *his*tory should be clear, that is, the stabilisation of identity through public performance in the form of self-glorification, it seems worthwhile to consider the question of why the contemporary press adopted Schliemann's life story so readily instead of critically evaluating and presenting it to the public. As demonstrated in the previous chapters, Schliemann's stories seemed to give the British audience what they were looking for, that is, stability and a confirmation of identity. This is underscored by the following reference to the public attention Schliemann found in Britain: »Dr. Schliemann then gave an account of his own exploration, *which was listened to with great attention since 1871*« (*Daily News*, Jun 25, 1875, emphasis mine), and furthermore, by the particular interest people in Britain showed in Schliemann's stories, as pointed out by the *Athenæum*: »The discoveries at Mycenæ have excited so much interest in England that many of your readers will, I think, be glad to hear what progress has lately been made there.« (*Athenæum*, Feb 2, 1878: 163) Apart from the relevance of Schliemann's discoveries for his popularity in Britain, however, his self-promotion and his life plays a crucial part, if not *the* crucial part, with regard to Schliemann's popular success in Britain.

This becomes particularly apparent if we consider the life and work of Frank Calvert, who had purchased parts of Hisarlık and, though a consular official, had profound knowledge of the archaeological sites in the Troad:[9]

»As Calvert became active in archaeology, he evolved from an antiquarian to an archaeologist, systematically confirming or disproving past identifications of ancient cities and filling in the blanks in the archaeological record of the Troad. By 1860, he had gained enough stature in the field to begin to tackle the big question – where was Troy?« (Allen 1999: 72)

Schliemann first met Calvert in 1868 on his journey through the Mediterranean and eagerly absorbed Calvert's archaeological opinion and information regarding the alleged site of ancient Troy. By this time Calvert was

9 Cf. Chapter 3.2.

convinced that Troy must lie buried under the surface of Hisarlık: »Trusting completely in Schliemann, Calvert shared his passionate conviction that Hisarlık was the site of Troy and volunteered not only this land and half of the artefacts found therein, but also his expert knowledge.« (Ibid.: 119) However, Calvert had been deceived by Schliemann, who immediately used the information he provided and declared it to be his own, first in his diaries and letters and soon also in his books and articles published in Germany and Britain:

»Schliemann's efforts to minimize Calvert's role in the discovery of Troy in order to maximize his own began immediately after their first meeting and continued throughout their relationship of over twenty years. [...] From the moment Schliemann left Calvert on 15 August, he deliberately began to rewrite the record. It is clear that he recast events in his diary to make it appear that he believed Hisarlık was the site of Homer's Troy prior to his influential meeting with Calvert.« (Ibid.: 120)[10]

Schliemann's subsequent popularity illustrates, however, that his strategies were successful and that the public eventually did not care about the fact that Schliemann had betrayed Calvert's trust and had achieved fame at his expense. In letters and articles published by the contemporary press, Schliemann did his best to either blank out Calvert's name completely or use it to promote his own name by slandering Calvert, as the following letter from Calvert, published in the *Athenæum*, illustrates:

»The public cannot, of course, be expected to care anything for certain private differences which have arisen between Dr. Schliemann and myself on the subject of a valuable monument discovered by him in a field belonging to me at Hissarlik. Since, however, the Doctor in one place, as will presently be seen, goes so far as to insinuate that I had purposely deceived him on a point of archæology, I consider it but due to myself to place the fact on these differences on record, and to observe that his imputation of bad faith on this small point seems to have suggested itself as the best available weapon for parrying complaints on my own part with regard to mat-

10 Susan Heuck Allen (1995a, 1995b, 1998, and 1999) and Marcelle Schidkowski Robinson (1994, 1995, and 2006) discuss Schliemann and Calvert's role with regard to Schliemann's decision to start his excavations at Hisarlık.

ters which are neither insignificant nor imaginary. Let me here point out that, whereas Dr. Schliemann has thought proper to represent me throughout his work as an adversary of his explorations and of the identity of Hissarlik with Troy, it was in truth I myself who first convinced him of that identity, and persuaded him to make the excavations which have yielded such interesting results. Having a turn for archæological pursuits, and as a resident of many years' standing in the Troad, I have made a special study of the topography and antiquities of this region. [...] My attention was then turned to Hissarlik (the ›Ilium Novum‹ of our maps, which therein adopt the view put forward by Demetrius of Scepsis and his copyist, Strabo) as the probable site of old Troy. [...] I suggested to the British Museum the advisability of making excavations in this promising field, but my proposal was declined. In 1868 Dr. Schliemann first visited the Troad. He asked me my opinion as to the true site of Troy, admitting that he had not as yet given any attention to the problem. I, on my part, frankly communicated to him the result of my researches, and the grounds on which I had arrived at the conviction, that if Troy ever existed, it must have been at Hissarlik. [...] In 1870, Dr. Schliemann commenced his excavations. A few insignificant walls were brought to light, which were at once pronounced by him to be the ruins of Priam's Palace.« (*Athenæum*, Nov 7, 1874: 610)

Despite Calvert's description here of his own archaeological work in the Troad, his relation to Schliemann, Schliemann's course of action at Hisarlık, and finally his betrayal by Schliemann, the broader public seemed to prefer Schliemann's »dreamy Romantic« (Allen 1998: 345) story of »rosy ruins of the past« (ibid.) to Calvert's (real) story,[11] as pointed out by Allen: »He [Calvert] was not just a dreamy Romantic looking for the rosy ruins of the past, for he knew the Plain of Troy well and had excavated more than a dozen archaeological sites throughout the Troad.« (Ibid.)

This does not mean that there were no critical voices in the British press regarding Schliemann's archaeological excavations and accounts. However,

11 The credibility of Clavert's story is supported by the objectivity characterising his account of the excavations in the *Levant Herald*: »Meanwhile Frank Calvert wrote up a summary account of Schliemann's excavations at Hisarlik, placing them in the larger context of the history of excavations in the Troad. His account, which appeared in the *Levant Herald* of 3 February 1873, is admirably objective in tone, entirely free from fanciful romanticism that sometimes colours Schliemann's reports.« (Traill 1995: 102)

it does underline the influence and power of Schliemann's self-promotion, as it dominated the presentation and reception of his image in the public: »Thus, Frank Calvert prepared the ground for Schliemann in Troy. The contemporaries could have known this, but Schliemann's biographical construction of the *self-made man* who makes a discovery that surprises everyone in fulfilment of a childhood dream covered up this view of things for a long time.« (Cobet 2006: 154, translation mine) The tragic example of Calvert, whose archaeological work was characterised by systematic and mature approaches to the historic sites, shows the relevance of public performance for popular acceptance as successfully realised by Schliemann. To the public, the *reality* behind the Schliemannian story was less interesting and entertaining than the story told by Schliemann – his opinions being more sensational than the *real* facts.

Subconsciously, people seemed to crave a storyteller to connect the mythical past to the present at a time when everyday life was dominated by rational thought and demystification. That archaeology as such appeared to be particularly suitable to these modern fairy tales is implied by the following statement in *The Times*: »We live in an age of archæological surprise, each more startling than its predecessor. Nineveh, Halicarnassus, Troy, Mycenæ, Cyprus, Thebes, in turn yield their wonders of sculpture, of arms, of jewels, of papyri, of long-buried dead.« (*The Times*, Dec 24, 1881: 3) The »wonders« (ibid.) of long-forgotten times and their »long-buried dead« (ibid.) lend themselves to fairy tales as much as the gold found at Mycenae.[12] However, as with every wonderful discovery or incident, these archaeological discoveries needed a storyteller or narrator to make them not only publicly known but also enthralling for the masses. The relevance of captivating storytelling with regard to the discoveries at Hisarlık can be observed in the two very different careers of Schliemann and Calvert. In order to grab public attention and maintain interest for a longer period of time, it is necessary to apply certain strategies, although these strategies are often far from being candid and unobscured. The popularity of Schlie-

---

12 As pointed out by Traill, »Schliemann sent a total of fourteen periodic reports of his excavations at Mycenae to the London *Times*« (Traill 1986b: 240). The relevance of gold is further emphasised by the public interest in the Great Zimbabwe legend associated with King Solomon's gold (cf. Chapter 7.5).

mann's stories emphasises the relevance of performing identity through the act of narration.

The relevance of storytelling for popular success had already been demonstrated earlier in the century by Layard. After his spectacular archaeological discoveries in Mesopotamia in the late 1840s, Layard was confronted with the problem that the British government was not particularly interested in financially supporting his explorations, while the public did not seem to support his venture either. With the intention of changing this, »[Layard] began to consider writing up his discoveries in a popular form so as to rouse public interest directly« (Damrosch 2003: 46):

»Encouraging this idea, his friend Sir Charles Alison shrewdly urged him to play up the biblical angel: ›Write a whopper with lots of plates, fish up old legends and anecdotes, and if you can by any means humbug people into the belief that you have established any points in the Bible, you are a made man‹ (quoted in Waterfield, 171). Home in England for several months in 1848, Layard did just this, rapidly writing a vivid account of his explorations. He made sure not only to play up the biblical angle but to tie his work into the British fascination of tales of travel and imperial outposts, a theme well indicated by his book's full title: *Nineveh and Its Remains: With an Account of a Visit to the Chaldaean Christians of Kurdistan, and the Yezedis, or Devil-Worshippers; and an Inquiry into the Manners and Arts of the Ancient Assyrians*. The book became a best-seller upon its appearance in 1849, and support for Assyriological work increased a modest but steady trickle.« (Ibid.)

Characteristically, the advice given by Alison to his friend Layard essentially comprises of what Schliemann later seemed to have taken to heart with regard to his own archaeological accounts. The stories of the Bible were to Layard what the stories of Homer were to Schliemann.[13] As the popular success of Layard's *Nineveh and Its Remains* (1848-1849) proves, both the elements of the story – in Alison's words the »old legends and anecdotes« (ibid.) and the connection to the Bible – and the newly evolving genre of the allegedly documentary excavation account were particularly effective in ensnaring the attention of a broad public. Although Schliemann

13 For a discussion of the parallel between Schliemann and Layard's excavations, cf. Arentzen (2001).

himself seemed to believe in *his*tory more than Layard did in his own,[14] as Schliemann's many demonstrably false claims and stories suggest he used the same dubious strategies to make *his*tory attractive and thus capture the public's interest, with an often flagrant disregard for more modest scientific enquiry: »The modern theory that the story of the *Iliad* is nothing more than a myth finds no entrance into the mind of Dr. Schliemann.« (*Illustrated London News*, Mar 24, 1877: 282)

In addition to the content and make-up of his stories, Schliemann was anxious to regularly have the writing and publication of his books announced in contemporary newspapers and magazines: »Dr. Schliemann has brought the long dead past literally out of the tomb for us to look at. He is working day and night just now at his book, and it will very soon be before the public.« (Ibid.) This announcement emphasises the function of Schliemann as a creator and storyteller, who not only excavates the »long dead past« (ibid.) »for [the public] to look at« (ibid.) but also makes efforts to present and interpret it for the public.

Another important factor that helped promote Schliemann's books was the fact that he had succeeded in having them published by John Murray in London:

»As scholars of the 19th-century publication practices have noted, books like *Mycenæ* were a major source of entertainment in Europe and America in the age before the radio, cinema, and television. Their imposing physical size, colorful decorated covers, and heavily illustrated texts conveyed an air of adventure and exciting discovery carefully crafted to appeal to the widest possible audience [...]. Schliemann recognized this and spared no expense in creating a series of books that was calculated to appeal to the largest possible number of readers in order to marshal public support for his discoveries and to silence his critics.« (Runnels 2002: 20)

14 Layard receives the advice to »›[w]rite a whopper with lots of plates, [to] fish up old legends and anecdotes, and [...] [to] humbug people into the belief that [he] [has] established any points in the Bible‹ (quoted in Waterfield, 171)‹« (Damrosch 2003: 46) from his friend, while right from the beginning Schliemann takes the initiative for the popularisation of his excavations and his career.

Murray had already published Layard's bestselling excavation accounts and was renowned for the publication of several other popular historical and scientific publications by authors such as Darwin and Lyell:

»By the time that Schliemann began to publish in archaeology in 1869, the conditions were in place to reach a larger audience with articles in newspapers and periodicals, and of course with books. Schliemann was arguably among the first archaeologists to make maximum use of new technologies in his publishing ventures. He was not the first: Austen Henry Layard's enormously popular books describing his adventures in the Near East and his excavations at Nineveh and Nimroud, which began to appear in 1849, must be accounted as among the first great popular successes in archaeology.« (Ibid.)

The fact that all of Schliemann's books were published by Murray was particularly important for their popularity, since Murray not only owned a renowned publishing house, but also had the financial means to announce and advertise Schliemann's forthcoming books, as the many examples in contemporary newspapers and magazines illustrate.

This meant that the public was not only regularly informed of the latest discoveries by Schliemann, but also referred to his forthcoming publications on the subject. Articles about and advertisements for Schliemann's work – both by himself and others – thus also helped promote Schliemann's works and popularity. The *Academy*, for instance, noted:

»Dr. Schliemann then referred to the unanimous opinion of Greek authors that Ilium Novum was built on the site of the ancient city, and proceeded to describe his own labours and their results, which are already familiar to the public from the translation of the Doctor's work published by Murray.« (*Academy*, Jul 1875: 20)

Through his many publications, the innumerable articles written by him and others about his life and work, and through his lectures and exhibitions, Schliemann created an interferential Schliemannian network which brought him glory and fame. The relevance and power of his self-established network is once more underlined by Calvert's less lustrous career, which, though Calvert continued his archaeological explorations in the Troad, remained unknown to the broader public because he lacked anything close to Schliemann's public presence. In this context, it is also crucial to note

Schliemann's strong presence in London, which – as opposed to Calvert, who remained isolated in the Dardanelles – put him right at the centre of public attention and thus was strategically important for his self-promotional efforts in Britain.

Another part of Schliemann's public network consisted of the many prominent men he won over and engaged to provide introductory notes and appendices to his books and write favourable reports on his archaeological work in the contemporary press.[15] In his article »Schliemann and His Academic Employees« (1990) Traill notes:

»Schliemann was commendably conscientious about writing up his reports promptly. That he employed people to help him in the production of these books is certainly not surprising nor should it be held against him. What is a little unusual, however, is the extent to which he was anxious to keep this assistance secret. He was generally successful in this endeavor. Among his contemporaries perhaps only one or two had any idea of the extent of this cooperation. The biographers have little to say on the matter. Dörpfeld's collaboration, of course, has always been well known.« (Traill 1999: 237)

Among the seven scientists Traill presents in his article as »those scholars who were actually remunerated by Schliemann for their services« (ibid.)[16] are three British intellectuals, the editor Philip Smith (1819-1885), the Irish classicist John Pentland Mahaffy (1839-1919) and the Oxford Assyriologist A. H. Sayce (1845-1933).[17] All three men were engaged by Schliemann to

15 Robinson points out the relevance of Schliemann's network for his archaeological career by comparing his situation to that of Calvert: »Calvert also lacked Schliemann's contacts with specialists in all aspects of the enterprise on whom Schliemann relied for their particular expertise.« (Robinson 1995: 336)

16 Schliemann's large network is also referred to by Schmied: »The man of the world knows how to seek help from American, British, German envoys by turns, telegraphs Bismarck and asks him to intervene, mobilises the world public with letters to the *Times*.« (Schmied 1990: 17, translation mine)

17 Sayce's friendship with Schliemann is described by Sayce in his *Reminiscences* (1923): »The letter brought me the friendship of Dr. Schliemann who was staying at De Keyser's hotel in London. He asked me to lunch with him there, and thus a friendship commenced which had a considerable influence upon my fu-

contribute to his publications by means of introductions, appendices, revisions, and editing. As Traill delineates in his article (cf. Traill 1990), Schliemann paid them well for their work but was anxious to keep the visible role played by them with regard to his own work to a minimum or even secret. Apart from their contributions to the English versions of Schliemann's books, Smith, Mahaffy, and Sayce also wrote articles and reviews about Schliemann's work, published in contemporary newspapers and magazines. These articles, which supported and promoted his work and interpretation of the archaeological sites, thus were an important part of his self-created promotional network:[18]

»Dr. Schliemann, it will be remembered, not only maintains that Hissarlik is the site of the Homeric Troy – which is now admitted – but refuses to allow that any other view was generally entertained in antiquity; and Prof. Mahaffy has written a short paper, printed as an Appendix to Dr. Schliemann's Ilios, in which he seeks to show

---

ture life. It also brought me into close contact with Sir Charles Newton, whom I had met several times at Dean Liddell's table, but with whom I now became very intimate.« (Sayce 1923: 150) That Sayce and Mahaffy's loyality to Schliemann was publically well known is further underlined by the following quote: »Injudicious partisans of Dr. Schliemann, such as Messrs. Sayce and Mahaffy, seek to represent criticism of his theories as if it were disparagement of his labours.« (*Fortnightly Review*, Apr 1884: 434)

18 Cf. Sayce's preface to Schliemann's 1884 *Troja: Results of the Latest Researches and Discoveries on the Site of Homer's Troy*, discussed in Chapter 3.2. Sayce's friendship with Schliemann is described by Sayce in his *Reminiscences* (1923): »The letter brought me the friendship of Dr. Schliemann who was staying at De Keyser's hotel in London. He asked me to lunch with him there, and thus a friendship commenced which had a considerable influence upon my fu-
ture life. It also brought me into close contact with Sir Charles Newton, whom I had met several times at Dean Liddell's table, but with whom I now became very intimate.« (Sayce 1923: 150) That Sayce and Mahaffy's loyality to Schliemann was publically well known is further underlined by the following quote: »Injudicious partisans of Dr. Schliemann, such as Messrs. Sayce and Mahaffy, seek to represent criticism of his theories as if it were disparagement of his labours.« (*Fortnightly Review*, Apr 1884: 434)

that Troy was not utterly destroyed, but continued to exist on the same site until it became the Aeolic ›Ilium‹.« (*Academy*, Apr 22, 1882: 290)

Since Schliemann paid men like Sayce, Mahaffy, and Smith for their contributions to his work, he could at least for a certain time be assured of having them on his side and verifying and legitimising his findings. This is illustrated by a statement of Sayce in the *Academy* in 1882:

»In the ACADEMY of October 28, in the report of the meeting of the Hellenic Society, it is recorded that, in place of a paper of mine which had been announced, ›Prof. Jebb gave an account of his recent visit to Hissarlik in company with Mr. Calvert, Prof. Goodwin, and others, and stated it as the unanimous opinion of the party that no such stratification of the ruins as is implied in Dr. Schliemann's theory of successive cities exists.‹ I read this with astonishment, as only three years ago, when I visited the excavations, the existence of the successive cities was visible enough to everyone who had been trained in practical archæology. I could only suppose that the section laid bare in the great trench which I explored in 1879 had been destroyed by Dr. Schliemann's recent diggings; but, even so, I could not understand how good archæologists like Prof. Goodwin and Mr. Frank Calvert, who had both been adherents of Dr. Schliemann's ›theory‹, and the latter of whom was my companion 1879, could have forgotten what had once been so plain, and have changed their views.« (*Academy*, Nov 18, 1882: 368-369)

The promotional effect this had on Schliemann's public image is reflected in the contemporary press, where Schliemann's name, as in the following example, is often mentioned next to Sayce's, Mahaffy's, and Smith's: »Dr. Murray's ›Handbook of Greek Archæology‹ and Professor Mahaffy's ›Problems of Greek History‹ are also largely based on Dr. Schliemann's latest discoveries.« (*Edinburgh Review*, Apr 1892: 434) In other words, the fact that Schliemann engaged men like Sayce, Mahaffy, and Smith also meant that he could rely on them and their reputation in general.

With regard to Schliemann's public reception in Britain, Schliemann's British supporters played an important role insofar as they not only promoted his name and work in Britain but also worked to assimilate it into British culture. This is reflected in the following comment in *The Times*: »We take leave of this volume [*Troy and Its Remains*] with a high sense of what it owes to the editorship of Mr. Philip Smith, both for the additions which are from his own pen and for the thoroughness with which he has *performed his task of rendering Dr. Schliemann's results easily accessible to the English reader.*« (*The Times*, Mar 31, 1875: 7, emphasis mine) His British supporters helped to turn *Heinrich* Schliemann into *Henry* Schliemann and thus made him ›one of them‹. In relation to Schliemann as a role model for Victorian society this is crucial, since in order to make identification possible an assimilation of the *per se* foreign Schliemann had to take place. As illustrated by his self-presentation in Britain, for example on the occasion of his address at an anniversary meeting of the Birkbeck Literary and Scientific Institution at Chancery Lane on December 7, 1877, Schliemann was aware of the necessity to build associations with his British audience: »Dr. Schliemann, whose rising was greeted with hearty cheers, spoke as follows: [...] ›But who knows how the ancient Greeks pronounced their language? Perhaps they used *our* English pronunciation‹ (Laughter.)« (*The Times*, Dec 8, 1877: 6, emphasis mine) Although intended as a joke, Schliemann's statement is illuminating considering the fact that he refers to the English pronunciation as »*our* English pronunciation« (ibid.). While the »our« (ibid.) used by Schliemann here could also be interpreted in terms of ›our *modern* pronunciation‹, in the first place it seems to suggest that Schliemann identifies himself with the British and thus considers himself to be part of them, which allows him to refer to the English pronunciation as »our English pronunciation« (ibid.) instead of using the more neutral ›*the* English pronunciation‹. The extent to which Schliemann used certain strategies to present himself to the British audience and promote his success among the British public will be discussed in more detail in the following chapter.

Apart from the more covert British supporters of Schliemann referred to above, one notable person who played a crucial part in Schliemann's popularity in England was the classicist and Prime Minister of the United Kingdom William Ewart Gladstone (1809-1898). Schliemann made contact with Gladstone through Charles Thomas Newton (1816-1894), who was keeper of the Department of Greek and Roman Antiquities at the British Museum. Newton had been friends with Calvert for some years when Schliemann appeared on the scene and had himself introduced to Newton.[19]

---

19 For Calvert's attempt to get the support of the British Museum for excavations at Hisarlık through Newton, cf. Robinson (1995): »Hisarlık might well have be-

Following the example of Calvert, who had sent Newton material on possible excavations at Hisarlık as early as 1863 (cf. Allen 1999: 97),[20] Schliemann contacted Newton with information about his planned excavation in the Troad hoping for Newton's support and influence at the British Museum.[21] Although Newton was at first sceptical with regard to Schliemann's plans to excavate at Hisarlık, this changed in 1873 after Schliemann »sent him plans and photographs of the excavations prior to publication« (Vaio 1990: 416). It was after receiving these new records that Newton introduced Schliemann's archaeological plans and hypothesis to Gladstone, who, »[l]ike Schliemann [...] was a ›true believer‹« (ibid.) as far as Homer was concerned: »Gladstone was a well-known, if not notorious, amateur Homerist, whose dedication and energy had so far produced four substantial volumes.« (Ibid.: 75) It was from that time onwards that Gladstone functioned not only as »Schliemann's prophet to the British public« (ibid.: 415) but also as a verifying and legitimising agent in relation to Schliemann's archaeological inferences:

---

come one of the most spectacular and significant excavations ever sponsored by the British Museum. Had the Trustees accepted Calvet's offer, Calvert would have shared with the Museum the world-wide acclaim Schliemann garnered for himself alone« (Robinson 1995: 331); cf. also Allen (1995) and (1998).

20 In the early 1870s Calvert tried a second time in vain to gain Newton's support for the excavation at Hisarlık.

21 Allen makes the scandal around Calvert's brother responsible for the failure of Calvert's request to the British Museum: »The only compelling reason for the postponement and subsequent failure of Calvert's proposal and thus of his archaeological career can be the widely publicized scandal caused by [his brother] Frederick's alleged fraud and subsequent bankruptcy. [...] The museum trustees would have been reading about the Calvert family's misfortune and Frederick's alleged felony only months before they received Frank Calvert's over generous offer. Still, the finality of Newton's language is curious, since the trustees simply had mentioned postponement. [...] Following this stunning rejection, Calvert withdrew. Just when he had the pleasure of seeing his refutation of Chevalier in print, he was rebuffed by the British Museum and his friend Newton.« (Allen 1999: 100)

»[Schliemann] went on to the Troad in 1870, and cut into the mound of Hissarlik, long marked by one school of topographical critics for the site of Troy, and actually opened first by Mr. Calvert. In the next two years Schliemann succeeded in arousing only sufficient interest to be accounted a spy by the Porte and a harmless enthusiast by Europe. But the year 1873 was to bring promise of greater things; for above a settlement of the stone age on the bed-rock Schliemann revealed a burned city with strong ramparts [...]. Who could doubt this was Priam's own treasure hastily concealed while Ilios burned? *The world was startled out of its apathy to archæological things, and England especially, led by Mr. Gladstone, was disposed to believe more than the half.*« (*Quarterly Review*, Jul 1898: 91, emphasis mine)

As this passage demonstrates, the British were particularly attracted by Schliemann's excavations and susceptible to his theories and inferences, which were supported and represented in Britain by Gladstone.[22] The importance of Gladstone for Schliemann's reception is also reflected by the many magazines and newspaper articles in which Gladstone comments on and is quoted in relation to Schliemann's excavations and publications. How much Gladstone's support meant for Schliemann's public image is also implied by the following review of Schliemann's *Mycenæ* (1877) in the *Academy*:

»The author of this book appears before the public under the disadvantage of having already announced his discovery with considerable detail, with every exaggeration of phrase and with surely enough of applause. By ›preserving request‹, however, he

22 Cf. also Gladstone's following statements in the *Contemporary Review*: »I am among those who have contended – 1. That the poems of Homer were in the highest sense historical, as a record of ›manners and characters, feeling and tastes, races and countries, principles and institutions‹, 2. That there was a solid nucleus of fact in his account of the Trojan War« (*Contemporary Review*, Jun 1874: 1-2); »Still more, and much more, important have been the excavations of Dr. Schliemann. His large collections have been inspected at Athens by Professor Burnouf of Athens, and by Mr. Newton of the British Museum. In this country we have had the opportunity of such examination as Dr. Schliemann's collection of photographs, in some instances rather imperfectly executed, would allow. Reviews of high authority have, within a few weeks of the publication of the ›Ausgrabungen‹, recognized their importance in elaborate essays.« (Ibid.: 3)

has obtained the novelty of a Preface by Mr. Gladstone, which will go far to furnish a new interest to this handsomely-published work.« (*Academy*, Dec 15, 1877: 558)

What is interesting here is that by engaging Gladstone to write an introduction to his book, which through his own premature announcements had lost its appeal of novelty, Schliemann was able to rehabilitate *Mycenæ* and thus make it attractive to the public again. As far as Schliemann's public image in Britain is concerned, this is crucial insofar as it represents the relevance Gladstone's name had for Schliemann's popularity, a fact also illustrated by the various examples of contemporary Schliemann criticism in which Gladstone is repeatedly mentioned to qualify certain points of criticism, as another review of *Mycenæ* illustrates:

»A disposition to smile at the simplicity with which Dr. Schliemann accepts poetry for history – apparently in unconsciousness how the poetical stories differ – is a little checked by finding that Mr. Gladstone gives him full encouragement in the Preface of some forty pages which he has contributed with liberal kindness.« (*Athenæum*, Dec 22, 1877: 819)

Thus, the fact that Gladstone had provided a preface to Schliemann's book is mentioned to make the book more valuable and serious.

Apart from his prefaces to Schliemann's books and articles published in newspapers and magazines, Gladstone contributed a large part to Schliemann's successful appearances in front of British audiences. This is documented by the various reports of Schliemann's appearances in public at the different societies and institutes discussing Schliemann's excavations. On most of these occasions Schliemann was either introduced or commented on by Gladstone, as a meeting of the Society of Antiquary attended by »[f]ellows and visitors« (*Illustrated London News*, Mar 31, 1877: 306) in 1877 demonstrates:

»Mr. Gladstone rose to address the meeting, and was *loudly cheered*. He said Dr. Schliemann had over-bountifully paid him for the little he had been able to do in the fields of Homeric inquiry. He felt the lecture's liberality must react in weakening the foundation of anything [Mr Gladstone] presumed to say on Dr. Schliemann's behalf, and must lead to the suspicion that he was only endeavouring to requite the person besides Dr. Schliemann had been mentioned on this occasion – he meant Mrs. Schliemann.« (*Illustrated London News*, Mar 31, 1877: 306, emphasis mine)

In all these accounts of Schliemann and Gladstone's public speeches, their popularity is supported by the positive reaction of the audience in the form of cheers and applause, as also reported for a speech by Schliemann at the Royal Institute of British Architects: »The attendance was very numerous, and included Mr. Gladstone and Dr. Schliemann, both of whom on making their appearance were cordially received.« (*Daily News*, May 1, 1877)

In addition to Gladstone, whom Schliemann continuously refers to as his close friend in public, Layard[23] and the geologist and archaeologist John Lubbock are repeatedly mentioned by Schliemann as friends: »›[I]n the direction of Bunarbashi, just before Hector, was, in October, 1872, excavated by *my esteemed friend Sir John Lubbock*, who found in it merely some painted Greek potsherds of the third century B.C..‹« (*The Times*, Aug 16, 1877: 3, emphasis mine)[24] Furthermore, Schliemann befriended Pedro II, Emperor of Brazil (1825-1891), who was extremely interested in Schliemann's excavations and visited his excavation sites. Schliemann would not have been Schliemann had he not also referred to this friendship

23 Cf. »›In conclusion, I here publicly most warmly thank my honourable friend, Sir A. Layard, the illustrious English Ambassador at Constantinople, for the powerful assistance he has lent me, and all the kindness he has shown me during the time of my excavation at Troy.‹« (*Athenæum*, Dec 14, 1878: 769)

24 Lubbock's reply was published the following day in a letter to the editor of *The Times*: »›Sir, – My friend Dr. Schliemann, in his letter on his extremely important and interesting excavations at Troy contained in *The Times* of yesterday's date, states that the so-called tumulus of Hector on Bunarbashi, which Mr. Grant Duff and I partially opened in 1872, contained some fragments of pottery which could not belong to an earlier date than 300 B.C.; and he therefore refers the tumulus itself to that comparatively recent period. Will you allow me to say that it would not be safe to infer the age of the tumulus from the pottery in question?‹« (*The Times*, Aug 20, 1877: 6)

in accounts and letters published by the contemporary press.[25] Eventually he even dedicated *Mycenæ* (1880) to Pedro II.[26]

Schliemann used prominent men not only to promote his excavations but also to defend himself against criticism, as illustrated with regard to the publicly reported dispute between him, the artist William Simpson,[27] and *The Times* correspondent William James Stillman (cf. Traill 1993c: 257).[28] Schliemann took any opportunity to publicly defend his archaeological theories and frequently referred to or quoted from other scientists to support his inferences, as the following example demonstrates: »›In fact, it sometimes took me several minutes to become convinced that I always played fair, that I had always stated the sacred truth, which only lately had, in so flattering terms, been publicly acknowledged by the high authority of Mr. Charles T. Newton.‹« (*The Times*, Aug 16, 1877: 3) Schliemann had a great

25 Cf. *The Times*, Nov 6, 1876: 4; *The Times*, Dec 22, 1876: 4; *The Times*, Jun 23, 1877: 114; *The Times*, Aug 16, 1877: 3.

26 »Dedicated to His Majesty Dom Pedro II Emperor of Brazil with the Profound Respect of the Author.« (Schliemann 1976b [1878]: n. pag.)

27 As a highly regarded special artist, William Simpson worked for the *Illustrated London News*: »There have been few scenes of great events in the world during the past quarter of a century that have not been visited and sketched by Mr. William Simpson: and there is perhaps no other man living with whom travellers generally have so many sympathies.« (*Academy*, May 4, 1878: 400) He provided the sketches of the excavational site at Mycenae. Simpson criticised and questioned Schliemann's inferences in regard to the excavations.

28 William James Stillman was an American journalist employed by the *London Times* as a correspondent in Athens (1876-1889) (cf. Chambers 1990: 409). Stillman referred to Schliemann's work as »›one of the most extraordinary hallucinations of unscientific enthusiast which literature can boast of‹« (cf. *Academy*, Jul 30, 1886). The controversy between Schliemann and Simpson and Stillman, which was carried out in *The Times* and *Fraser's Magazine*, was also referred to in the contemporary press: »We do not wish here to revive the controversy that went on, last July and August, in *Fraser's Magazine* and the *Times*; but as Dr. Schliemann's case rests partly on his collection of portable relics from the Troad, now on view at the South Kensington Museum, our readers must be warned of the conflicting arguments for and against the Hissarlik site of Troy.« (*Illustrated London News*, Jan 5, 1878: 11)

ability to get in touch with prominent men like Gladstone, Newton, Lubbock, Layard, Sayce, and Mahaffy and use them for his own ends. His popularity in Britain shows how those contacts soon became self-generating and helped promote Schliemann to the British public.

5.6 ENTERTAINING THE MASSES: FROM BURLINGTON HOUSE TO SOUTH KENSINGTON

> As already briefly stated in the *Mercury*, an extraordinary meeting of the Institute of British Architects was held on Monday evening at their rooms, in Conduitstreet, for the purpose of granting to Dr. H. Schliemann the explorer of Troy and Mycenæ, his diploma of election as honorary and corresponding member. There was a large gathering of ladies and gentlemen.
> (*LEEDS MERCURY*, MAY 4, 1877)

So far the focus has been on Schliemann's *publications* and their discussion in contemporary newspapers and magazines. In the present chapter, the emphasis is on Schliemann's *public performances* in Britain.[1] As already mentioned, there were various occasions on which Schliemann was invited to present his discoveries to the British public. In addition, Schliemann exhibited his Trojan Collection at the South Kensington Museum in London for four years starting in December 1877.[2] Both Schliemann's public lectures and his museum exhibition will subsequently be analysed in relation to their effect on the contemporary audience and visitors. Reports and reviews published in newspapers and magazines will once more be the primary source for this analysis.

---

1 Schliemann's ability to entertain his audience is also pointed out by Traill's reference to Schliemann as a tourist attraction in Athens: »Schliemann was by now almost as much of a tourist attraction in Athens as the Parthenon itself. He catered to the stream of visitors who wanted to see him by inviting all those who had left their cards to the literary salon held at his house every Thursday evening during the winter.« (Traill 1995: 216)

2 For a description of the exhibition, cf. Easton (1994: 230-232).

Schliemann gave his first public speeches in London as part of a series of lectures delivered in various European cities from June to December 1875. At this point Schliemann was already well known through the publication of his books and the innumerable British newspaper and magazine articles reporting and discussing his excavations and accounts. Although Schliemann's London lectures had already been followed with great interest before, the climax of his public appearances in Britain was to take place on his second England tour in 1877, which Cottrell views as the beginning of Schliemann's public career: »But what years they had been! First there had been his triumphal tour of England in 1877, when thirty learned societies had vied with each other to honour him, and he renewed acquaintance with Gladstone, whom he had first met in 1875.« (Cottrell 1984: 224) For Schliemann himself his success in Britain must have been the realisation of a dream. In a statement in which he retrospectively describes his experiences of 1877 in Britain he notes: »›In London last year I was received for seven weeks as if I had discovered a new part of the globe for England. How very different it is in Germany. There I met only with abuse.‹« (Schliemann, quoted in Ludwig 1931: 180) Schliemann's comment not only confirms Cottrell's about the former's extraordinary popularity and success in Britain but also contrasts it with the disregard he met with in Germany addressed above.[3]

Most of the lectures given by Schliemann were very well documented in contemporary newspapers and magazines, as the prominent example of lecture he held before the Society of Antiquaries on March 22, 1877, illustrates (cf. fig. 4):

»Last night a meeting of the Society of Antiquaries was held at its rooms, Burlington-house; [...] There was a *crowded attendance of Fellows and visitors*. [...] Mr. Watson [secretary] then said that they had now arrived at the real business of the meeting. After eight days and nights of incessant travelling, Dr. Schliemann arrived in this country only this morning with the object of reading his paper before the society on his discoveries at Mycenæ. (*Cheers.*) In illustrations of these discoveries photographs and plans were exhibited on the walls. Upon the last occasion Dr.

3 The difference in the reception of Schliemann has to be read in connection to the standing ›amateurs‹ had in Britain, which originated in the long tradition of the *Dilettanti* (cf. Chapter 3).

Schliemann was present, the late President, who then occupied the chair, said he was certain that if the lecturer again visited this country he was sure to have a good reception. The *crowded meeting* that night verified the late Lord Stanhope's prophecy. (*Cheers.*) On rising to read his paper the illustrious discoverer was greeted with the heartiest of welcomes.« (*The Times*, Mar 23, 1877: 10, emphasis mine)[4]

Fig. 4: »Dr. Schliemann Giving an Account of His Discovery at Mycenæ before the Society of Antiquaries at Burlington House«

Illustrated London News, March 31, 1877: 301. Reproduced with kind permission of Frankfurt University Library Johann Christian Senckenberg.

4 The identical report on Schliemann's lecture at the Society of Antiquaries was also included in an article on Schliemann published by the *Illustrated London News* (Mar 31, 1877).

The number and variety of contemporary newspapers and magazines reporting and commenting on Schliemann's speech at the Society of Antiquaries further underlines the popular interest people showed in Schliemann's works, which Traill acknowledges in his comment:

»Schliemann arrived in London on 22 March. [...] Photographs were posted for the audience to examine. The talk was extremely well received and at its conclusion Schliemann was made an honorary member of the Society by acclamation. This was the first of many talks that Schliemann gave before the learned societies in the spring and summer of 1877. [...] Schliemann was the lion of London society.« (Traill 1995: 166)

Schliemann's popular appeal was marked in addition by the many people attending his speeches and their fervid reaction, referred to in reports as marked with ›cheers‹ and exclamations of ›hear, hear‹.

Schliemann's lecture at the Society of Antiquaries on March 22, 1877, takes on a special position in this respect, since it was his first and at the same time most widely received public appearance in Britain. Although Schliemann's excavations and publications had been discussed by the contemporary British media right from the beginning of his career and his first excavations in the Troad, Schliemann had his public breakthrough with his lecture at the Society of Antiquaries in Britain:

»The arrival of Dr. Schliemann in London, and his address to be delivered here to the Society of Antiquaries, *must increase the amount of public interest* already felt in his successful explorations of the sites of ancient classic history, or early traditions of romantic events in Greece, celebrated by the epic and tragic poets of that highly gifted nation.« (*Illustrated London News*, Mar 24, 1877: 281, emphasis mine)

This breakthrough was promoted by the recent discoveries at Mycenae, which because of the gold finds were even more popular than Schliemann's Trojan excavation had been. 1877 was the year in which the *Illustrated London News*, one of the leading illustrated magazines of the time, published more reports and sketches on Schliemann and his exploration than ever before or after.[5] Two days after Schliemann's speech at the Society of Antiquaries, the *Illustrated London News* published a multi-page account of Schliemann's excavations with a portrait of Schliemann subtitled »Dr. Schliemann, the Explorer of Troy and Mycenæ« (ibid.). This article is interesting with regard to Schliemann's public reception, if we consider that it does not mention his lecture at the Society of Antiquaries although it was published shortly afterwards. Since this was presumably due to editorial deadlines, it is even more interesting to consider the way in which the *Illustrated London News* prepares its readers in its issue of March 24 for the report of Schliemann's lecture to be published by the magazine one week later:[6]

»The portrait of Dr. Schliemann, which accompanies the other Illustrations given in this Number, is offered as a tribute of personal respect, in recognition of his laudable endeavours and large pecuniary sacrifices, at his own private risk, to carry on these laborious and costly researches. His example at the present time appears more worthy of note, from the circumstances that he is not a man trained to the profession of literary and academic scholarship; that he has never been a professor of any of the German or other Universities: but that his youth and part of his manhood, in Hamburg and in London, as we understand, were incessantly occupied with commercial business. Since his retirement, within the last few years, from an active mercantile career, he has devoted nearly all his time, and great deal of his money, to the self-imposed task of examining the places associated with the chief actions related in Homer's ›Iliad‹, and those of collateral importance described in the narratives or dramatic compositions of other Greek poets.« (*Illustrated London News*, Mar 24, 1877. 281)

This passage sums up all the aspects of Schliemann's career discussed in Chapter 5.5 as being particularly relevant to British society in general: Schliemann as the self-made man who through hard work and perseverance succeeded in spite of his poor educational background. What Schliemann meant to the British public is once more reflected in the fact that the maga-

5 »We gave some account of these discoveries, with many Illustrations, in several Numbers of our Journal published in 1877.« (*Illustrated London News*, Jan 3, 1891: 5)

6 *Illustrated London News*, Mar 31, 1877.

zine honours Schliemann with a portrait »as a tribute of personal respect and in recognition« (ibid.) of his achievements and refers to the fact that »[h]is example at the present time appears more worthy of note, from the circumstances that he is not a man trained to the profession of literary and academic scholarship« (ibid.). Furthermore, Hamburg and London are mentioned as places where Schliemann supposedly spent »his youth and part of his manhood« (ibid.). Interestingly, Schliemann never really lived in London and only started to visit it regularly after beginning his archaeological career. For this reason, it seems likely that the reference to London is used here to make Schliemann even more attractive to a British readership, who thus are not only invited to see Schliemann as an example of a self-made man, but also as a person with a deep affinity with their own country. There could have been no better way to prepare the report on Schliemann's lecture at the Society of Antiquaries in the *Illustrated London News*' next number, where the famous illustration, subtitled »Dr. Schliemann giving an account of his Discoveries at Mycenæ before the Society of Antiquaries at Burlington House« (*Illustrated London News*, Mar 31, 1877: 301), shows Schliemann holding his speech surrounded by a large British audience (cf. fig. 4).

This illustration would most certainly have satisfied Schliemann's desire for public attention, since it presents him not only surrounded by prominent men of science but also *as part of* a British collective. His effort to be accepted and thus become part of his British audience is emphasised by the many references to Britain he makes in his public speeches, as demonstrated by the following example: »Dr. Schliemann concluded by saying that his labours were the object of both praise and obloquy; but he was willing to submit to the candid judgement of an English audience (Cheers.)« (*Daily News*, Jun 25, 1875) As exemplified by one of Schliemann's first lectures at the Society of Antiquaries in 1875, he skilfully uses the audience by pretending to submit to their »candid judgement« (ibid.), while at the same time he adulates their Englishness and is thus indirectly able to control their reaction. As suggested by the »[c]heers« (ibid.) following his statement, Schliemann achieved the desired affirmative reaction from his audience, confirming his success in subtly establishing a connection between the British audience and himself.

His ability to win the audience over is also illustrated by another instance reported in *The Times*, when Schliemann concludes a paper on Troy and Mycenae at the British Archaeological Society in 1877: »In the midst of their suffering and privations in the depths of Ilium, it would always be to Mrs. Schliemann and himself a great consolation to think that they had the kind sympathies of this great and noble nation (Cheers.)« (*The Times*, Apr 12, 1877: 8)[7] Here again, after receiving a diploma from the Society for his archaeological achievements, Schliemann gives the audience the feeling that their »great and noble nation« (ibid.), that is, Britain, is through its support indirectly part of his success and glory.[8] As these examples show, Schliemann used his audience as a means of self-affirmation in much the same way as he used prominent men of the time, like Gladstone and others, to promote his own public success in Britain.

As much as anything else in his life, Schliemann knew how to stage his gratitude to the British public. This is made explicit by the following passage from a letter written by him[9] and published in *The Times*:

7 Schliemann's ability to connect with his audience is further underlined by people's reactions to his speeches, as documented by *The Times*: »In conclusion he said that if they thought Mrs. Schliemann and he had by their disinterested labours contributed a little to show that Homer did not describe myths, but real events and tangible realities, this would be to them most flattering acknowledgment and a greater encouragement in the continuation of their works in Troy, which they would resume very soon, for they had the necessary Firman of the Turkish Government in their hands. (Loud Cheers.)« (*The Times*, Mar 23, 1877: 10)

8 Another example of Schliemann's strategies to gain acknowledgement and appreciation in Britain is his dedication of *Troja* (1884) to Queen Victoria: »Dedication to Her Imperial and Royal Highness Victoria, Crown-Princess of the German Empire, Crown-Princess of Prussia, Princess Royal of Great Britain and Ireland, Duchess of Saxony, the Illustrious Patron of Art and Science this Work is Dedicated, with the most Profound Respect, by the Author.« (Schliemann 1972 [1884]: n. pag.) Schliemann dedicated *Ilios* to Layard, who, as the British Ambassador at Constantinople, had helped Schliemann gain a permit for his excavations at Mycenae.

9 »Dr. Henry Schliemann writes to us from Boulogne-sur-Mer, where he is now staying.« (*The Times*, Aug 16, 1877: 3)

»›I have much pleasure in informing you that, in order to show my gratitude to the English people for the warm reception I have found with them during my three months‹ stay at London, I have resolved to bring my Trojan collection, including the treasure in gold and silver, to England, and to exhibit it provisionally at the South Kensington Museum. Of course it is not for sale. [...] [T]he Trojan antiquities have no resemblance whatever to anything yet found elsewhere; they are, consequently, a new world of archæology, and I have therefore no doubt that they will be duly appreciated by the intelligent English public, and that they will encourage with them the classical studies and increase their love for divine Homer.‹« (*The Times*, Aug 16, 1877: 3)

Here Schliemann characteristically combines praise for Britain with self-praise by referring to the uniqueness of his Trojan antiquities and offering them to an extraordinarily »intelligent English public« (ibid.). By referring to »treasure in gold and silver« (ibid.), Schliemann emphasises not only the antiquarian but also the material value of his collection. Against this background, Schliemann's act of offering his Trojan collection in order »to show [his] gratitude to the English people for the warm reception [he] [has] found with them during [his] three months' stay at London« (ibid.) can be read as a financial transaction, which confirms his basic strategy of purchasing – both in a figural and a literal sense – his own glory. As indicated above, Schliemann in some cases actually did engage and pay people for their support, while in others he found support by flattering them and seemingly putting his own glory in their hands.[10] The combination of both is further illustrated by another announcement by Schliemann with regard to the collection he exhibited at the South Kensington Museum the following year:

»After an interruption of five years and a half – not idly spent, but fruitful in results with which the public are already acquainted – Dr. Schliemann has resumed his search for the buried Troy. In a series of communications to *The Times*, commencing October 14 and coming down to the 11th inst., he describes the progress of his work and the discoveries he has made with great minuteness. Having, through the friendly

10 Schliemann proceeded in a similar way by publicly announcing the donation of his collection to Greece, France, and Britain before eventually presenting it to Germany (cf. Allan 1999: 181-184).

interposition of Sir A. Layard, obtained a firman from the Porte authorizing him to proceed, he opened his trenches on Mount Hissarlik on the 30th of September, and has since prosecuted his labours with the greatest energy and a confident expectation of success. [...] To show his gratitude to the English people for the valuable assistance which their Ambassador, Sir A. Layard, has rendered in procuring the firman and for the protection ever since afforded, Dr. Schliemann promises to send to South Kensington for exhibition that portion of the Trojan treasure which may fall to his share.« (*The Times*, Nov 27, 1878: 4)

Here again, Schliemann's staged gratitude for British support, in particular for that of Layard, is presented as the reason for his promise to send another portion of the Trojan treasure to the South Kensington Museum.[11] The fact that he regularly reported on the excavations he had resumed in communications with *The Times* once again emphasises his consciousness of public relations. The close link between his style of self-presentation and his popularity in Britain is further emphasised by another of Schliemann's statements about these antiquities, namely that »›[w]hatever may fall to my share in the ensuing division with the Turkish Government, I shall exhibit in the South Kensington Museum‹« (*Graphic*, Apr 19, 1879).

Against this background, Schliemann's exhibition of his Trojan collection at the South Kensington Museum was a very effective advertising strategy that surpassed the momentary popularity he had achieved in 1877. Whereas previously Schliemann had had to travel to London in order to make himself the talk of the town and refresh people's interest in his person and excavations, through his exhibition at the South Kensington he was omnipresent in Britain. In November 1877 Schliemann finally brought his Trojan collection to London, where it was open to the public until he took it

11 Further references to Schliemann's commitment with regard to the public display of the treasures in Britain were made as follows: »Dr. Schliemann has to make over to the Imperial Museum at Constantinople, and the remaining one-third, his portion, he intends to deposit in the South Kensington Museum. Our metropolitan friends would do well to make a visit thither.« (*After Work*, Feb 1879: 36); »They are now in the museum at Constantinople; but he had been able to save for himself the third treasure of gold and silver ornaments, goblets, and vases, now on view at the South Kensington Museum.« (*Illustrated London News*, Jan 12, 1878: 45)

to Berlin in January 1881, as the press subsequently gushed:[12] »Apart from the absence of any ground for questioning Dr. Schliemann's graphic account of the discovery of the famous ›Treasure‹ – which was so long open to inspection in the South Kensington Museum* – no candid or competent judge ever doubted its genuine and high antiquity.« (*Quarterly Review*, Jul 1881: 212)[13]

When Schliemann's collection was opened on December 28, 1877, the public had already been informed by innumerable announcements about both Schliemann's decision to present his collection in London and the whole process of transportation, arrangement, and presentation of the collection prior to the exhibition opening:[14]

»London was the obvious place because he had been well received by the Society of Antiquaries and because he thought Gladstone could intervene on his behalf. His first hope was for a couple of rooms in the British Museum, but the trustees decided that they had no available space. A suggestion from Gladstone, however, took him to the South Kensington Museum – now the Victoria and Albert.« (Easton 1994: 231)

Shortly before the exhibition opened to the public, a private viewing was given, as reported in the *Manchester Times*: »The collection of antiquities

12 »Dr. Schliemann has come over to England for the purpose of removing from South Kensington Museum his Trojan collections, which have been exhibited there for the last three years. A few days, therefore, only remain for any persons who have not yet seen them to satisfy their curiosity.« (*Athenæum*, Dec 25, 1880: 872)

13 *»The collection exhibited at South Kensington has been presented by Dr. Schliemann to the Berlin Museum; but a still greater number of the objects dug up at Troy remains at his disposal. Could they find a fitter home than the British Museum?« (*Quarterly Review*, Jul 1881: 212)

14 »The antiquities found at Hissarlik, on the plains of Troy, by Dr. Schliemann during his remarkable excavations, have been arranged for exhibition, and are now on view, in one of the courts of the South Kensington Museum.« (*Illustrated London News*, Dec 29, 1877: 627) Various illustrations of Schliemann's exhibition were also published in the daily press (cf. *Illustrated London News*, Jan 5, 1878: 13).

found at Troy by Dr. Schliemann during his remarkable excavations has been arranged for exhibition at the South Kensington Museum, and a private view has been given, at which a large number of visitors were present.« (*Manchester Times*, Dec 29, 1877) Reports in various newspapers and magazines about this private preview stirred public interest in the exhibition even more, which upon its opening was commented on as follows:

»An exhibition of extraordinary interest was opened to the public yesterday at South Kensington Museum. We refer to Dr. Schliemann's collection of antiquities from Troy, the arrangement of which in one of the courts has just been completed by the learned and enthusiastic explorer. Archæologists may differ as to the antiquity of the articles comprised in this collection, but there can be no second opinion as to their exceeding value as memorials of a very remote age, or of the enterprise, perseverance, and ability displayed by Dr. Schliemann in their discovery. Many of our readers have seen the Doctor's book on Troy and its Remains, published about two years ago, in which he gives a history of his exploration of the Hill of Hissarlik, and describes minutely the objects which he found. A perusal of his work will enable the visitor who desires to make a careful examination of the collection to do so with increased pleasure.« (Ibid.)

In terms of promotion and publicity, this passage is interesting for various reasons, since it not only announces the opening of the exhibition but also includes references to Schliemann's character, excavations, and recently published book *Troy and Its Remains* (1875). Considering the fact that most people only knew about Schliemann and his excavations from the contemporary press and from his books, while some of them might actually have been able to attend one of his speeches, the fact that the objects excavated by him were now displayed at the museum could be seen as material confirmation of what until then had remained ›abstract‹ and ›fictional‹. Schliemann's exhibition gave people the chance not only to get *in touch with* what had been at best a *story* narrated by Schliemann and the media, but also to come face to face with the past: »Already the collection has a symmetrical appearance, and promises, when Dr. Schliemann and Mr. Streatfield have concluded their labors, *to enable the spectator to turn over at his ease page after page of the history of man, as written in the mound of windy Troy.*« (*Daily News*, Dec 10, 1877, emphasis mine) Instead of the rather abstract accounts on Schliemann's excavations and his antiquities, people were now not only directly confronted with them but also got a chance to almost literally *grasp* them.[15]

In June 1877 the *Graphic* had published an illustration of Schliemann's exhibition of Mycenaean antiquities in Athens entitled »Exhibition at Athens of the Mycenæ Treasures Discovered by Dr. Schliemann« (*Graphic*, Jun 16, 1877).[16] The illustration showed visitors to the museum inspecting Schliemann's collection of Mycenaean antiquities. Now, only five months later, the British public had the archaeological objects excavated by Schliemann before their very eyes. However, the disappointment expressed in many reports on Schliemann's exhibition in London implies that the collection lacked the entertaining character of Schliemann's own archaeological accounts:

»The collection of antiquities from Hissarlik now on view at the South Kensington Museum is not so large as we had expected; and, from the fact that a considerable number of the objects do not vary in any essential particular from their neighbours, it might have been smaller. [...] The whole thing is disappointing to the last degree. Yet, when *the collection fails as a show*, it begins to be interesting to the student; not, however, the student of Homer, who, should he take with him a text of the poet in the hope of obtaining illustrations of it, will find himself in a worse position than the inexperienced playgoer who takes a common version of a play which he is to

15 Cf. also Chapter 5.3. This is underlined by the following comments in the *Illustrated London News* and the *Academy*: »Dr. Schliemann has brought the long dead past literally out of the tomb for us to look at« (*Illustrated London News*, Mar 24, 1877: 282), as well as: »[T]here is in the museum [in Athens] a model of the tomb, with the body of Agamemnon in it.« (*Academy*, Apr 29, 1882: 300)

16 According to *The Times*, the Athens collection was also seen by various British people (cf. *The Times*, Mar 17, 1877: 12), while the *Illustrated London News* emphasised its own concerns to provide the British reader with immediate information and illustration of Schliemann's discoveries: »Our readers are fully aware of the direct efforts made by the proprietors of this Journal, in the employment successively of two Special Artists on the spot, as well as at Athens, to provide complete and accurate Illustration of Dr. Schliemann's remarkable discoveries.« (*Illustrated London News*, Mar 24, 1877: 281)

see, hoping to be able to follow the actors.« (*Academy*, Dec 22, 1877: 581, emphasis mine)

Unlike his books, which were commented on as »entertaining and instructive« (*Leisure Hour*, Jul 1877: 444) both for the »learned and even the popular reader« (*Edinburgh Review*, Apr 1892: 429), Schliemann did not succeed in captivating the masses in Britain with his Trojan collection at the South Kensington Museum, as stated by Easton: »There were neutral or favourable reviews in The Builder and The Times. Bertram Hartshorne wrote a long piece in The Archaeological Journal, and The Illustrated London News published some engravings of its own. A.S. Murray, writing in The Acadamy, found the whole thing ›disappointing to the last degree‹.« (Easton 1994: 231) The reason for this most likely was the fact that the objects exhibited by Schliemann essentially lacked the spectacular character of Egyptian and Assyrian exhibitions, which included such sensational public displays as the unwrapping of mummies. The lack of spectacular objects in Schliemann's collection is implied by the following comment in the *Manchester Times*: »The most valuable part of Dr. Schliemann's relics is the great ›Trojan Treasure‹. Most of the articles which it comprises are of silver and gold, and independently of their historic association, must be of great intrinsic value.« (*Manchester Times*, Dec 29, 1877) For many visitors the ›Trojan Treasure‹ must have been the highlight of the exhibition, considering the fact that the other antiquities exhibited were more common objects people were likely to have seen before or even knew from their own everyday lives. It is due to this lack of *extra*ordinariness which Egyptian and Assyrian exhibitions achieved that »the collection fail[ed] as a show« (*Academy*, Dec 22, 1877: 581), in other words, that it was unable to enthral and entertain the masses. However, this does not mean that Schliemann's exhibition was not popular. As the many contemporary newspaper and magazine articles announcing and discussing the exhibition illustrate, it was widely received and many people went to see the objects excavated by the famous explorer.

The reason for this popularity in spite of the fact that the collection ultimately turned out to be disappointing for many, certainly lies in the *idea* of ancient Troy which Schliemann had been successfully popularising since his early excavations in the Troad. What the exhibition of Schliemann's Trojan collection meant for many was material proof of Schliemann's *his*tories, which they had been reading about and seen illustrations of in his books and in newspapers and magazines. With his exhibition of ancient Trojan objects in London, Schliemann had brought antiquity as the *most familiar strangeness* to Britain and *into* British society. The *materiality* of Greek antiquity, as the alleged familiar, was thus imported into actual British culture to meet with the *idea* of Greek antiquity:

»Above all else, of course, the material increase in evidence of a completely lost historical period [...]. After all, the vividness of the historical memory created by Schliemann also includes the myth of the pioneer who promised to break free of the no longer transparent trivialities of science, a myth not without a methodological – propaedeutic, as it were – implication. Thanks to Homer, Troy is a place of great European tradition. It seems that we have Schliemann to thank for being able to assure ourselves of that in a real way.« (Cobet 1992: 360, translation mine)

Schliemann's excavations and the exhibition of his Trojan antiquities suddenly made ancient Greece very palpable to a broad public as a ›defining culture‹ for Britain.

As has been shown, Schliemann had a keen sense of how to make himself and his work public and popular. He knew how to use people and the media for his own ends and fascinate people with the announcements and presentation of his excavations and life story in his accounts, public speeches, exhibitions, and reports in newspapers and magazines. Similar to the earlier examples of Belzoni and Layard, Schliemann was prone to dispense with the truth in order to make *his*tory more attractive to the broader mass of people and present himself in a better light. Schliemann the archaeologist was still the businessman aiming at the greatest possible profit, both with regard to his personal achievements and his archaeological pursuits. He was the populariser, the showman, and the storyteller who consolidated his identity and entertained the public with his popular *his*tories of Troy Divine at a time of alienation and demystification in Britain – this was what made him so popular, even when the material reality failed to live up to the public's expectations, as Schliemann the man remained larger than life.

5.7 The Fall of the Mighty: Troy, Mycenae, and Britain

> At Mycenæ we know certainly that we are standing at the very centre and citadel of the powerful kingdom which engaged the interest of Greek poetry and poetical tradition for centuries.
> (*Athenæum*, Dec 22, 1877: 819)

As has been shown, Schliemann's Trojan excavation was associated, as a part of Greek history, with the *most familiar strangeness* and thus served as a means of creating and sustaining individual and collective identity. Moreover, however, the remains brought to light by Schliemann also gave rise to fears of decline and degeneration. In this context, the remains of the ancient cities that he excavated become symbols of the downfall of once powerful and prosperous empires and cultures. In addition, the different layers of cities and cultural remains found at Hisarlık raised the question of whether one could really speak of a cultural development in the sense of an increasing degree of ›civilisation‹ over the centuries, or whether the contrary might actually have been the case, as Glover points out with regard to Bram Stoker's texts:

»The theoretical vocabulary which attempted to make sense of the conviction that much of the social world was slipping regressively into a more and more pathological condition was a kind of rear-mirror Darwinism whose central concept was that of ›degeneration‹. As several recent historians have painstakingly demonstrated, this type of thinking was widely diffused across the whole political and intellectual spectrum during the late Victorian and Edwardian periods, providing a meeting ground for socialists, conservatives, and liberals old and new.« (Glover 1992: 986)

Both of these aspects, the decline of the once powerful and prosperous and the degeneration of the once civilised and cultivated, were themes that deeply fascinated nineteenth-century British society due to the subtle threat people felt from them in relation to their own situation (cf. Arata 1996: 27).

This underlying threat can be seen as the result of an ultimate identification on the part of the British with the ancient remains excavated abroad.

When Schliemann announced his excavations at Hisarlık, the world – and in particular the British public – was hoping to be presented with material proof of that great and ancient city of Troy once ruled by the mighty King Priam. They were not disappointed when Schliemann in 1872 first proclaimed to have excavated the »Pergamus of Troy« and one year later reported having discovered Priam's Treasure. Yet, in spite of these promising discoveries at Hisarlık, which in terms of material value were later surpassed by the treasures excavated by Schliemann at Mycenae, people had to recognise that instead of a story of continuous wealth and progress the various layers at the site of Hisarlık seemed to tell a different tale:

»In short, not to multiply details, the newly opened mound of Hissarlik […] stands henceforth as *a lasting witness to a progressive decay of civilisation and industry and wealth among the successive races of its inhabitants;* and it *completely overturns*, for that part of the world at least, *the hasty assumption of the progress of mankind through the ages of stone, bronze, and iron,* which has been derived from the regions that may well have been the last retreats of degraded races, rather than the first abodes of primitive men.« (*Quarterly Review*, Apr 1874: 544, emphasis mine)

With regard to contemporary British society, this passage is significant, as it stands in sharp contrast to the Victorian belief in progress and continuous development by pointing out that the site of Hisarlık proves the exact opposite, that is, the »progressive decay of civilisation and industry and wealth among the successive races of its inhabitants« (ibid.). In addition, the »hasty assumption of the progress of mankind through the ages« (ibid.) was being questioned, as was the theory that these places were »the first abodes of primitive men« (ibid.). The vocabulary used here in relation to the site at Hisarlık is significant insofar as it also represents discourses prevelant in the Victorian Age. In particular, concepts of progression, civilisation, industry, wealth, succession, and race were determining parameters of Victorian society and crucial for its self-identification as distinguished from the uncivilised and primitive ›other‹.

While people seemed to firmly believe in these concepts, there also existed a subtle fear of decay despite the power and stability Britain experienced at the time, which was to become a dominant theme in the reception of Egyptian and Zimbabwean archaeology towards the end of the century. In relation to this subtle fear, the sites at Hisarlık and Mycenae become relevant in two respects. First of all, as underlined by the passage published in the *Quarterly Review*, rather than representing *progressive development* as one of the catchwords of Victorian Britain, they seem to prove *progressive decay*. This is further illustrated by various articles in magazines and newspapers of the time, as the following two examples from the *Edinburgh Review* and the *Academy* illustrate. The *Edinburgh Review* speculates: »The civilisation of the ›Iliad‹ is doubtless higher; in respect of age, however, it may have been, not ›centuries later‹, but centuries earlier« (*Edinburgh Review*, Apr 1881: 527), while the writer of a letter to the editor of the *Academy* regrets that Schliemann does not show any interest in layers not attributed to ancient Troy: »In the first place Dr. Schliemann finds under his supposed Troy a city of a *more ancient, yet more civilised people*, of which *most unfortunately* he takes little note, so full was his forgone conclusion that he had found the end of his quest.« (*Academy*, Apr 11, 1874: 402, emphasis mine) In a lecture on Schliemann's discoveries held by the archaeologist Charles Newton for the Society of Antiquaries on April 30, 1874, Newton gives voice to the following:

»On a comparison of the pottery and disks from Hissarlik with the pottery and other antiquities of the very earliest period which we can connect with the Hellenic race, we find that the rudeness of Dr. Schliemann's antiquities far transcends the rudeness of all previously-known archaic art. *Are we then justified in assuming that, because the Hissarlik antiquities are ruder, therefore they are earlier*; that because their rudeness is non-Hellenic, therefore it is pre-Hellenic? That is the question really at issue in regard to Dr. Schliemann's discoveries.« (*Academy*, May 2, 1874: 496, emphasis mine)

What is significant with regard to the three passages quoted is that they all question the *a priori* assumption of a progressive development by pointing out that in accordance with Schliemann's findings certain ›civilised‹ people had existed *earlier* than certain less ›civilised‹ people. Against the background of Victorian society, the fact that rather than an endless progressive development Schliemann's findings seemed to prove that progressive degeneration was significant as it corresponded to theories of degeneration prominent at the time. Contemporary degeneration theories assumed that progressive development as brought about by industrialisation and modernisation would finally, after reaching a climax, result in degression and decline. The subtle fears stirred by these theories were enforced by Darwin's theories about evolution and human descent. Although generally assuming development, these theories also involved ideas of degeneration in that they connected the human species to the primitive and uncivilised by tracing its descent back to prehistoric apes. The primitive and uncivilised thus seemed in some way inherent in human nature, and while the present state of civilisation and cultivation promised stability, this was no safeguard against the uncontrollable destructive forces of degeneration. In this context, the different archaeological layers brought to light by Schliemann at Hisarlık and Mycenae seemed to show that this was actually possible and that eventually no civilisation was immune to degeneration and decline.

In addition to these underlying fears of the degeneration of Victorian society, which seemed to be justified by the fact that earlier, more civilised cultures followed by less civilised cultures had been discovered at the sites excavated by Schliemann, the decline of once powerful and prosperous cities represented by the remains at Hisarlık and Mycenae are significant in the context of contemporary British society. The remains found at Mycenae impressed people in particular, as suggested by the following passage from the *Athenæum*:

»The discoveries of Dr. Schliemann himself in the Troad are far less important; Hissarlik may probably enough have been the site of a city destroyed by confederated Greeks, but too many cities have been captured and only imperfectly sacked in the confusion of battle and conflagration for the mere discovery of the abandoned treasure to prove that the city was necessarily Troy. *At Mycenæ we know certainly that we are standing at the very centre and citadel of the powerful kingdom which engaged the interest of Greek poetry and poetical tradition for centuries, and transmitted a strong political impulse to historical times.*« (*Athenæum*, Dec 22, 1877: 819, emphasis mine)

The vocabulary used here to describe Mycenae as the centre of a »powerful kingdom« (ibid.) suggests a potential for identification insofar as it could well be transferred to nineteenth-century Britain and certainly can be said to correspond to the self-perception of Victorian society. Just as the ancient society at Mycenae centuries ago, Britain had also become the centre of a »powerful kingdom« (ibid.) controlling a vast territory in the realm of the British Empire. Apart from this political power and influence, which is further emphasised by the »strong political impulse to historical times« (ibid.) attributed to Mycenae, Britain would also have considered itself to be as highly educated and cultivated as the ancient Mycenaean society with its »interest in Greek poetry and [the] poetical tradition« (ibid.). Furthermore, the fact that the »interest in Greek poetry and poetical tradition« (ibid.) had been carried on in Mycenae »for centuries« (ibid.) might well have been transferred by contemporaries to British culture, which saw itself as continuing the tradition of classical learning and education.

Although Schliemann's discoveries in the Troad were, according to the *Athenæum* article, »far less important« (ibid.) than those made by him at Mycenae, there were also contemporary voices that described ancient Troy similarly to Mycenae:

> »[Schliemann] has found, indeed, no records and scarcely any certain inscription [...]. But he has found monuments which place beyond doubt the existence of flourishing and civilised inhabitants on the spot that has always, within historic memory, borne the name of Ilium, and which proves the real existence of a pre-hellenistic city, small but strong, civilised and wealthy, and having some most striking points of correspondence with the Troy of which Homer sang.« (*Quarterly Review*, Apr 1874: 558)

The Ilium portrayed here, if less distinguished than the ancient civilisation at Mycenae, is depicted as »strong, civilised and wealthy« (ibid.) with equally »flourishing and civilised inhabitants« (ibid.).

As illustrated by these passages, the ancient civilisations at Hisarlık and Mycenae are associated with power, prosperity, and cultivation, which were all characteristics attributable to Britain in the second half of the nineteenth century. It therefore stands to reason that people actually identified with these ancient civilisations, and also with the terrible fate and eventual doom these civilisations suffered – a fate so powerful that it overtook even the strongest and fittest, as the remains at Hisarlık and Mycenae indicated: »We must not forget to notice the curious confirmation afforded by the Hisarlik remains. Older for the most part than the Mycenaean civilization, *they show its influence at the time when the golden age of the city was about to end in the destruction which overtook it.«* (*Pall Mall Gazette*, Oct 3, 1892, emphasis mine) Although this quotation was published in 1892, a time when fears of destruction and decline of the »golden age« (ibid.) of Britain had already found expression in the Decadent movement, underlying fears of an abrupt ending of Victorian progress and prosperity had been in existence well before that time. In this context, the paper Schliemann read before the Society of Antiquaries on his recent discoveries at Mycenae in 1877 is instructive. The report given by the *Illustrated London News* on this occasion runs as follows:

»[Schliemann's finds] further proved that Homer had lived in the Mycenæ's golden age, and at or near the time of the tragic event by which the inmates of the five sepulchres lost their lives, because *shortly after that event Mycenæ sank by a sudden political catastrophe to the condition of a poor powerless provincial town, from which it had never again emerged.* They had the certainty *that Mycenæ's flourishing school of art disappeared, together with its wealth*; but its artistical genius survived the destruction, and when, in later centuries, circumstances became again favourable for its development it lifted a second time its head to the heavens.« (*Illustrated London News*, Mar 31, 1877: 306, emphasis mine)

The assumptions presented by Schliemann with regard to the downfall of ancient Mycenaean civilisation are significant since they again underline the fact that eventually no civilisation is immune against an unforeseen, »sudden [...] catastrophe« (ibid.). Interestingly, Schliemann does not go into detail as far as the character of the catastrophe is concerned but refers to it in general terms by presuming it to have been of a »political« (ibid.) nature. The fact that the catastrophe is described as sudden and that no precise information is given about its cause turns it into something universal. In relation to British society this is significant, since in the late 1870s many a contemporary might have felt the same subtle fears, in spite of – or rather because of – the powerful position occupied by the British Empire. As the remains excavated by Schliemann at Mycenae suggested, even the most powerful nations were doomed to fall, with their »flourishing school[s] of art« (ibid.) and their »wealth« disappearing forever.

Fears of doom through an unforeseen catastrophe were also enforced by the fact that the remains excavated at Hisarlık were ascribed by Schliemann to the *last* monarch of Troy, »[t]he chief or king at the time of Troy's tragic end […] called Priamos by Homer and by tradition« (*Academy*, Feb 14, 1874: 175). Furthermore, »[Schliemann] professes to have found the ruins or palace of its *last monarch*, and has extracted from them, and safely recovered to Athens, a vast accumulation of articles […] consisting of gold, silver and copper vessels and ornaments« (*Leisure Hour*, Jul 1874: 438-439, emphasis mine). Priam had no successor, and therefore a »great or enduring« power was denied to Troy: »But the Troad had never become the seat of a state which attained to great or enduring power. Priam has had no successor; the true heroes of the land are still Hector and Achilles.« (*Fortnightly Review*, Apr 1883: 514) The question of succession might also have troubled contemporary Victorians, in particular since Victoria had completely withdrawn from public life after Prince Albert's death in 1861.

In addition to the underlying fear of an unexpected catastrophe, however, the ancient remains at the sites of Hisarlık and Mycenae seemed to address another anxiety displayed by human societies in general and British contemporary society in particular, namely the fear of not being remembered and thus of being excluded ad infinitum from historiography. This is implied by Gladstone's comment on Schliemann's discoveries at the occasion of the aforementioned meeting of the Society of Antiquaries:

»Now, thanks to splendid munificence, unwearied perseverance, discernment, [Mr and Mrs Schliemann] seemed to have attained to a great accession to the antiquarian wealth of the world. (Cheers.) They were told there were great men before Agamemnon who remained unknown because they had no sacred poet to sing their praise.« (*The Times*, Mar 23, 1877: 10)

As Gladstone here mentions with regard to those »great men before Agamemnon who remained unknown« because their names did not become part of the collective memory and thus were lost over the centuries, greatness is no safeguard against oblivion, just as power is no safeguard against doom. The importance of archaeological excavation in this light is further emphasised in the following passage, presenting Schliemann's sites at Mycenae:

»No one would have supposed that the soil around had ever been anything else than its present condition indicated; but we have of late seen how explorations have exposed to our view long buried cities, how temples long lost have been brought to light again, and the ashes of the dead have been revealed to us after long ages of repose. The spade used by the excavator and explorer is now the chief instrument among archæologists; and here we had an evidence of it. We noticed some men digging a little distant from the road, and on inquiring what they were about, the answer was that they were ›tomb-hunting‹, and so it turned out.« (*Illustrated London News*, Apr 14, 1877: 358)

As underlined by this passage once again, the archaeological spade became the key to the past by retrieving greatness long forgotten, concealed underneath the featureless soil.

However, in analogy to the fact that other »great men before Agamemnon [...] remained unknown because they had no sacred poet to sing their praise« (*The Times*, Mar 23, 1877: 10), the possibility that a long lost civilisation could be discovered and excavated also pointed out the possibility that it also *would not be* found and would remain literally buried in oblivion. In relation to this, Schliemann's excavations are ambivalent insofar as on the one hand they entail the chance to retrieve former greatness, while on the other they raise awareness of the risk of being eternally lost. The possibility of a civilisation's sudden downfall, brought about by an unforeseen catastrophe *in combination* with the complete negation of its existence from history, might well be considered one of society's deepest anxieties. To be forgotten is possibly worse than to be destroyed, as oblivion involves the deletion of memory, which means the final destruction of a civilisation and its cultural practices.[1] The tragic combination of destruction and oblivion is also emphasised in Sophia Schliemann's talk before the Society of Antiquaries:

»But, alas! Greek books have had a like fate with Greek works of art, and I make bold to say that not even one-thousandth part of our ancient classics has escaped destruction. But I must not forget that my ancestors have also distinguished themselves by their heroism and military skill, and that our Greek history is full of names such as Agamemnon, Achilles, Diomedes, Ulysses, Aristodemos, Miltiades, Themistocles [...] [and] Alexander the Great, whom the mightiest of the mighty and the proudest of the proud warriors of later ages took as ideals of military virtues.« (*The Times*, Jun 9, 1877: 9)

1 Cf. Aleida Assmann *damnatio memoriae* (cf. Assmann 2006: 105).

The juxtaposition of Sophia's assumption that the vast majority of Greek books had been destroyed with the enumeration of the names of Greek heroes emphasises the power of both memory and oblivion. The destruction of Greek books and art results not only in their elimination from cultural heritage but also in the elimination of all the names, places, and stories contained in them. That we have knowledge of all the famous men named by Sophia is also due to the same tenuous archival process, that is, that their names have been passed on by historical records and legends. Consequently, where historical records are missing, the retrieval of the past at first seems impossible. However, it is at this point that archaeological excavation becomes a theoretical chance to retrieve parts of the past which have been eliminated from cultural memory:

»[A]rchaeology does not take the same view of history as do the historians of older or of modern days. Far from recording only wars, she has by far the most to say about peace and the manners and customs of time of quiet. Far from telling us of government and the balancing of political forces, she tells us rather fact about the external life of men, of the houses in which they lived, the weapons they used, what deities they adored, and in what light they regarded the life beyond the grave. From the outward facts revealed by spade and pick, we are enabled to judge of the degree of civilization attained by a nation, of its commerce and its art, perhaps of its ethnical affinities, but not of its laws or government, nor of its moral and intellectual condition.« (*Quarterly Review*, Jan 1886: 116)

As this passage from a review of Schliemann's book *Tiryns – The Prehistoric Palace of the Kings of Tiryns* (1886) underlines, archaeology has the potential to reveal certain aspects of the past which would otherwise remain in darkness. However, it is also admitted that there are other aspects of the past about which archaeology would not be able to make any statements.

In spite of the fact that archaeology has no power to retrieve the past as a whole, Schliemann's excavations at Hisarlık, Mycenae, and elsewhere were still highly popular due to the fact that archaeology seemed to be a promising means of stopping oblivion. This is made explicit by Gladstone's comment on Schliemann's excavations in *The Times*: »The right hon. Gentleman concluded by congratulating the Society that light was now beginning to pierce the thick mist which had hitherto obscured such remote subjects of antiquarian researches and by a renewed eulogy of the labours of Dr. Schliemann.« (*The Times*, Jun 26, 1875: 9) At a time of an underlying fear of doom in Britain in spite of its political and economic power, the retrieval of the once powerful civilisations at Troy and Mycenae seemed to offer comfort insofar as it suggested the possibility that greatness endures after all and can be regained, even though it might have disappeared from the surface of the earth for centuries.

Interim Findings: Heinrich Schliemann's Troy as the *most familiar strangeness*

On the assumption that the archaeological venture can be read as the symbolical desire to retrieve Lacanian wholeness, the present chapter has looked at how Heinrich Schliemann's excavations and their popularisation as associated with Greek antiquity as the *most familiar strangeness* were used as a means of creating individual and collective identity in Britain. As Schliemann's case has shown, according to what I have termed the *Frankenstein Discourse*, identity was created through the juxtaposition and junction of heterogeneous – fictional and factual – fragments derived from individual and collective memory and material remains. Schliemann's (auto)biography and his reception as a popular role model in Britain have been introduced as examples of the close interconnection between archaeology and the search for identity and the public interest in both at the time. In this context, I analysed Sophia Schliemann's image as promoted by Heinrich in public against contemporary gender roles. I then elaborated on Heinrich Schliemann's popular reception by reading his search and excavation of the alleged ancient Troy as a symbolical excavation of individual and collective identity, functioning as a prosthetic memory in Britain at the time. Moreover, I analysed the search for individual and collective identity with regard to alleged racial similarities associated with the ancient remains at Hisarlık. The prominent interrelation of prosperity, archaeology, and identity creation in relation to Schliemann's career was examined in a separate chapter. On the assumption that identity is consolidated and perpetuated through performance, this chapter further focused on Schliemann's self-promotion and popularisation in Britain. It then showed how the remains excavated by Schliemann also stirred fears of decline and degeneration through an extreme identification with the lost ancient Greek civilisation. This points to underlying Victorian uncertainties and instabilities, which become even more prominent in the reception of Egyptian archaeology, which will be discussed in the following chapter.

6. The Mummy as the *less familiar strangeness*

Man in short, is an animal who, like every other animal, is finally subdued by his environment and takes his color for his surroundings, as cattle do from the red soil of Devon.
(*SMITH AND THE PHARAOHS*: 138)

In the previous chapter I showed how the popular archaeological discourse functioned as a space of constructing, performing, and consolidating identity by the example of Greek archaeology as the *most familiar strangeness*. Focusing on Egyptian archaeology as what I have termed the *less familiar strangeness*, the present chapter illustrates how the archaeological space functioned not only as a space for (re)creating and completing identity in Victorian and Edwardian society, but also as a counter-discourse for subverting, inverting, and (re)negotiating it. Unlike the reception of Greek archaeology, which, as I have illustrated in regard to Schliemann, was consistently positive, the reception of Egyptian archaeology was characterised by a strong ambivalence oscillating between affiliation and repulsion, causing comfort and fear at the same time:

»Archaeology is about desire: and Egyptian archaeology more so than any other. Egypt is the East; to many it is erotic, shameless, liberating, sexually threatening, fascinating. It was more alien to the Victorians than Greek or Roman civilization.

The archaeology of Egypt was more often than not associated with romance and excitement, with erotic desire – and also with death.« (Pearson 2010: 223)

On the basis of what I have introduced as the *Jekyll and Hyde Discourse*, I will show how frequently a Hyde-like subliminal ›other‹ of what in *Dr. Jekyll and Mr. Hyde* is referred to as part of the »duality of man« (*Jekyll & Hyde*: 82) prominently surfaces in the characters of archaeological texts, disturbing their Jekyll-like smooth and socially accepted facade which represented the norm in contemporary Victorian and Edwardian society. The relevance of the *Jekyll and Hyde Discourse* in this context becomes clear if we consider the coexistence of ›good‹ and ›evil‹ in the figure of Jekyll/Hyde. This coexistence of these allegedly exclusionary categories eventually become exchangeable and, in accordance with what Butler points out in regard to sex and gender, underlines the arbitrariness of identity attributions and emphasises its dependence on performativity. Since we are both ›good‹ and ›evil‹, as much as we are ›male‹ and ›female‹, whether we appear as Dr Jekyll or Mr Hyde is a matter of performance and not of nature. In all these cases the boundaries are fluid and the alleged ›other‹ can at any point turn into the self, since these distinct categories are nothing but cultural constructions. In the present chapter I will show in regard to mummy fiction how conventional notions of identity are deconstructed through the emergence of the Hyde part in the form of deviation and dissidence from the norm within the archaeological space. Significantly, with regard to the texts looked at this is particularly prominent in regard to gender roles, concepts of sanity and madness, and the spatial and temporal ›other‹. I will further demonstrate how, as a consequence of the emergence of this Hyde part, identities are destabilised, subverted, and inverted in the archaeological space in spite of the characters' desperate attempt to maintain and regain stability and continuity. As I will show in the first three chapters, in Theo Douglas's *Iras*, Austin Freeman's *The Eye of Osiris*, and Bram Stoker's *The Jewel of Seven Stars* the ›archaeological‹ search is characterised by the struggle to recover and confirm identity, a process which is, however, continually threatened by the power of the past. The last two chapters will focus on the increasing failure to use the past in a constructive way, which climaxes in Stoker's *Jewel of Seven Stars* and Boothby's *Pharos*, where forces of the past infiltrate the present and lastingly disturb traditional notions of identity.

6.1 Narrating *History*: Memory, Fantasy, and Madness

›Curses on thee, fiend!‹ cried Father Morris, rushing from the grove, and pressing his cowl round his head with both his hands, as though he feared the horrid laugh of Cheops should echo in his ears, and sting him to madness.
(*Mummy*, Vol. II: 81)

The example of Heinrich Schliemann has shown in regard to Greek archaeology how both individual and collective identity were created through the construction of biographical, cultural, and racial continuities. The present chapter shows that the underlying structures and mechanisms of identity formation are essentially the same for both factual and fictional texts. In both cases the final product of this construction of continuities is an identity-forming narrative. I will further demonstrate how this process of creating a life continuum by means of narration on the basis of a remembered[1] past functions as a displacement strategy reflecting the Lacanian desire to compensate for a feeling of lack.

A novel in which the archaeological discourse is closely connected to memory and desire is Theo Douglas's *Iras*: *A Mystery*, where the autodiegetic narration becomes a means of restoring meaning and thus identity. In *Iras* the amateur archaeologist Ralph Lavenham acquires the past by buying an ancient mummy from a friend at Alexandria. With the intention to investigate »certain secrets both of ingredient and methods« (*Iras*: 9) of embalmment, Lavenham has the mummy sent and imported to Britain under a false declaration:

»The true enclosure was a mummy of superior class, still sealed in its original casings and unviolated by the wholesale system by which the dead have been despoiled

1 This includes everything that is conceived as memory by the narrating subject, regardless of whether it is ›true‹ or ›false‹.

– not only of such valuables as it was the practice to bury with them, but of the very spices with which their bodies were embalmed – stolen in former times to figure again in the market, and take a fresh turn of services with the modern dead.« (Ibid.)

When acquiring the mummy, Lavenham disguises it as a batch of »[e]xpert sponges« (ibid.:10). His contact with the mummy and his later experiences with it are thus, right from the start, characterised by (false) pretence and feigned identity. The only one who knows the true contents of the expected delivery in Britain besides Lavenham is his friend Knollys, who is supposed to make a sketch of the mummy prior to its examination by Lavenham. Nonetheless, Lavenham's relationship to the mummy, which is at first purely scientific and commercial, is solitary and subjective in that both his knowledge and his expectations of it are not shared by anyone else: »I had certainly awaited the intimation [in regard to the expected mummy] with some anxiety, and was gratified by its reception; but the singular excitement which agitated me from that time forward was quite unexpected and wholly annoying.« (Ibid.) Thus, like Frankenstein, Lavenham, through his ›secret‹ intention to examine the mummy, isolates himself from society. In both cases the self-chosen isolation is necessary due to the moral and ethical boundaries Frankenstein and Lavenham are aware of trespassing through their acts, which turns them into outsiders in society.[2]

While both Frankenstein and Lavenham's ventures seem to be motivated by an initial thirst for knowledge, a closer look at Lavenham's case reveals an underlying conflict in regard to his own identity. At the beginning of the novel Lavenham introduces himself as a man without family or friendly connections:

2 Although Frankenstein's actions have to be considered as more severe as far as moral and ethical issues are concerned, the reaction of Lavenham's landlady, who eventually finds out about the mummy in Lavenham's flat, underlines the breach of taboo Lavenham commits by bringing the mummy into a British house: »What my landlady had called to ask I never discovered; the sight before her seemed effectually to scatter her ideas. […] ›Why, Mrs Mappinbeck, this is what is called a mummy-case, and it has been sent to me direct from Egypt by a friend‹. […] I was going on to descant on the rock tombs and their contents, when the indignant horror deepened on the face of my auditor. ›Are you meaning to tell me, sir, as how there is *a body* inside of this?‹« (*Iras*: 61)

»I am singularly devoid of near ties. I have no certain knowledge whether the name I bear is mine by right or mine by adoption: I never knew father or mother. I was placed at school by a guardian who showed no disposition to cultivate a nearer acquaintance; and from school I passed to college, where I made few friends and took only a moderate degree.« (Ibid.: 2)

Lavenham informs the reader that after earning his college degree he had spent »the next ten years [...] aimlessly enough – partly in foreign travel, and in writing one or two descriptive books which met with some degree of ephemeral popularity« (ibid.). However, it is only afterwards that he claims to have »found at last the absorbing interest of [his] life in Egyptian exploration« (ibid.): »From 1877 to 1882 I labored in the Nile Valley, working with my hands as well as with brain, uncovering temple sites, opening tombs, deciphering hieroglyphics, driving fellaheen, and building up theories like the rest.« (Ibid.: 2-3) Unfortunately, Lavenham's happiness was not to last, since a sunstroke followed by a severe fever eventually forced him to stop his archaeological work in spring of 1882 and to return to Britain (cf. ibid.: 3). As underlined by Lavenham's account, for a long time his life had been characterised by isolation and disorientation. He had lacked both a foundation and a *telos*, which, following Storey's idea of the formation of identity on the basis of memory and desire, can be read as indicative of a destabilised identity. Considering the stabilising effect of archaeological sites and excavation displayed in previous chapters, the fact that Lavenham eventually finds the »absorbing interest of [his] life in Egyptian exploration« (ibid.: 2) may be seen as further supporting this idea, as it suggests that by performing the archaeological work Lavenham is finally able to realise his true identity. The analogy of archaeological excavation and the search for identity is supported by other mummy texts, in which archaeology works as a catalysing force that finally makes possible the formation and confirmation of identity.[3]

Often this process of self-discovery is initiated in these texts by characters finding their true disposition in archaeology after having spent a

3 One of these texts is Haggard's short story »Smith and the Pharaohs« (1913), in which the protagonist, James Ebenezer Smith, through Egyptian archaeology is able to find the passion and love of his life, who finally helps reveal his true identity.

disoriented life.[4] Most of the characters get involved with archaeology accidentally before it turns into their passion, which eventually leads to self-awareness. Frequently the contact with Egyptian archaeology is initiated through an inherited archaeological collection, artefact, or interest in the subject from another family member, as in *Pharos*,[5] which once more underlines the affinity of archaeology to individual history and identity.

In *Iras*, the stabilising effect archaeological work has on Lavenham is underlined by the positive memories he has back in Britain of the time he spent at the excavation site in Egypt:

»As I looked out over the heaving of plain green water, I heartily wished myself back again, *spade in hand, among my comrades* in the East – *battling with our enemy the sand*, at once the great obliterator and the great preserver of those relics of a dead past. Welcome the hot wind like the breath of a furnace, the torrid sun, the plague of flies beyond the power of even a Moses to remit, could I have exchanged for them England and inaction such as weighed upon me now.« (*Iras*: 24, emphasis mine)

In contrast to the situation Lavenham is in at the beginning of the novel, which can essentially be characterised as ›lack‹, this passage is marked by a strong sense of community, a clear intention, and a definite aim. On a more abstract level, the community experienced by Lavenham can be understood as providing him with the social frame of reference lacking in his previous life and connecting him not only to the present but also rooting him in the past. The »battling with [their] enemy the sand« (ibid.) in this context can be seen as motivated by the desire to achieve completion in the Lacanian sense, represented by the possibility to symbolically recreate wholeness through archaeological excavation. Lavenham's positive experience with archaeological excavation can thus be interpreted as originating in the

4 Cf. »The Ring of Thoth« (1890), *Iras* (1896), *Pharos* (1898), and *It Happened in Egypt* (1914).

5 Cf. Chapter 7.4. Other mummy texts in which an archaeological object/mummy is inherited or connected to the family history in some way are *The Ring of Amasis* (1890), »The Story of Baelbrow« (1898), *The Mummy and Miss Nitocris* (1906), *Jewel of Seven Stars* (1903), *Nemesis of Fire* (1908), *The Green Mummy* (1908), *The Eye of Osiris* (1911), and »The Cat« (1914).

continuous illusion that the Lacanian state of the Real can be retrieved. This illusion, however, is radically destroyed by Lavenham's forced departure from the archaeological site to his books in Britain due to the fever he contracted. Although the fever seems to be the cause for his departure, it can actually be read as a reaction to Lavenham's anagnorisis/recognition that wholeness can never – or only momentarily – be found and that the archaeological search will thus forever remain unsatisfying. However, whereas Lavenham still felt close to the realisation of his longing for wholeness when he was engaged in archaeological work at the excavation site, which was experienced by him as something *real*, the return to his books essentially marks the departure from the real. Lavenham's intellectual occupation with Egyptology far from its geographical and material origins in this context is thus only a distorted image of the real/reality and in terms of Walter Benjamin bereaved of the aura of the original[6] with its promise of Lacanian completion.[7]

---

6 »The situations into which the product of technological reproduction can be brought may leave the artwork's other properties untouched, but they certainly devalue the here and now of the artwork. And although this can apply not only to art but (say) to a landscape moving past the spectator in a film, in the work of art this process touches on a highly sensitive core, more vulnerable than that of any natural object. The core is its authenticity. The authenticity of a thing is the quintessence of all that is transmissible in it from its origin on, ranging from its physical duration to the historical testimony relating to it. Since the historical testimony is founded on the physical duration, the former, too, is jeopardized by reproduction, in which the physical duration plays no part. [...] One might encompass the eliminated element within the concept of the aura, and go on to say: what withers in the age of the technological reproducibility of the work of art is the latter's aura. The process is symptomatic; its significance extends far beyond the realm of art. *It might be stated as a general formula that the technology of reproduction detaches the reproduced object from the sphere of tradition. By replicating the work many times over, it substitutes a mass existence for a unique existence. And in permitting the reproduction to reach the recipient in his or her own situation, it actualizes that which is reproduced.*« (Benjamin 2003: 254)

7 This is also implied by Knollys, who refers to Lavenham's appearance as follows: »›You look like a ghost, Lavenham.‹« (*Iras*: 6) Thus, a solely intellectual

Against this background, Lavenham's return to Britain marks the continuation of the disenchantment suffered through his anagnorisis at the excavation site in Egypt. The consequence is an even deeper feeling of disintegration and isolation than before:

»No, there was no one. I had forgotten my former associates, and had been forgotten by them; and there was not a single intimacy I felt disposed to revive. I had been away too many years, and since my return had inclined only to bury myself more and more remotely, like a hermit-crab in its shell.« (Ibid.: 6)

Significant for the further development of the novel, it is in this state of mind that Lavenham receives the mummy he had secretly exported from Egypt, which consequently becomes his final attempt to stabilise his identity by achieving completion.

Lavenham's first encounter with the mummy, to whom he later gives the name Iras, takes place in the middle of the night and thus serves as another example of a nocturnal mummy encounter:[8]

»As I sat at the table with the inscription before me my back was turned to the coffin. I pushed the chair sideways and glanced round at it, when all power of movement was arrested by surprise. Thrown carelessly out of the disturbed wrappings, and hanging over the edge, was a woman's arm – slender, exquisitely rounded, warm with life. […] I knew my heart's one love when I saw her face to face. *All the aching loss of my solitary life – all I had lacked hardly knowing – was present to me in that moment, as I recognized a need filled, an incompleteness suddenly made whole.*« (Ibid.: 77, emphasis mine)

As Lavenham's strong reaction towards the reawakened mummy makes explicit, the mummy is seemingly able to satisfy Lavenham's underlying feeling of ›lack‹, experienced as an »incompleteness suddenly made whole« (ibid.). It thus assumes the function of a substitute for the wholeness in the Lacanian sense that Lavenham failed to achieve at the archaeological site.

---

occupation with archaeology beyond the actual historical site is only a shadow of the real and also leaves its traces on the archaeologist.

8 As mentioned in Chapter 4.2, ›archaeological encounters‹, particularly those with mummies, are prone to occur at night.

Even though Lavenham has his own initial doubts in regard to Iras's awakening, asking himself whether »it [was] believable that a human creature could have existed under such conditions and in such suspension for three thousand years« (ibid.: 77), his alleged love at first sight makes him almost immediately accept Iras as a flesh-and-blood woman: »The face I looked upon was beautiful, but it was a marvel the more that I did not regard it in the least as one looks upon the beauty of a stranger. I knew my heart's one love when I saw her face to face« (ibid.: 78). Thus, Lavenham does not perceive Iras's face as the »beauty of a stranger« (ibid.) upon first looking at it, but as someone, or rather some*thing*, familiar. The reason for this feeling of familiarity can be seen as pointing towards the real cause of Lavenham's immediate love for Iras, a foreigner of both time and space to him. Right from his first encounter with Iras, Lavenham uses her as the alleged answer to his inner longings for completion and consequently constructs and forms her according to the idea he has of an ideal woman *prior* to meeting her for the first time: »Speechless with the wonder of it, new born into joy, and into a rarer atmosphere where it was difficult at first to breathe, *I stood and looked upon the sleep which I alone from the beginning of the ages had been ordained to break.*« (Ibid., emphasis mine) Here, the re-creation of identity is directly connected to archaeology, which invests Lavenham with the power to possess and control. This puts Lavenham close to Frankenstein, who through creating the creature becomes its sole possessor and the master of its life.

To what extent Lavenham turns Iras into a part of him is made explicit by the fact that he wants her to have his name: »›I will give you my own name, if you will take it‹ [...]. She looked at me with a sweet half-comprehension, like a child at once perplexed and confiding.« (Ibid.: 81) That Iras belongs to Lavenham is not only implied by the fact that she henceforth carries his name but also by the fact that it is Lavenham who finally comes up with a surname for her. The novel is already well advanced when Lavenham, who is about to marry Iras, remembers the necessity that Iras must have a name of her own, a name that distinguishes her from him, before she entirely becomes his:

»It was necessary for me *to invent* a maiden surname as well as the familiar prefix; and probably some boyish recollection of Shakespeare grafted itself on to the idea of her nationality, by suggesting the names of the great queen's handmaidens in ›Antony and Cleopatra‹. She would be Iras Charmian as preliminary to her conversion into Iras Lavenham.« (Ibid.: 103, emphasis mine)

That Lavenham has to »invent« (ibid.) a surname as well as a »familiar« (ibid.) forename for Iras is significant in several respects. It suggests that as a stranger Iras has to be domesticated by Lavenham before becoming his legitimate wife. This Lavenham accomplishes by giving her a (fake) British identity, that is, a »familiar prefix« (ibid.), instead of a foreign one. That the British name and identity Lavenham imposes on Iras are purely fictitious and a sham is suggested by the fact that by association he takes her names from Shakespeare's *Antony and Cleopatra.* That Lavenham draws on Shakespeare in choosing a name for Iras is also significant in that Shakespeare, being part of the British tradition and canon, represents a crucial part of the collective British culture and identity. In other words, before Iras, as the embodiment of a foreign culture both in a spatial and temporal sense, is allowed to enter British culture thoroughly through the marriage to Lavenham, it is necessary for her to undergo certain formal adaptations in the form of naming. Once this has been accomplished, there is nothing left coming from Iras to disturb the Victorian middle-class discourse after Lavenham has also provided her with a British outfit: »The things once there, it seemed to come to her by nature how to wear them, there was nothing, save her uncommon beauty and grace, to distinguish her from an ordinary English girl.« (Ibid.: 110) Neither her name nor her appearance will now betray Iras as a foreigner, since both are brought into line with British culture, and consequently nothing stands in the way of the perfect union between Iras and Lavenham.

That Lavenham still projects this union into the future is underlined by the fact that he does not want to kiss Iras before they are married: »I kissed her hand as I took the pen from it – I would not touch her lips till she was mine.« (Ibid.: 111) The fact that Lavenham emphasises that he would not kiss Iras on the lips until she is his (possession) suggests that he obviously feels the need to verbally assure both the reader and himself that he will not act against the convention of the time even though he *could* if he wanted to, since as a spatiotemporal foreigner and woman Iras is at his mercy. What this underlines is that Iras as much as Lavenham's relationship with her is determined by Lavenham's application of the contemporary gender-specific code of conduct. In other words, Iras functions as a projection screen for

Lavenham's male idea of the ideal nineteenth-century woman and wife that would make him whole again. Lavenham has isolated Iras from her original historical context and her social frame of reference, which lies in the past, and thereby robbed her of her own identity – her personal life story – to then use her as a blank page on which he writes *his*tory in order to complete his identity. In this context, Lavenham's love for Iras can be described as what Storey sees as the »classical substitute object« (Storey 2008: 88) for the Lacanian state of wholeness, that is, romantic love: »In the west we live in a world in which romantic love is held up as the ultimate solution to all our problems. Love is a narrative compelled by lack. If we find love it will make us whole again. The discourse of romance is the most fundamental displacement strategy.« (Ibid.) It is due to this subjective and isolated character of Lavenham's encounter with Iras that in the further course of the novel it remains essentially unclear what is real and can be objectively attested and what is illusionary and thus only part of Lavenham's fantasy. This is also pointed out by MacFarlane when she states that »Everett's novel constantly draws the reality of Iras's reanimation into question [because] Lavenham has been physically and mentally incapacitated by his work in Egypt« (MacFarlane 2010: 19). As a result of this, the story that is being narrated by Lavenham becomes essentially *his*tory characterised by ambivalence, oscillating between fact and fiction, sanity and madness, and motivated by the underlying desire to sustain the wholeness momentarily experienced through his love for Iras.[9]

The momentariness of Iras and Lavenham's union is underlined by the fact that Iras's signature on the marriage certificate has already started to fade shortly after her death: »While my signature and those of the witnesses remained plain to read in all the ordinary blackness of ink, that of Iras was faded so as to be barely legible. [...] [A] few more months and the paper will be blank. It matters little; there is one place where her name is written indelibly, and that is my heart.« (*Iras*: 241) Tellingly, the signatures of Lavenham and the other (male) witnesses are still »plain to read in all the ordinary blackness of ink« (ibid.), while that of Iras, the only woman among the signers, is almost invisible and about to disappear completely to leave no more than a blank. On the one hand this can be read as pointing

9 In that sense Iras, very much like Frankenstein's creature, becomes the realisation of a wish or mental conception of her creator, i.e., Lavenham.

towards the discrimination against women in general and their exclusion from collective memory, while on the other it emphasises again the instability of Iras's identity, which is essentially dependent for its formation on Lavenham. Consequently, even Iras, whose mummy has outlasted three thousand years, has no chance to overcome these patriarchal mechanisms represented by Lavenham's longings to stabilise his own identity. To be a woman in the traditional sense means to be invisible and forgotten. The only way to be remembered is through the name or hand of a man, as Lavenham's story about his relationship to Iras emphasises:

»If this narrative should ever fall into other hands than mine, it may be wondered why I should dwell on so trivial a matter as the purchase of the outfit when so dark a cloud hung in the future. But *I write to preserve the memory* of our pleasure in it, and the joy it was to me to see it used and worn, as used and worn it was, despite after-evidence.« (Ibid.: 110-111, emphasis mine)

However, were it not for Lavenham's records there would be nothing left in the modern world to remember Iras with except for her mummy, which can only serve as a proof of her existence in the remote past of Egypt but not in the more recent one of contemporary Britain. That women take no active part in the making of world history is further hinted at by Lavenham's following remark:

»I had inquired about one or two expeditions into the interior of the island, and, as we paced up and down on the sea-wall, I told her what we were going to see, and why these places were thought of interest – *some fragment of the long roll of the world's history which had written itself in achievement and suffering while she lay asleep in her rock-tomb in the cliffs beyond Luxor.*« (Ibid.: 141-142, emphasis mine)

The very fact that Lavenham tells Iras about the history she has missed by lying passively in her tomb can be read as pointing towards the lack of female involvement and influence in the making of history in general. Consequently, the statement of the clairvoyant Madame St. Heliers, whom Lavenham meets at a London party at the beginning of the novel, is true in more than one sense: »›But I was attracted to you directly I saw you, and determined I would have a talk with you. You are a man with a history – it is either made or yet to make.« (Ibid.: 19) Lavenham is not exactly a man with a history, but he is a man and is therefore able to make *his*tory, whereas Iras, whose experiences have to be considered as much more spectacular than those of Lavenham, only becomes *part of* Lavenham's *his*tory. In this respect history does not write itself, as remarked by Lavenham (cf. ibid.: 142), but rather is written by the hand of men about men.

In what way Lavenham's retrospective account of the development of his relationship with Iras is an attempt to recreate this feeling of wholeness, and thus Iras's and his identity after Iras's death, is emphasised by the affinity of Lavenham's narration presented in retrospect to the archaeological work as such. In that sense Lavenham's narration can be understood as a last means of trying to recreate identity, which he failed to achieve/sustain twice before: first, through his aborted archaeological excavations in Egypt and, second, through his tragic love relationship with Iras. Lavenham's narration thus becomes an archaeological venture with the aim of recreating completion of a past identity he experienced through the union with Iras. In this process, the constant threat of oblivion and insanity that characterises Lavenham's narration reflects his instability and his subconscious knowledge of the futility of his desperate endeavour to sustain identity. This is further reflected by various other mummy texts in which a character's journey to modern Egypt already marks the passing of the border separating reality from dream/fantasy, rationality from irrationality, and sanity from madness.[10]

---

10 The mystery and romance associated with (ancient) Egypt is illustrated by the following choice of quotations from different mummy texts: »[Vyning:] ›I dare say Egypt hypnotises one.‹« (*Tombs of the Kings*: 440); »[Smith:] ›Perhaps [the face of the sculptured head] was not really beautiful save for its wondrous and mystic smile; perhaps the lips were too thick and the nostrils too broad.‹« (*Smith and the Pharaohs*: 140); Sir Marcus Auckland's thoughts on Inez Durward: »[T]onight the mystic phase of her nature asserted itself – the phase he did not understand, had never understood.« (*Cat*: 79); »[A] man [Forrester's father] whose whole mind and being was impregnated with a love for that ancient country and its mystic past.« (*Pharos*: 6); »I [Lord Ernest Borrow] paced up and down, acutely conscious of my great secret, the secret inspiring my voyage to Egypt. For months it had been the hidden romance of life; now it began to seem real.« (*Happened in Egypt*: 3)

To what extent reality and fantasy, sanity and insanity, are interlinked with concepts of memory and desire in the novel is underlined by the introductory sentence of Lavenham's narration:[11] »›I have become subject to a failure of memory, the partial consequence of recent severe illness. Already, as I look back, a confused mist obliterates certain portions of the past; and names and dates have a trick of deserting me at the moment of requirement.‹« (*Iras*: 1) Since, as pointed out above, Lavenham's »severe illness« (ibid.), which developed as a result of the sunstroke and fever he contracted at the Egyptian excavation site, can be seen as the reaction to his traumatic anagnorisis that wholeness cannot be attained, the fact that he makes his »severe illness« (ibid.) responsible for his »failure of memory« (ibid.) links it directly to this anagnorisis. Significantly, however, it is only in retrospect, after Iras's death, that Lavenham is openly confronted with the fact that his sanity is in question in regard to what he claims to be the recent past, as becomes explicit in his friend's comment on Lavenham's situation:[12]

»›I feel sure of this – that you entirely believed what you have told me. But the view the doctor takes seems to be that your illness has been a long time threatening – possibly ever since the sunstroke in Egypt; and when this is so it is not at all unusual to find the patient subject to what he calls persistent hallucination. […] [U]nder his theory you were hallucinated when you left London believing you had a travelling companion, and the same impression has continued since in your wanderings.‹« (Ibid.: 193)

Although Lavenham seems convinced of the reality of Iras's existence, his effort to ascertain this for himself and the world is undermined by the incompleteness of his memories, and thus the process of recollections becomes a search for evidence in the past. What follows is Lavenham's investigation into this recent past and the presentation of evidence to prove

11 Other stories in which madness/insanity play a role are *Pharos*, »Lot No. 249«, »The Miraculous Explorer«, »In the Tombs of the Kings«, »A Professor of Egyptology«, »The Lost Elixir«, »Smith and the Pharaohs«, »The Case of the Headless Mummies«, and »The Strange Discovery of Dr. Nosidy«.

12 Cf. also the comment of the landlady at the mansion in Scotland when Lavenham inquires after his wife: »›Dear, dear! he's off again for sure!‹« (*Iras*: 184)

the truth of *his*tory to the world: »I had totally forgotten Iras's belongings. Surely here would be another *fragment of evidence for her reality*!« (Ibid.: 220, emphasis mine) This is significant, since both the language used to describe Lavenham's approach as well as the process itself turn Lavenham's search in his own past into an archaeological venture, whose aim it is to reconstruct the recent past and fill certain gaps by »piec[ing] together the *broken chain* of memory‹« (ibid.: 184, emphasis mine) by means of material evidence such as Iras's clothes. As an archaeologist, Lavenham is aware that making a theory of the past plausible to the world and thus turning it into ›factual‹ history requires material evidence in the form of relics from that past. If one fails to provide this, however, the theory remains nothing but a theory, a ›fancy‹, of the archaeologist, since unlike the extraordinary case of Iras, normally there are no »eye-wittness[es]« (ibid.: 119) to confirm or correct an archaeological theory of the remote past: »She was never weary of hearing how we labored in the excavations, and of the treasures of antiquity restored to light, and the theories deduced from them. *Touching these she – an eye-witness of that remote past – had much to say, both in correction and confirmation; and I began to note down various matters to be incorporated in the book.*« (Ibid., emphasis mine) However, as with an archaeological theory, there is more at risk if Lavenham fails to provide the required evidence to *his*tory, since the lack of material proof threatens not only Iras's but also his own identity: »›I will prove it if I live – both her existence and my sanity; for the two stand or fall together.‹« (Ibid.: 196) Similar to the situation at an archaeological site, Lavenham is confronted with fragments of memories of the recent past which he has to join together in order to reconstruct what he has experienced and thereby provide both for himself and for his surroundings the material proof of his sane existence, that is, his proper identity.[13]

13 Another prominent example in which the protagonist's identity is negotiated along the lines of sanity/insanity is Grant Allen's short story »My New Year's Eve among the Mummies« (1878). In Allen's short story, the *nameless* – this can again be seen as pointing towards the lack of identity – protagonist accompanies his fiancée and her family to Egypt, and, being unable to sleep, gets up one night to go for a walk around the »great unopened Pyramid of Abu Yilla« (*New Year's Eve*: 34): »I feel half awake, half asleep, and altogether feverish: but I poke about the base [of the pyramid] in an aimless sort of way, with a

That Lavenham is continuously close to losing himself in this process of stabilising his identity in the archaeological sphere is made explicit right from the beginning of his narration, when he refers to his exited and agitated state upon the arrival of Iras's mummy (cf. ibid.: 10). However, at the time, Lavenham, according to his friend Knollys, makes his overwork responsible for his curious feelings in regard to the expected arrival of the mummy and decides to distract himself by following his friend's invitation to join him at a party at Mrs Payne's. Ironically, however, it is at Mrs Payne's party, which is arranged in accordance with the hostess's latest interest in palmistry and clairvoyance[14] with a clairvoyant present, that

vague idea that I may perhaps discover by chance the secret of its sealed entrance, which has ere now baffled so many pertinacious explorers and learned Egyptologists.« (Ibid.: 34) The young man is finally able to enter the pyramid and inside meets the living queens and kings of ancient Egypt, who »[o]nce in every thousand years [...] wake up for twenty-four hours [to] recover [their] flesh and blood, and banquet once more upon the mummied dishes and other good things laid by for [them] in the Pyramid« (ibid.: 41). The protagonist falls in love with Hatasou and agrees to his own mummification in order to be able to stay with her. When he wakes up, however, he finds himself in a bedroom of Shepherd's Hotel at Cairo with a ring presented by Hatasou in his pocket. In spite of the ancient Egyptian ring the young man is able to produce, people do not believe his story and suggest that he »must have gone to the Pyramid already in a state of delirium« (ibid.: 45). The fact that people ascribe his story to his delirious state of mind is crucial as it points towards the theme of sickness and madness prominent in many mummy stories, which is also strong in *Iras*, and marks the traumatic recognition of the protagonist that wholeness is forever lost.

14 This reflects the Victorian fascination with the occult in general and in combination with ancient Egypt in particular: »[O]ccultists were also exploring the hidden reserves and dissociated states of consciousness revealed during magical rites. One of the commonest ways of making sense of such phenomena was to conceive the mind as a series of levels or strata, an analogy legitimated by the scientific prestige of Victorian archaeology or geology.« (Glover 1996: 85) This is further emphasised by the popular Victorian medium Elizabeth D'Esperance alias Elizabeth Hope (1855-1919), whose spirit-guide was, *inter alia*, Nepenthes, an Egyptian beauty. D'Esperance published her autobiography *Shadow Land of Light from the Other Side* in 1897.

Lavenham has the first vision of Savak, the ancient Egyptian priest who is eventually going to destroy Lavenham's union with Iras: »He came into view from behind a knot of people near the doorway, some twenty feet away, across the width of the room, and stood for a few seconds in a vacant space, his eyes blazing into mine and his hand clinched, while I remained transfixed with astonishment.« (Ibid.: 14) Significantly, Lavenham is the only one who sees the Egyptian and first believes him to be a disguised companion of the clairvoyant, Madame St. Helier. Yet, St. Helier assures Lavenham that even though she has some knowledge of the Egyptian, he is in no way associated with her: »›Beware of him; it is he that is your enemy.‹« (Ibid.: 21) As a consequence, instead of being enlivened and entertained by the party, Lavenham leaves it deeply confused, and only in the cold air outside the house does he find some refreshment and relief: »I think it was that cold breath about me which cleared my brain and brought me back to myself, for I felt as a man might *who is drunk with the inhalation of a narcotic.*« (Ibid.: 22, emphasis mine) Although Lavenham holds the atmosphere of the »psychological entertainment« (ibid.: 11) at the house responsible for his strange condition, considering the nervous and strained state he has been in ever since his illness broke out in Egypt the happenings at the party have to be seen as the culmination of Lavenham's inner crisis and at the same time as the beginning of his growing isolation in the attempt to regain wholeness in the Lacanian sense.[15]

In the further course of the novel, Lavenham's experiences can be depicted as oscillating between sanity and insanity, reality and fantasy, and thus point towards the *Jekyll and Hyde Discourse* as characterised by the »duality of man« (*Jekyll & Hyde*: 82).[16] On a symbolic level this duality of man can be seen as reflecting the Janus-faced quality of both Victorian

15 This essentially marks Lavenham's trespassing into the sphere of the ›other‹ beyond Victorian discourse, as is prominent in the *Dracula*, *Frankenstein*, and *Jekyll and Hyde Discourse*.

16 The *Jekyll and Hyde Discourse* also represents the potential for subversion of traditional norms and constructs. Although in Stevenson's *Jekyll and Hyde* the »duality of man« (*Jekyll & Hyde*: 82) focuses on the categories of ›good‹ and ›evil‹, I want to use this concept to also refer to connotations of these categories which include what is conventionally conceived as ›normal‹ and ›abnormal‹, respectively, e.g., ›sane‹ and ›insane‹.

society and archaeology in general. Victorian society and archaeology was/is very much concerned both with rational thought and with materialism in the wider sense. However, beyond that rationality and materiality there was/is also an irrational, fantastic, and vague side to both that often cannot be grasped.[17]

In *Iras* this is further emphasised by the fact that in order to secure credibility it is always necessary to incorporate theory into reality, since a theory lacking support might also cause a character's sanity to be called into question. Knollys reminds Lavenham of this as follows: »›You must have patience with me, old fellow, but I want you to see the other side and what will occur to people as the natural explanation of all this.‹« (*Iras*: 193) What Knollys here refers to as »the natural explanation« (ibid.) is people's reaction to Lavenham's story as that of a madman. Accordingly, in order to prove his sanity, and with it his and Iras's identity, once he is able to leave his sickbed Lavenham insists on seeing the remains of Iras, depicted by Knollys as that of

> »›a mummy – [a] mummy of a woman; a thing swathed and bandaged in cements and dry as a stick, which had been dead for hundreds – nay, thousands – of years.[…]. [A] mummy [that] had been wrapped from head to foot in a fur coat, and had a modern gauze veil tied over its own bands of long hair.‹« (ibid.: 194)

When Knollys leads Lavenham to the granary of the farm Lavenham was brought to after his breakdown in the Scottish mountains and shows him the mummy, Lavenham has to admit that even for him it is hard to recognise his wife's complexion in the mummy's face, which is »changed and perished, as if from contact with the air« (ibid.: 211). However, in spite of the mummy's obvious traces of old age and decay and Knollys's attempt to convince him of the opposite, Lavenham begins to unwrap the mummy in order to prove the truth of his account by means of a former injury only someone who has spoken to the living Iras could know of: »It was the left wrist I wanted to examine, but Knollys's exclamation drew my attention to

17 Furthermore, it might also be argued that the destructive power of Savak threatening Lavenham and Iras's happiness really is part of Lavenham himself or contemporary society in general and thus points towards the double-standard of British society, which is a major aspect of the *Jekyll and Hyde Discourse*.

the hand. There on the attenuated finger, still a finger and not a bone, was a new broad wedding-ring of shining gold.« (Ibid.: 213) Even more valuable than traces of an old injury is the fact that underneath the seemingly undisturbed bandages appears a »Victorian ring on the hand of [the] mummy three thousand years old« (ibid.), which serves as material evidence for Lavenham's *his*tory and thus confirms his past union with Iras, representing the completion of his identity. It further emphasises that even though things seem to appear impossible and are judged accordingly, it does not mean that they are ultimately not true. On the contrary, in analogy to archaeological excavation there is always the need for deep investigation and examination which penetrates the surface of things to finally realise that there potentially exists another materiality, reality, and history beyond first sight.[18]

---

18 This is also made explicit in George Griffith's story »The Lost Elixir« (1903), in which the British archaeologist Professor Hessetine meets an Egyptian, named Pent-ar, who claims to be the brother of Ramesses II. When Hessetine makes clear that he has reason to doubt this, Pent-ar tells him the following: »›The matter is in this wise. Thou art going to Susa, the city of my youth and happiness, and the scene of the crime against the High Gods which made the one unfading and destroyed the other for ever. At Susa thou wilt seek to clear the dust of ages from the house in which I and mine dwelt, the temples in which we worshipped, and the tombs where the mummies of my dear ones are resting [...]. Now, what I ask is this: that thou shouldst make me one of thy company, [...] and take me to Susa, and there I will show thy workmen where to dig that they may find that which thou seekest. I will draw thee pictures of the temples and the theatres and the tombs, and mark out the streets and the squares, until all Susa in its ruins shall be as plain to thee as it was in its glory to me.‹ I don't suppose that any archaeologist had ever had such an astoundingly tempting offer made to him, and I candidly admit that I was not only tempted – I fell.« (*Lost Elixir*: 134-135) As Lavenham in *Iras*, Pent-ar here is able to offer proof of his extraordinary and incredible story beneath »the dust of ages« (ibid.), i.e., beneath the surface of things. In the further course of events, Pent-ar's story and identity are confirmed by material evidence unearthed in the subsequent excavation at Susa. In addition, the mummy of his long-lost love, Amaris, is rediscovered and the two lovers are finally reunited in death: »At the end of a week we had laid bare a small pyramid, the apex of which, only showing a couple of

This brings to mind figures like Dracula, which, very much like Iras,[19] are, as »possible impossibilities« (*Dracula*: 193), embodiments of the idea of a parallel reality beyond everyday reality, as Knollys tries to make clear: »›You know, my old fellow, it is *impossible* to expect [the doctor] to believe that the dried-up corpse in the granary, centuries old as it is, could have been alive a month ago and your wife.‹« (*Iras*: 208, emphasis mine) Against the background of contemporary British culture, however, the idea of a parallel reality beyond everyday reality seems plausible when one considers the general social, political, and economic instabilities of the time. This is also what Lavenham tries to suggest when he says: »›Have you convinced [the doctor] there may by chance *be a thing or two in the universe not included in his knowledge-box*?‹« (*Iras*: 207, emphasis mine) Here the »knowledge-box« (ibid.) referred to by Lavenham points towards the power structures of Victorian discourses with the including and excluding strategies by which they regulate and protect contemporary society. And even though Lavenham is eventually ready to accept the fact that there seem to be things which cannot be explained by the doctor's »knowledge-box« (ibid.), at the beginning of the novel Knollys's »cheerful conviction that all lay within the realm of the physical and the reach of a doctor's prescription *comforted [him] in spite of [himself]*« (ibid.: 43, emphasis mine).

Although Lavenham is eventually able to prove his marriage to Iras by means of the wedding ring, there always remains the danger that people will still not accept the truth after all and, in order to protect the contemporary civil discourse, will brand people like Lavenham as ill and insane. Ultimately, this is indeed what happens to Lavenham, who has to accept

feet or so above the sand, he had found with unerring instinct or memory after an hour's survey of the wilderness of ruins amidst which it stood.« (Ibid.: 142) Without the search beneath the »dust of ages« (ibid.: 134-135), which in *Iras* is symbolised by the unrolling of Iras's mummy, both Pent-ar's identity and story as much as that of Lavenham would have been judged by those in his surroundings as the fantasy of a madman or swindler. Pent-ar's account thus can be read as a further hint at the interdependence of conceptions of narration, memory, and identity which play a prominent role in *Iras*.

19 Unlike *Pharos, the Egyptian*, and *Dracula*, the title of the novel, *Iras: A Mystery*, already suggests a certain uncertainty in regard to Iras's existence.

that his doctor, representing Victorian matter-of-fact society, is convinced that *his*tory is nothing but a fantasy: »I know of the great man's theory about me, and how he disposes of my ›delusion‹ on the score of previous entanglement and conspiracy.« (Ibid.: 249) Lavenham knows that even his friend Knollys, although he is willing to believe in Lavenham's story, is unable to accept it as thoroughly *real*: »Knollys hangs suspended between us like Mahomet's coffin – he can neither quite credit me nor wholly believe the doctor; the *narrowing faculty which he calls common-sense is with him a stone of stumbling and rock of offence, as it is to so many*.« (Ibid.: 250, emphasis mine) The very fact that even Knollys, who would like to believe his friend is controlled by what Lavenham refers to as »common-sense« (ibid.), points towards the discursive power of a culture in which rationalism and materialism are the determining parameters, as is emphasised by the fact that Lavenham diagnoses this for »so many« (ibid.). Furthermore, it is instructive that Lavenham declares »common-sense« (ibid.) to be a »narrowing faculty« (ibid.), since unlike in the Victorian discourse this suggests that common sense is not *per se* a positive faculty. Rather than leading to truthful insight, it might sometimes even prevent people from seeing the truth as it really is, as a truth beyond the rationally explicable. That there is a reality beyond the rationally explicable and the visible is emphasised by the behaviour of Lavenham's dog Nell on several occasions:

> »Nell had been sleeping, to all appearance, but now roused herself and looked eagerly at the door, as if at an entering figure […]. She followed with her eyes invisible movement across the room, dropped from the bed, and seemed to approach what excited her curiosity […] […] Knollys and I witnessed the little pantomime of what we called in jest Nell's ghost, but it was of frequent after-recurrence.« (Ibid.: 200)

Here, the dog's behaviour seems to prove the existence of what remains invisible to Lavenham *and* Knollys. Yet, what they are both able to perceive is the dog's exceptional behaviour, which points towards an *im*material existence the men refer to as »Nell's ghost« (ibid.). Later in the novel, the reality of this immaterial sphere seems to be confirmed by material evidence when Lavenham finds a branch of palm in Iras's empty sarcophagus upon the dog's characteristic behaviour.[20] In spite of the fact that Lavenham, by means of ›archaeological‹ *re*-membering, succeeds in proving his past union with Iras, his attempt to achieve Lacanian wholeness through his love for archaeology and his love to Iras fails, as symbolised by his approaching death at the end of the novel: »This wasted body of mine is to be taken the long journey northward; the grave at the Hill of Sorrow will be opened once more. Knollys has promised to lay me besides her, and to set up a white cross over us, writing upon it, plain for all to read, the names of Ralph Lavenham and Iras Lavenham, his wife.« (Ibid.: 250-251) Ironically, as made explicit by Lavenham's last will in regard to his shared grave with Iras, symbolising a reunion after death, the only thing ever possible for Lavenham is an unsatisfactory substitute completion – real wholeness is never attained.[21]

20 At the same time, the two poles of materiality versus spirituality can also be read as representing the Victorian period on the one hand and the beginning of modernism on the other, with the former's emphasis on materiality and the latter's emphasis on subjectivity and impressionism. In this context, the novel can be read as emphasising the fact that, rather than judging one of these poles as positive or negative, an integration of both is necessary. In this regard, *Iras* can be read as an attempt to reconcile the seemingly contradictory trends between what Virginia Woolf later refers to as the conflict between materialists and modernists (cf. »Mr. Bennett and Mrs. Brown« (1923) and »Modern Fiction« (1925)).

21 Note that once more Lavenham's wish to have his and Iras's names written on the cross can be read as Lavenham's desire to achieve eternal union, i.e., wholeness, with »Iras, his wife« (*Iras*: 251).

6.2 Reconstructing the Past: Search for Evidence in the Present

›I am a simple sort of chap, and have no use for mysteries which beat all the detective stories I have ever read. That sort of thing is all very well in fiction, but in real life – humph.‹
(*Green Mummy*: 270)

So far I have looked at different ways in which archaeology was used as a means of forming and constructing continuity and eventually identity for both the past and the present. A genre that demonstrates the actual process of identity (re)creation in combination with archaeology is the archaeological detective story, which plays with the affiliation of the two fields and strongly evokes the *Sherlock Holmes Discourse*. This has been pointed out by Wieczorkiewicz in regard to her interpretation of museum narratives:

»The detective narrative is often used to present scientific procedure as if they were detective procedures. [...] The detective figure can also appear in museum rhetoric in a more obvious way. This is the case when, for example, methods of criminal investigation are applied to the scientific study of museum objects, and when the mutual benefits of cooperation between scientists from both fields are reported.« (Wieczorkiewicz 2005: 57)

In both detective and archaeological work the detective and archaeologist, respectively, is concerned with the past by searching for evidence in the present, which he or she collects and arranges in order to recreate a past chain of events, that is, a narrative of the past, to then interpret and understand the present in relation to that past:

»The fragmented historical becomes a complete piece of historical evidence by virtue of its concrete presence and additional historical knowledge acquired through investigations. The historical object also made its arrival in the historical novel at the same time in a similar way, where it conveyed the authenticity of the myth in the spirit of ancient times. Consequently, crime fiction and archaeology are linked in two different ways. On the one hand, they both study causes, that is, they investigate a sequence of events that lies in the past at the time of investigation. On the other hand, they both involve tracing or predicting past events, in that the detective, like the archaeologist, can bring a past world back to light by means of research or investigation and discover something new in it.« (Patzek/Hauses/Dudde 1999: 395-396, translation mine)

The affinity of detective and archaeological work is also mirrored in the numerous archaeological detective stories, which started to appear at the end of the nineteenth century and flourished during the first half of the twentieth century with stories by prolific authors such as Agatha Christie.[1] In the present chapter, I will analyse the parallels between detective and archaeological work by focusing on Austin Freeman's detective novel *The Eye of Osiris* and Bram Stoker's *Jewel of Seven Stars*.

In *The Eye of Osiris*, the forensicist John Thorndyke[2] confronts his students after one of his popular lectures with the unresolved case of the mysterious disappearance of John Bellingham, a passionate Egyptologist, reported on in the morning papers:

»›The circumstances in this case are very curious; in fact, they are highly mysterious; […] The gentleman who has disappeared, Mr John Bellingham, is a man well known in archaeological circles. He recently returned from Egypt, bringing with him a very fine collection of antiquities – some of which […] he has presented to the British Museum, where they are now on view – and having made this presentation, he appears to have gone to Paris on business. I may mention that the gift consisted of a very fine mummy and a complete set of tomb-furniture.‹« (*Osiris*: 2)

1 The 1920s and 1930s are also referred to as the golden age of the detective story. Barbara Patzek (1993), Charlotte Trümpler (1999), and Barbara Korte (2000a) discuss the affinity between the detective story and archaeology by looking at the prolific writer and amateur archaeologist Agatha Christie. Trümpler's book was published to accompany the eponymous travelling exhibition *Agatha Christie und der Orient. Kriminalistik und Archäologie*, shown from 2000 to 2002 in Essen, Vienna, Basle, Berlin, and London.

2 Donaldson refers to Freeman as having introduced Thorndyke as »the first real scientific investigator into the pages of detective fiction« (Donaldson 1971: 79).

It turns out that after his departure for Paris, Bellingham has disappeared without a trace, and neither his family nor his friends have any clues about his whereabouts. Eventually, they expect the worst when Bellingham remains missing for a longer period of time. In the aftermath of Bellingham's disappearance, his family does not only have to deal with his probable death and their loss, but also with an extremely complicated will, which leaves the family empty-handed until Bellingham *or* his remains surface. Consequently, the brother of the missing John Bellingham, Godfrey, and his niece Ruth are extremely anxious to solve the mysterious disappearance of their relative. It is by accident that one of Thorndyke's former medical students, Paul Berkeley, who is also the narrator of the story, is called to a home visit at the Bellinghams's house and there meets Godfrey and Ruth for the first time. Berkeley has to return a couple of times to the house and gradually befriends Godfrey and Ruth and also learns more about the circumstances of Bellingham's disappearance. At this point in the novel the first of several human bones are discovered in the process of cleaning a watercress-bed in Kent, as reported by the newspapers:

»›The operation began two days ago. A gang of three men proceeded systematically to grub up the plants and collect the multitudes of water-snails that they might be examined by the expert to see if any obnoxious species were present. They had cleared nearly half of the beds when, yesterday afternoon, one of the men working in the *deepest part* came upon some bones, the appearance of which excited his suspicion. Thereupon he called his mates, and *they carefully picked away the plants piece-meal, a process that soon laid bare an unmistakable human hand lying on the mud amongst the roots*. Fortunately they had the *wisdom not to disturb the remains*, but at once sent off a message to the police. Very soon, an inspector and a sergeant, accompanied by the divisional surgeon, arrived on the scene, and were able to view the remains lying as they had been found. [...] After a *thorough examination on the spot, the bones were carefully collected and conveyed to the mortuary, where they now lie awaiting further inquiries*.‹« (Ibid.: 44, emphasis mine)

The scene described here with its initial search for water snails very much resembles that of an archaeological excavation, which culminates in the accidental discovery of bones »in the deepest part« (ibid.) of the watercress-bed.[3] The men's procedure of »carefully pick[ing] away the plants piece-meal« (ibid.), which leads to the discovery of a human hand as well as their »wisdom not to disturb the remains« (ibid.), links them to excavators at an archaeological site, digging for bones and artefacts of the past to then present them to the site director responsible for their identification.[4] In this case the function of the site director is fulfilled by the inspector, the sergeant, and the divisional surgeon, who examine and collect the bones, which are then brought to the mortuary for further investigation. Significantly, the bones found by the workmen in the watercress-bed, in analogy to archaeological relics, are only fragments of a whole, that is, a human body, of which the other remains are still missing. Thus, in order to identify the person whose hand has been found in the watercress-bed, it is first necessary to reconstruct the human skeleton, which constitutes the physical and material rudiment of a human identity. Accordingly, as put forward in one of the newspaper articles, the authorities' next step is likely to be the search for the other missing fragments of the skeleton: »›Also we believe that a systematic search is to be made for further remains.‹« (Ibid.: 45) The very fact that a »systematic search« (ibid.) for the remains is required, with the ultimate aim of first recreating the human skeleton and then determining his or her identity, once more emphasises the analogy to an archaeological venture. Similar to the situation at an excavation site, the discovery of the fragment of a skeletonised human body leads to an attempt to restore the complete whole by means of a systematic search for other fragments, which eventually make possible the reconstruction of a past entity.

---

3 The presentation of the discovery is reminiscent of a newspaper report on a new archaeological find.

4 The similarity in the methodological work of the archaeologist and the detective is also pointed out by Volker Neuhaus (1999) and Barbara Korte (2000a) in regard to Wilkie Collins's *The Law and the Lady* (1875): »Like in Poe's *The Purloined Letters*, the goal in *The Law and the Lady* from 1875 is to find a compromising letter. But all effort is in vain until the searchers strike on the systematic excavations in Pompeii, which extend even to the cesspits. Accordingly, they then proceed to search through the household rubbish heaps, even orienting themselves toward archaeology in that they end up abandoning tools and work exclusively with their hands, like at a delicate excavation.« (Neuhaus 1999: 429, translation mine)

A further instructive passage in regard to the interweaving of detective and archaeological work in the novel is the following scene, where members of the police are digging for further bones of the missing man:

»Suddenly the man bent down by the side of a small pool that had been left in one of the deeper hollows, stared intently into the mud, and stood up. ›There's something here that looks like a bone, sir‹, he sang out. ›Don't grub about then‹, said the inspector. ›Drive your shovel right into the mud where you saw it and bring it to the sieve‹. The man followed out these instructions, and as he came shorewards with a great pile of the slimy mud on his shovel we all converged on the sieve, which the inspector took up and held over the tub, directing the constable and labourer to ›lend a hand‹ [...]. [...] Presently the inspector raised the sieve from the water and stooped over it more closely to examine the contents.« (Ibid.: 126-127)

Here the actions of digging, sieving, and examining the findings performed by the police in search for human bones are the same as those undertaken at an excavation site with the aim to reconstruct the past. The difficulty of bringing together various pieces of evidence to recreate a past scenario is illustrated in the text by the continuous search for further human remains. This renders the past a puzzle which has to be put together in order to make this evidence in the present readable as a story of the past, that is, a history. It is only through the recreation of the past that both the detective and the archaeologist are able to make statements concerning bygone events and times and thus to *explain* the past and through it also the present.

It becomes clear in the further course of the novel how this process of searching for evidence results not only in the reconstruction of the past events that finally lead to Bellingham's death but also in the recreation of Bellingham's alleged identity through the assembly of his fragmented skeleton, fulfilling what Wieczorkiewicz describes as follows: »Detective procedures lead in these cases toward a twofold reconstruction: the reconstruction of past events and processes [...]; and the reconstruction of individual biographies [...]. The reconstruction of individual biographies is the task of a creative Storyteller [sic].« (Wieczorkiewicz 2005: 57) Before the completion of ›Bellingham's‹ skeleton, he remains essentially absent and abstract, even though he appears in the stories told by his family and his will. This is underlined in the question Berkeley addresses to Ruth: »›What sort of man was your uncle? [...] I hope you don't think me inquisitive, but, to my mind, he presents himself as a *kind of mysterious abstraction.*‹« (*Osiris*: 55, emphasis mine) All that we know of Bellingham is through other people or other things, such as his writings and his collection of antiquities, while he himself remains absent until the case has been solved at the end of the novel. This is significant since it once more shows strong parallels to archaeological research. The stories and memories the Bellinghams relate and have of their missing relative can be seen in terms of Aleida and Jan Assmann's communicative memory, in which memory is perpetuated orally by the living members of a group. In this sense, the accounts given by the Bellinghams on John Bellingham function as an oral history, as a commemorated narrative of the past shared by a group of people, which serves as a written record of past peoples and events: »Deep excavations and collection of remembered stories allow some reconstruction of what these remains, themselves already signifying yet baffling images, commemorate.« (Pollock 2006: 9) Both written and oral records, however, are also always characterised by a certain tentativeness as far as their historical ›truth‹ is concerned. Accordingly, the task of the archaeologist as much as of the detective is to provide material proof for what otherwise only exists in the form of abstract knowledge and hypotheses or is stored as oral or written memory in order to eventually create meaning and identity.[5]

5 In this context it is interesting to note that in several mummy texts the characters are fascinated by the idea of obtaining *first-hand* historical information through a character of the past. Under normal circumstances this would be impossible. Not so, however, in mummy fiction, where the past awakens in the present and agents of the past provide first-hand accounts of historical events. As Bridges notes in regard to *Jewel*: »Yet the Egyptologist's dream of *intense, personal communion with antiquity* is realized to most horrifying effect when his scientific (or is it occult?) experiment succeeds and Tera is resurrected but then decimates his household.« (Bridges 2008: 150) While in *Jewel* the »personal communion with antiquity« (ibid.) ends fatally for the archaeologist and his friends and thus essentially marks a miscommunication with the past, attempts to communicate with the past in other texts are more successful, although not always satisfying. In *Iras* Lavenham finds Iras's first-hand knowledge about the Egyptian past she supplements the stories about his excavations with quite useful and adopts them into his own narrative: »[Iras] was never weary of hearing how we

In *Osiris*, the combination of material proof and historical record for the recreation and interpretation of the past is illustrated by the Bellingham case, in which the oral history of John Bellingham is gradually filled in with factual proof provided by the various detective and archaeological findings. However, although an increasing number of bones are found of the skeleton, as summed up by Thorndyke (cf. *Osiris*: 103-105), it remains impossible to identify the bones as the remains of John Bellingham, since the idiosyncratic parts of the skeleton are still missing:

---

labored in the excavations, and of the treasures of antiquity restored to light, and the theories deduced from them. Touching these – she an eye-witness of the remote past – had much to say, both in correction and confirmation; and I began to note down various matters to be incorporated in the book.« (*Iras*: 119, quoted above) Two other examples in which the characters try to gain first-hand information from historical eye-witnesses are the short stories »The Strange Discovery of Doctor Nosidy« (1886) by Ernest Suffling and »The Mummy of Thompson-Pratt« (1904) by C. J. Cutliffe Hyne. In both of these stories a scientist successfully reanimates the brain of a mummy, which renders an interview with the latter possible. However, in both cases the scientists are disappointed by what they hear from the mummies: »Of course the thought-words I had read, by means of the wire around my head, were in the *Egyptian* tongue, hence the reason for my not understanding them.« (*Doctor Nosidy*: 24) Even though in Hyne's story the scientist is able to *understand* what the mummy has to tell, he is deeply disappointed by *what* he is told by the mummy: »But Gargrave cursed as he listened. He had looked for a dissertation on history, and he was getting *chroniques scandaleuses*; he had expected talk about Kings' policies, and he was hearing the tales of flirtations with their housemaids. [...] It was a gorgeous piece of description. But as it was given to us it would not publish; no, not even inside a yellow cover.« (*Thompson-Pratt*: 230-231) On a more abstract level, this also reflects the common difference in an appreciation of a people's social history rather than world history, underscored by the actuality that an interest in the former has only been developed more recently. Perhaps most notable in regard to these stories of mummy-communication in general, aside from their entertaining character, is the fact that they reflect a contemporary longing for a *first-hand* – as quasi-experienced – affirmation of historical records usually associated with material remains, functioning as dumb witnesses of the past.

»›Now, the question is, Are these bones the remains of my brother John? What do you say, Doctor Thorndyke?‹ ›I say that the question cannot be answered on the facts at present known to us. It can only be said that they may be, and that some of the circumstances suggest that they are. But we can only wait for further discoveries. At any moment the police may light upon some portion of the skeleton which will settle the question definitely one way or the other.‹« (Ibid.: 104-105)

The bones of the skeleton necessary for identifying the remains as that of John Bellingham besides the teeth are the left ankle and the knee-caps, since they, according to the accounts of his family and acquaintances, bear the marks of former injuries: »›What accident was that?‹ I asked. ›Oh, hasn't my father told you? It occurred while he was staying with us. He slipped from a kerb and broke one of the bones of the left ankle – somebody's fracture –‹ ›Pott's?‹ ›Yes; that was the name – Pott's fracture; and he broke both his knee-caps as well.‹« (Ibid.: 173) As the impossibility of identifying the discovered bones as that of John Bellingham without having corresponding evidence in what is remembered and communicated by his social environment illustrates, the (re)construction and formation of identity is a joint process of memory, communication, and substantial remains. This is true of both archaeology and forensic sciences, as formulated by Thorndyke in regard to the Bellingham case:

»›You are asking me for an incriminating statement [...] delivered in the presence of a witness too. But, as a matter of fact, there is no use in speculating *a priori*; we should have to reconstruct a purely imaginary situation, the circumstances of which are unknown to us, and we should almost certainly reconstruct it wrong.‹« (Ibid.: 111)

Thus, in order to be able to reconstruct the past as ›true‹ and as precisely as possible, a combination of abstract knowledge, in the form of memory and narration, and concrete material evidence is necessary. Consequently, origin, continuity, and finally personal identity are formed through a complex interaction between communicative and cultural memory on the basis of material objects, which assume the function of sites of memory and the link between present and past.

The intricate interrelation between memory and material remains can also be found in Ruth's remark on her uncle's obsession with Egyptology:

»›He knew the contents of every museum in the world, in so far as they were connected with Egyptian antiquities, and had studied them specimen by specimen. Consequently, as Egyptology is largely a museum science, he was a learned Egyptologist. *But his real interest was in things rather than events*. Of course, he knew a great deal – a very great deal – about *Egyptian history*, but still he was, before all, *a collector*.‹« (Ibid.: 54, emphasis mine)

In this passage, John Bellingham is introduced as a person with a strong interest in and fascination for »things« (ibid.), that is, objects rather than »events« (ibid.), which eventually are responsible for the formation of history. This is significant since, as implied by the fact that »before all, [he was] a collector« (ibid.), people like Bellingham are responsible for isolating a ›thing‹ from its original context, and thus the ›thing‹ as such becomes more important than its history. As mentioned in connection to the character of Iras, however, this also means that the ›thing‹ is bereaved of its genuine meaning and its ›aura‹ (cf. Benjamin 2003: 254) and, imported to a foreign cultural context, loses its original identity. This does not suggest that an object isolated from its original context cannot be given another meaning. Yet, to use the words of Thorndyke just quoted in regard to the reconstruction of the Bellingham case, »›we should have to reconstruct a purely imaginary situation, the circumstances of which are unknown to us, and we should *almost certainly reconstruct it wrong*‹« (*Osiris*: 111, emphasis mine). Consequently, in analogy to the reconstruction of a crime case, it is particularly crucial for an archaeologist to read a ›thing‹ in accordance with its original context, as bequeathed by both written and oral historical sources.

As already mentioned, the relevance of history in concurrence with an object of the past for its identification and (re)familiarisation is also the main underlying theme of the Bellingham case. This is emphasised when Thorndyke asks Godfrey Bellingham to write down the medical records of his brother:

»›What I want you to do is this: Write down a full description of your brother, including every detail known to you, together with an account of every illness or injury from which you know him to have suffered; also the names and, if possible, the addresses of any doctors, surgeons, or dentists who may have attended him at any time. The dentists are particularly important, as their information would be invaluable if the skull belonging to these bones should be discovered.‹« (Ibid.: 105)

As it eventually turns out, this medical record becomes essential for the identification of the skeleton as the remains of John Bellingham and thus underscores the importance of both material and ›historical‹ evidence for the successful work of the detective *and* the archaeologist. As in archaeology, there has to be some historical record in order to make the reconstruction of certain historical events possible, which does not imply that a statement on the past is impossible as long as no historical records exist. However, in that case the histories remain essentially unspecific, ambivalent, and impersonal:

»The story of a buried person gradually changes into another story, which conveys information about (say) beliefs, everyday life and craft. The burial place is deconstructed and its contents classified according to subject and information value. A dead person can also be deconstructed: every bone can be made to tell a different story.« (Wieczorkiewicz 2005: 54)

The same is true for the Bellingham case, as the hypotheses presented by Thorndyke in regard to the case illustrate:

»›There are five conceivable hypotheses […]. First, he may still be alive. Second, he may have died and been buried without identification. Third, he may have been murdered by some unknown person. Fourth, he may have been murdered by Hurst and his body concealed. Fifth, he may have been murdered by his brother.‹« (*Osiris*: 108)

Although at that point the skeleton found scattered throughout the whole county is nearly complete, the only thing that can be ascertained is that it is a male skeleton, which allows for the conclusion that at some point in the past a man died. However, there is no certainty as to the cause and time of his death, let alone the identity of the dead man. Alongside these uncertainties in regard to the material remains of the past, as in archaeology there are certain hypotheses as to what might have happened that would *explain* the material findings in the present. However, as the bones and Thorndyke's hypotheses demonstrate, as long as these things cannot be brought together by means of some kind of historical record, a specific as opposed to a general interpretation and reconstruction of the past remains impossible. In other words, even though the reconstruction of a past scenario might be possible, the identity of the agents involved will remain unspecific:

»The bones were apparently those of *a man* of fair though not remarkable muscular development; over thirty years, but how much older I was unable to say. His height I judged roughly to be five feet eight inches, but my measurement would furnish data for more exact estimate by Thorndyke. *Beyond this the bones were quite uncharacteristic.*« (Ibid.: 122, emphasis mine)

Thus, as the Bellingham case illustrates, a precise and specific reconstruction of the past is necessary for an interpretation of the present, which can also be transferred to historical and archaeological work.

In the novel, the importance of reconstructing the past by combining material remains, historical records, and hypotheses culminates in an experiment conducted by Thorndyke and Dr Norsbury, an employee at the Egyptian Department of the British Museum. Following Thorndyke's invitation, Ruth and Berkeley join Thorndyke and his assistants at the British Museum without knowing anything of Thorndyke's intention: »Clearly an X-ray photograph was being made; but of what? I strained my eyes, peering into the gloom at the foot of the gallows, but though I could make out an elongated object lying on the floor directly under the bulb, I could not resolve the dimly seen shape into anything recognisable.« (Ibid.: 203) Although Berkeley and Ruth presently learn from Dr Norsbury that the »elongated object« (ibid.) on the table is a mummy, namely the mummy of Sebek-hotep, imported to London by John Bellingham, they still have no idea about the purpose of the experiment being conducted by the scientists. Hence, together with the others they curiously watch the development of the X-ray photograph taken of the mummy:

»There is something singularly fascinating in the appearance of a developing photograph; in the gradual, mysterious emergence of the picture from the blank, white surface of plate or paper. But skiagraph, or X-ray photograph, has a fascination all its own. Unlike the ordinary photograph, which yields a picture of things already seen, it gives a presentment of objects hitherto invisible; and hence, when Polton poured the developer on the already wet paper, we all craned over the tray with the keenest curiosity.« (Ibid.: 204)

Gradually, the outlines of a skeleton start to appear on the plate, and the watchers are granted a view into the inside of the mummy:

»The developer was evidently a very slow one. For fully half a minute no change could be seen in the uniform surface. Then, gradually, almost insensibly, the marginal portions began to darken, leaving the outline of the mummy in pale relief. The change, once started, proceeded apace. Darker and darker grew the margin of the paper until from slaty grey it had turned to black; and still the shape of the mummy, now in strong relief, remained an elongated patch of bald white. But not for long. Presently the white shape began to be tinged with grey, and, as the colour deepened, there grew out of it a paler form that seemed to steal out of the enshrouding grey like an apparition, spectral, awesome, mysterious. The skeleton was coming in view.« (Ibid.)

The process described here is interesting since a view beneath the surface of things is made possible by means of modern technology, that is, through deep investigation.[6] However, it turns out that what the onlookers finally perceive on the X-ray photograph is more than just a look beyond the mummy wrappings, since the apparition of the body and skeleton also reveals the mummy's history and its identity. What Dr Norbury initially misinterprets as an extraordinary Egyptian mummy showing gold and tin dental fillings Thorndyke reveals to be a modern specimen by pointing out a tattoo mark showing the eye of Osiris on the chest of the body and »pieces of silver wire« (ibid.) in the knees of the skeleton. Although Dr Norbury is still unable to interpret the X-ray photograph correctly, Thorndyke eventually solves the mysterious case by presenting his reading of it, which identifies the body in the photograph as that of John Bellingham. This he does by combining both *oral* and *written history* in the form of Bellingham's medical records and the description of a distinguishing mark, the Osiris tattoo, produced by his family: »›John Bellingham is here, Doc-

6 This underlines again the Victorian fascination with a materiality beyond the commonly visible, pointed out above in regard to Lavenham's endeavour to prove Iras's existence in Victorian England.

tor Norbury. This is John Bellingham.‹ Dr Norbury started back and stared at Thorndyke in speechless amazement. ›You don't mean‹, he exclaimed, after a long pause, ›that this mummy is the body of John Bellingham!‹ ›I do indeed. There is no doubt of it.‹« (Ibid.: 206) In other words, the body of John Bellingham was swapped with that of the mummy Sebek-hotep and was smuggled to the British Museum under a false identity, while the remains of Sebek-hotep »›are at present lying in the Woodford mortuary awaiting an adjourned inquest‹« (ibid.: 207).

What is interesting in regard to Thorndyke's solution of the case is his means of solving it, namely the use and interpretation of the X-ray photograph. In this context, the process of developing the photograph can be read as the process of visualising the past or at least a momentary state of the past. What becomes visible on the photograph is the conserved body of a dead person at the point of its embalmment, which virtually turns the body into the representation of a moment frozen in time. Thus, the development of the X-ray photograph functions as the recreation of the past, which, as shown by Thorndyke's presentation, if interpreted properly, allows the reconstruction of a personal history *and* identity. In this way, the X-ray photograph becomes the missing link which connects history with the material remains. Significantly, as a picture of the past the X-ray photograph also *proves* that the common assumptions in regard to the recently discovered bones being the remains of John Bellingham were simply wrong, as Berkeley, who has analysed the bones, is forced to admit: »Now that the explanation was given, how obvious it was! And yet I, a competent anatomist and physiologist and actually a pupil of Thorndyke's, had mistaken those ancient bones for the remains of a recent body.« (Ibid.: 207) This underlines the necessity of both detective and archaeological work to incorporate hypotheses *and* material evidence to achieve a veritable reconstruction of the past, which emphasises at the same time the constructed character of a past reality as such and its dependence on the present.

As Dr Norbury's initial misinterpretation of the X-ray photograph illustrates, without specific knowledge in the form of a historical record the interpretation of material evidence might be similarly misleading. Rather than reading the tattoo of the eye of Osiris as a mark of modernity, Dr Norbury reads it spontaneously as proof of the body's affiliation with ancient Egyptian culture. Only through Thorndyke's explanation of the untypical marks on the body and skeleton of the alleged Egyptian mummy is the real identity of the body revealed. This demonstrates how evidence of the past is prone to misinterpretation, as its meaning changes depending on the cultural context it is attributed to. The tattoo on Bellingham's chest as such is already a reproduction of a past symbol with an intrinsic meaning, which, however, is reproduced once more through the X-ray photograph and thus becomes a reproduction of a reproduction. This invests the tattoo with an ambivalent meaning, since isolating the symbol and transferring it from its original cultural context, that is, from ancient Egypt into modern Britain, changes its meaning and makes it arbitrary.

This is also illustrated in Bram Stoker's *Jewel of Seven Stars*, where the meaning of the past in the form of the diffuse Egyptian smell and the innumerable antiquities is first of all perceived as indefinite and thus disturbing in the present, which the unsettling effect the Egyptian smell on the characters' minds discussed earlier underlines. Yet, once a fragment is inserted into the whole of the semantic field and thus interrelated with other objects, it can be identified and interpreted to some extent. This process of joining together separate pieces in order to create a coherent whole mirrors not only the necessary steps and preparation finally leading to the Great Experiment in *Jewel* but also the stong archaeological and detective theme in the novel, as seen in the reconstruction and interpretation of the nightly attacks on Mr Trelawny, and thus evokes the *Sherlock Holmes Discourse*.

In all these cases, the characters create meaning and identity (or attempt to create them) by assembling various pieces and facts which have to be searched for in the past and the present as much as in Britain and abroad, while each thing for itself essentially seems illegible. Before these isolated facts and things can be put together to create a semantic whole, however, it is necessary for the characters to first ›discover‹ and then examine them. This becomes clear in *Jewel* when Ross and Margaret explore curiosities spread over Mr Trelawny's house:

»Together we went round the various rooms and passages, examining and admiring the magnificent curios. There was such a bewildering amount and variety of objects that we could only glance at most of them; but as we went along we arranged that we should take them *seriatim*, day by day, and examine them more closely.« (*Jewel*: 91)

The way Ross and Margaret approach the variety of objects accumulated by Mr Trelawny resembles both an archaeological and a detective quest. By virtually following the traces of the objects, Ross and Margaret examine them »*seriatim*« (ibid.) and »day by day« (ibid.) »more closely« (ibid.), as an archaeologist or detective would do with remains and evidence of a more or less recent past to eventually piece them together in the fashion of Sherlock Holmes and form a coherent reconstruction of their meaning in relation to each other. This is emphasised by the fact that in this process Ross suddenly becomes aware of the changed meaning of the ancient relics at Trelawny's house:

»From what I had heard from Mr Corbeck I began to have some idea of the vastness of his enterprise in the world of Egyptian research; and with this light everything around me began to have a new interest. As I went on, the interest grew; any lingering doubts which I might have had changed to wonder and admiration.« (Ibid: 90)

Ross's changed perception of the Egyptian relics, his nascent interest in Egyptology, and his and Margaret's subsequent exploration of Mr Trelawny's collection resemble the behaviour of young children discovering the world around them step by step to finally comprehend and organise it cognitively and thereby create meaning and identity. This is underscored by Ross's feeling during the exploration of the house: »[Margaret] was so fresh in her views and ideas, and had so little thought of self, that in her companionship I forgot for the time all the troubles and mysteries which enmeshed the house; and I felt like a boy again …« (Ibid.: 91) Initiated by Corbeck's accounts on Egyptology, a new world has opened up for Ross and Margaret in which each object has a yet undefined meaning and is wait to be discovered, examined, identified, and finally fitted into the conception of a new old world. This underlines once more how the archaeological space represented by the Egyptian objects in Trelawny's house thus becomes a sphere of construction and creation in which the isolated objects are rediscovered and – endowed with meaning – rearranged, which emphasises the creative and liberating potential of the past.

However, it also implies that the conception of the past essentially depends on the recovery of facts *and* their interpretation by a person, such as the archaeologist or detective, in the present.[7] This interdependence of past and present becomes particularly clear in connection to the resurrection of Queen Tera's mummy in *Jewel*. In the preparation of what is referred to as the Great Experiment, Mr Trelawny and the others accumulate ancient Egyptian objects, assuming that they are part of a complex mechanism devised by Tera in the past for the future, that is, their present: »›This woman seemed to have had an extraordinary foresight. Foresight far, far beyond her age and the philosophy of her time. She seemed to have seen through the weakness of her own religion, and even prepared for emergence into a different world.‹« (Ibid.: 166) As the dramatic consequences of the Great Experiment suggest, however, the past has been essentially misinterpreted through its rearrangement in the present by Mr Trelawny and the others. Ironically, however, during the time before the experiment Mr Trelawny is sure that he is acting according to the plans and wishes of Queen Tera:

»›I have come to the conclusion that Queen Tera expected to be able to effect her own resurrection, when, and where, and how, she would. […] *Now comes the crown of my argument*. The purpose of the attack on me was to get the safe open, so that the sacred Jewel of Seven Stars could be extracted. […] But despite all its power, the astral body could not remove the Jewel through the chinks of the safe. The Ruby is not astral; and it could only be moved in the ordinary way by the opening of the doors. To this end, the Queen used her astral body and the fierce force of her Familiar, to bring to the keyhole of the safe the *master key* which debarred her wish.‹« (Ibid.: 176-177, emphasis mine)

Even though Mr Trelawny here points out that Tera intended her own resurrection, the very fact that *he* is in possession of both the jewel, the essential object for the resurrection, and the »master key« (ibid.) underlines that he is the one who is in control of the situation and thus that the conception of the past is essentially dependent on the present in the Halbwachsian sense. This further suggests that Mr Trelawny's interpretations of Tera's plans are coloured, if not determined, by his own motives and ambitions, as the overconfident presentation of his argument implies. Thus, rather than respecting Tera's alleged wish in regard to her own resurrection, Mr Trelawny pre-

7 Cf. Halbwachs, Aleida and Jan Assmann, Gehrke.

vents her from carrying it out independently by interposing himself. Accordingly, Mr Trelawny's plan to resurrect Tera is *his* venture right from the start, which at first can be seen as motivated by a drive for higher knowledge[8] in the manner of Victor Frankenstein:

»Margaret accuses the men of indecency in unwrapping the mummy of the queen, but by imaging Queen Tera's resurrection as a birth – and one, moreover, from which the woman is to be excluded – Trelawny is perpetrating more than mere indecency: he, like Victor Frankenstein before him, is proposing to act as a male midwife at an act of life creation which radically pre-empts the participation of the female. By subjecting the birth-chamber to the exclusive control of male science, however, all he produces is an abortion.« (Hopkins 1998: 137)

At second glance, however, it can also be read as originating in Mr Trelawny's urge to recreate a past identity in its wholeness, originating in his desire for Lacanian completion. The impossibility of the recreation of wholeness is underlined by the dramatic failure of the resurrection, which can be seen as paradigmatic for mummy resurrections in fiction of that time in general.[9]

In *Jewel* the failure of Tera's resuscitation is already foreshadowed by Doctor Winchester's small experiment with Margaret's cat Silvio, in which by means of behavioural sciences he successfully proves that Silvio's aversion to the cat mummy, Tera's familiar, is restricted to that specific mummy, while Silvio shows no interest in another cat mummy provided by Winchester from the British Museum:

»He had a large parcel with him, which, when unwrapped, proved to be the mummy of a cat. With Miss Trelawny's permission he placed this in the boudoir; and Silvio was brought close to it. To the surprise of us all, however, except perhaps Doctor Winchester, he did not manifest the least annoyance; he took no notice of it what-

8 »»If we are successful we shall be able to let in on the world of modern science such a flood of light from the Old World as will change every condition of thought and experiment and practice.«« (*Jewel*: 200)

9 Even if mummies are successfully reanimated, they frequently become a threat to the present, or if that is not the case, have a destabilising effect on the present (cf. Lavenham's relationship with Iras).

ever. […] Then, following out his plan, the Doctor brought him into Mr Trelawny's room, we all following. […] The moment Doctor Winchester got into the room, Silvio began to mew and wriggle; and, jumping out of his arms, ran over to the cat mummy and began to scratch angrily at it. Miss Trelawny had some difficulties in taking him way; but so soon as he was out of the room he became quiet. When she came back there was a clamour of comments. […] Then by common consent we dropped the theme – for the present.« (Ibid.: 99-100)

Interestingly, although Winchester's experiment provides a clear outcome, none of the party is able to come up with an interpretation, apart from the fact that they all find it strange (cf. ibid.: 100). Instead of further discussing the remarkable behaviour of Silvio, they »drop[] the theme« (ibid.) »by common consent« (ibid.) and thereby simply exclude the inexplicable from their discourse, refusing to acknowledge its existence. In other words, *because* there is no explanation at hand for Silvio's »strange« (ibid.) and »odd« (ibid.) behaviour, it is simply ignored in the further conversation. This ignorance and inability to understand the past essentially marks the misinterpretation of the past through the present in the novel in general and at the same time underlines the dependence of the conception of the past on the present.

Thus, the cat experiment underlines not only the group's ignorance but also their blindness in regard to an immanent danger and their own limits, as becomes clear later in the novel in regard to the Great Experiment. Very much like in *Frankenstein*, this turns their decision to undertake the Great Experiment into a case of hubris, which eventually culminates in Ross's anagnorisis *ex post* when he finds himself the only survivor in the cave at the house in Cornwall after the Great Experiment: »I found them all where they had stood. They had sunk down on the floor, and were gazing upward with fixed eyes of unspeakable terror. Margaret had put her hands before her face, but the glassy stare of her eyes through her fingers was more terrible than an open glare.« (Ibid.: 244) Although Ross is the only one who has the chance to fully realise the horrible dimensions of the miscarried experiment, the fact that the dead »were gazing upward with fixed eyes of unspeakable terror« (ibid.) suggests that there must have been some recognition prior to their death of the terrible thing conjured by them. The direction of the gazes further implies that the group was looking up to something/someone, which also implies that ultimately they have realised that they have actually been dealing with a force larger than life.

Apart from the fatal misinterpretation of the past and Tera's identity,[10] there is another crucial underlying aspect connected to the misinterpretation of the past that helps explain the failure of the Great Experiment. This becomes clear when we read Mr Trelawny's urge to resurrect Tera in the context of the Lacanian desire to retrieve wholeness, as suggested above. First of all, Mr Trelawny's initial situation is identical to that of Shelley's *Frankenstein*, since as in *Frankenstein* Mr Trelawny's wish to resurrect Queen Tera is motivated by his pursuit of higher knowledge and scientific reputation:

»I am about to make an experiment; the experiment which is to crown all that I have devoted twenty years of research, and danger, and labour to prepare for. Through it we may learn things that have been hidden from the eyes and the knowledge of men for centuries; for scores of centuries. [...] As to myself, I am willing to run any risk. For science, and history, and philosophy may benefit; and we may turn one old page of wisdom unknown to this prosaic age.« (Ibid.: 163)

Similar to Frankenstein, Trelawny also considers the possible consequences and dangers of his (re)creation. However, he plays them down by delivering extensive scientific explanations on ancient Egyptian occultism and mythologism in front of the others, as emphasised by Bridges:

»Beyond the archaeologist's conventional aspiration of unearthing material remains and thereby piecing together the history of the ancient cultures from which those relics survive, Trelawny betrays a dangerous desire to learn metaphysical truths, mystical wisdom, and otherworldly systems of practice. Like that other Faustian necromancer, Victor Frankenstein, the Egyptologist-as-Egyptomaniac wants to overreach the bounds of nature by attempting to accrue ancient knowledge that is otherwise secreted from modern human comprehension.« (Bridges 2008: 156)

10 As the tragic course of the Great Experiment suggests, Tera has been misinterpreted, Mr Trelawny and the others having presupposed that her intentions were good and conciliatory.

If Trelawny has any doubts and fears at all regarding his planned experiment, they are projected onto Ross: »Then and there, with the feeling strong upon me, I determined to warn Margaret and her father of dire possibilities; and to await, as well content as I could in my ignorance, the development of things over which I had no power.« (*Jewel*: 212) Although the obvious motivation for the Great Experiment seems to be the pursuit of higher knowledge, in analogy to *Frankenstein*, and on a more abstract level, the wish to (re)create life is strongly characterised by fantasies of omnipotence. Accordingly, in *Jewel* the desire to resurrect the past can be read as the desire to recreate a lost state of wholeness. This is underscored by the events surrounding Mr Trelawny's excavation of Tera's mummy, which coincided with the birth of Trelawny's daughter Margaret. Apart from the fact that several people died in connection to Queen Tera's mummy and her tomb, Mr Trelawny had lost his young wife in childbirth while he was exploring Tera's burial chamber for the first time. In addition, Margaret is said to look like Queen Tera rather than her mother: »›She is unlike her mother; but in both feature and colour she is a marvellous resemblance to the pictures of Queen Tera.‹« (Ibid.: 136) Both the fact that Mrs Trelawny's death coincided with Mr Trelawny's advance into Tera's tomb and Margaret's resemblance to Tera suggest that Margaret, rather than being the daughter of her mother, is the daughter of Tera (cf. Byron 2007: 60; Hopkins 1998: 136). Furthermore, it seems crucial in this context that Mr Trelawny's penetration into Tera's burial chamber seemingly results in the birth of Trelawny's daughter, who then turns out to look like the ancient Queen. In other words, Margaret's birth seems to be the consequence of Mr Trelawny's visit to Tera's (burial) chamber, where some kind of creational process was carried out.[11] As a result, Trelawny's wish to (re)create Tera by resurrecting her can also be read as his unconscious craving to restore his dead wife, whom he lost through symbolical ›adultery‹ while visiting the burial chamber of the ancient Queen.[12] Thus, Mr Trelawny's desire to exhume Tera and to later resurrect her can be read as his desire to regain completeness in the Lacanian sense, which dramatically fails. This is emphasised by the feeling of isolation and loneliness Ross experiences as the

11 Also note that Queen Tera is wearing a wedding dress under her mummy wrappings, as Trelawny discovers upon unrolling her mummy.

12 For the sexual connotation of archaeological excavation, cf. also Chapter 4.1.

only survivor of the Great Experiment: »I did what I could for my companions; but there was nothing that could avail. There, in that *lonely house, far away from aid of man*, naught could avail. It was merciful that *I was spared the pain of hoping.*« (Ibid.: 244, emphasis mine)

The failure of the Great Experiment can thus in general be read first as underlining the ambivalence of the past,[13] which is likely to result in its misinterpretation through the present, and second as emphasising the fact that the past can never be retrieved in its completeness. Rather than restoring identity by offering completion and fulfilment, it results in fragmentation, disintegration, and destruction of identity,[14] essentially reflecting the feeling of alienation of the Victorian and Edwardian subject: »The feeling of impending disintegration seized not only those who thought in political and social terms. There was a sense of dissociation between the outer and the inner worlds, an anticipation of the alienation which was to become almost a cliché of the next century.« (Chapman 1986: 189)

13 This is reflected by the description of the Great Experiment: First something strange seems to happen (cf. *Jewel*: 241), then something wonderful (cf. ibid: 242), and finally something horrible seems to have occurred (cf. ibid.: 244), as both the facial expressions and gestures of the bystanders reflect.

14 What is interesting in regard to Tera's miscarried resurrection is the fact that the four characters in *Jewel* have to die at the end because they have tempted the past. This means that the forces of the past have been provoked by modern men of the present, which finally brings about the lethal consequences. This can also be found in other mummy texts such as Loudon's *The Mummy!* and Doyle's »Lot No. 249«, where the resuscitation of Egyptian mummies becomes a threat to the present. In a broader sense this is true for many other mummy narratives, since, as in Shelley's *Frankenstein*, the curse or evil brought about by a mummy is in most cases nothing but a reaction to an initial provocation or misdeed by modern men. Often this curse or evil from the past can be read as a feeling of guilt which, in combination with superstition, culminates in a feeling of threat from the past (cf. Chapter 7).

6.3 VICTIM AND PERPETRATOR: EXCHANGING ROLES

> No sooner had their lips met than the fair flesh of the mummy grew dark and shrivelled into a thousand wrinkles.
> (*LOST ELIXIR*: 144)

While in *The Eye of Osiris* the characters are preoccupied with the reconstruction of identity, ironically spatial and temporal identities eventually turn out to be inverted when the alleged body of an Egyptian mummy at the British Museum is identified as that of the Egyptologist John Bellingham, and the human remains, which have been scattered all over the British countryside by Bellingham's friend Jellicoe in an attempt to conceal any suspicion of him having murdered Bellingham, are recognised as actually being those of the mummy Sebek-hoptep (cf. Chapter 6.2).

This is particularly significant considering the circumstances that lead to this confusion of identities, as they once more emphasise the need for contextualisation of material evidence from the past. The only reason why the human bones found in and around London are assumed to be the remains of Bellingham is the fact that the context they are found in suggests this. The same is true of the mummy at the British Museum, which is recognised as the mummy of Sebek-hotep. The context of a museum evokes antiquity, particularly so in regard to the exhibition of human remains. It seems completely absurd and therefore out of the question that the remains of a modern man should become part of a museum's stock. The reason for this can be found in the fact that to exhibit a modern Western human corpse in a museum was beyond the discourse of the time and hence simply impossible.[1] Thus, the mummy *is* an Egyptian mummy *because* it is tagged with the name Sebek-hotep and *because* it is exhibited at the British Mu-

1 Interestingly, this has changed if we consider the ›Body Worlds‹ exhibition of Gunther von Hagens which has travelled the world since the late 1990s (cf. Chapter 3.1).

seum.[2] The same applies, even though in a less pronounced form, to the remains of the Egyptian mummy found in ponds and soils of modern Britain, which are naturally taken to be those of a modern person. What this illustrates once more is the need for a historical record that, in combination with material evidence, allows one to contextualise, categorise, and thus identify the non-specific human remains, as is finally achieved by Thorndyke in the novel.[3] If either historical records or material evidence are missing, only general and impersonal statements are possible in both forensic investigation and archaeology, since otherwise, as the Bellingham case illustrates, misinterpretation and false identification are the result.

Apart from these practical questions in regard to the identification of human bones, there is a further, more symbolic aspect to the confusion of identities of Bellingham and Sebek-hotep. It is significant that Bellingham not only assumes the identity of the mummy Sebek-hotep, but that his body is actually turned into a mummy by Jellicoe after Bellingham's accidental death at his own house.

»›I had, of course, some knowledge of the methods of embalming, but principally of those employed by the ancients. Hence, on the following day, I went to the British Museum library and consulted the most recent works on the subject; and exceedingly interesting they were, as showing the remarkable improvements that modern knowledge has effected in this ancient art. I need not trouble you with details that are familiar to you. The process that I selected as the simplest for a beginner was

---

2 As pointed out by Leishmann, the exhibition of ›foreign‹ human remains was quite common at the time: »Many collections in museums in the Western world include human remains such as bones, skin, hair and teeth. The Pitt Rivers Museum in Oxford shows several shrunken heads, or *tsantsas*, as part of a display on ›The treatment of dead enemies‹. […] This is a vivid illustration of how cultural attitudes towards death and dead bodies can vary markedly across societies, and reflect the tension that still exist as a result of colonialism and the colonial mindset that dominated, especially during the 20th century, and with the advent of late modernity, when the importance of positivist science became a key force especially in the West. Different cultures hold different paradigmatic representations and value the dead in different ways.« (Leishmann 2009: 45)

3 This once again underlines the constructed and arbitrary character of the past and its dependence on the present.

that of formalin injection, and I went straight from the Museum to purchase the necessary materials. I did not, however, buy an embalming syringe: the book stated that an ordinary anatomical injecting syringe would answer the same purpose, and I thought it a more discreet purchase.‹« (*Osiris*: 239)

This is only the beginning of Jellicoe's detailed description of the preparation of Bellingham's corpse, a preparation which appears strange and particularly shady and unethical from the perspective of the Christian Western tradition. The decisive factor for this evaluation, however, seems not to be the process of embalmment as such but the object of embalmment. What from the perspective of contemporary British society seems to be acceptable for a human corpse of another cultural area, that it, that of ancient Egypt, seems to be objectionable for a corpse of one's own culture.[4] This assumption is indirectly supported by another passage in the novel in which Berkeley appears at the morgue to inspect the discovered bones:

»The bones were laid out on a large table and covered with a sheet, *which the sergeant slowly turned back, watching my face in*tently as he did so *to note the impression that the spectacle made upon me.* I imagined that *he must have been somewhat disappointed by my impassive demeanour*, for the remains suggested to me nothing more than a rather shabby set of ›student's osteology.‹« (Ibid.: 119-120, emphasis mine)

First of all, the situation described here, though the context is a different one, is very similar to that of a museum, where, amongst other things, ancient human remains are exhibited to be looked at by the visitors. However, the strict admission restriction at the morgue as well as the sheet covering the bones stand in stark contrast to the public masses at a museum and the revealing exhibition of ancient human remains.[5] Consequently, the

4 Although there was indeed a certain interest in embalmment in Victorian Britain (cf. Scandura 1996), it was still rather uncommon.

5 A further interesting parallel of the Bellingham case to archaeology is the public interest people take in the case, as reflected both by the various newspaper articles quoted in the novel and the popular occupation of searching the region for Bellingham remains. This is mentioned by Thorndyke in the following: »›The original discovery was made, apparently quite accidentally, at Sidcup on the

sergeant's way of presenting the bones seems out of place and tasteless, which is evident when Berkeley compares the sergeant's presentation to that of a showman: »The sergeant was in the act of replacing the sheet, *with the air of a showman who has just given an exhibition*, when there came a sharp rapping on the mortuary door. The officer *finished spreading the sheet with official precision*.« (Ibid., emphasis mine) The change in the sergeant's behaviour from his showman-like presentation to the final covering up of the bones with »official precision« (ibid.) *upon* the »sharp rapping« (ibid.) at the door emphasises the fact that the sergeant has temporarily left the discursive structures of the morgue and is now brought back by the officially sounding knock at the door. This change in the sergeant's behaviour is crucial since the contrast between the two modes of dealing with the human remains emphasises the crudity of the initial sensation-seeking presentation of the human skeleton. What is interesting is that both in the morgue and in the museum human bones are laid out to be looked at by other people. The crucial difference, however, is that at the morgue the purpose of gazing at and examining the bones is supposedly a scientific one, while at the museum the official emphasis is on educating, informing, and entertaining the public. This is emphasised by Jellicoe's reference to the collection of Egyptian tomb furniture John Bellingham brought home from his last expedition to Egypt: »›And a very *instructive collection* it is, in a *popular sense*; *very suitable for exhibition in a public museum*, though there is nothing in it of unusual interest to the expert. The tomb furniture is excellent of its kind and the cartonnage case of the mummy is well made and rather finely decorated.‹« (Ibid., emphasis mine) Although Jellicoe here underlines the »instructive« (ibid.) character of

---

fifteenth of July. It consisted of a complete left arm, minus the third finger and including the bones of the shoulder – the shoulder-blade and collar-bone. This discovery seems to have set the local population, especially the juvenile part of it, searching all the ponds and streams of the neighbourhood –.‹« (*Osiris*: 103) Significantly, this is reminiscent of the popular interest stirred by archaeological excavations both in Britain and abroad in the nineteenth and twentieth centuries, as discussed above. Hence, both the coverage of the latest findings in the Bellingham case as well as the public involvement in the form of searching for the missing bones of the unidentified skeleton reflect the interest in popular archaeology of that time.

Bellingham's collection for the public, the fact that the objects in question are a mummy and part of its tomb furniture suggests yet another, less explicit aspect to the collection's potential for public interest. In more general terms, this aspect might be identified as a potential to entertain the masses, as indicated by Jellicoe's comments that the collection is instructive »in a popular sense« (ibid.) and that »there is nothing in it of unusual interest to the expert« (ibid.). The question that now arises is what makes the collection attractive to a broad public but less appealing »to the expert« (ibid.). One of the main reasons for this public interest is, I would argue, the subtle voyeuristic and sensation-seeking atmosphere at museums in general, and more particularly if the exhibition is in any way connected to taboos excluded from contemporary discourse, as emphasised by Wieczorkiewicz in regard to the exhibition of mummies: »[O]ne feature does remain constant: if death enters the discourse, it concerns Others; our existence is directed towards life.« (Wieczorkiewicz 2005: 56) Thus, the exhibition of a collection presenting human remains and tomb furniture, especially if they are conceived of as part of the spatial or temporal ›other‹ as in Bellingham's case, is particularly attractive and suitable in this regard. The interest in the ›*ab*normal‹ in connection with the human body is illustrated by the popularity of the so-called freak shows particularly loved in the nineteenth and early twentieth centuries:[6] »However, when buried people belong to past or remote cultures, the validity of ceremonies is cancelled out and the dead body may be incorporated into museum's messages. The line is thus drawn between those (living and dead) who belong to our society, and those excluded form it.« (Ibid.) The popularity of these kinds of exhibitions and presentations of the human body, both dead and alive, is documented by the many reports on public mummy unwrappings as well as the exhibition of ›*extra*ordinary‹ human beings and allegedly ›primitive‹ people at fairs, ethnological expositions, and world exhibitions.[7]

What all these various shows and exhibitions have in common is that they appeal to a voyeuristic and sensationalist desire. While the nature of this lust seems to be universal and temporally transcendent, the way people

6 For a discussion of nineteenth- and early twentieth-century ›freak shows‹, cf. Nadja Durbach (2010), Marlene Tromp (2008), Robert Bogdan (1990), Lillian Craton (2009), and Robert D. Altick (1978).

7 Cf. Chapter 3.3.

deal with it, however, seems to be dependent on a particular discourse, namely on what is accepted or not by the dominant power structures of a society. This is illustrated on the one hand by a diachronic change as to what is commonly accepted as moral and immoral in a community and on the other by the synchronic coexistence of different discourses with their respective inclusion and exclusion principles.

In the novel, the coexistence of two different discourses with the same underlying principle though diverse codes of behaviour is evoked by the morgue scene. While in this passage the sergeant's attempt to create an atmosphere of sensationalism seems tasteless, a similar way of presenting human bones at a museum would probably seem completely acceptable. Thus, as long as the human remains exposed are those of a fellow being from the same temporal and regional context, their exhibition will be judged as morally wrong, while the more strange the remains are, the less immoral their exhibition will seem. This certainly results from the fact that in the former case a strong identification with the exhibited body is likely, while this will decrease the more the displayed body is perceived as different from the self. In other words, the identification with the human remains becomes weaker the further we move from the self towards the ›other‹.

In *Osiris*, however, this underlying principle is essentially disturbed by the confusion of Bellingham and Sebek-hotep's identities, underlining the potential for an inversion of the archaeological space. What for a long time are considered the remains of Bellingham turn out to be those of Sebek-hotep, whereas Bellingham's corpse, concealed in the mummy case of Sebek-hotep, is taken for the mummy of the latter. This correspons to what Daly refers to as the »objectification of the self« (Daly 1994: 39): »Exhibits take on a life of their own, and collectors face their own objectification.« (Ibid.: 35-36) The irony of this confusion lies on the one hand in the inversion of the principle described above, that is, that instead of the remains most like the self, those most *un*like the self function as the object of identification. On the other hand, the confusion of the two men's identities is also deeply ironic in regard to the inverted roles of Bellingham and the mummy of Sebek-hotep. While at first Bellingham has to be seen as a perpetrator who has violated the peace of the dead by invading the ancient tomb of Sebek-hotep and taking his mummy and grave goods away from Egypt to exhibit them in Britain, he eventually becomes the victim himself, exhibited at the British Museum. Considering the chain of events that lead to Bellingham's death and are presented by Jellicoe in retrospect, the mummy can be seen as being indirectly responsible for it:

»›I had left John on the third floor opening some of the packing-cases [which he had brought with him from Egypt] [...] with a tool somewhat like a plasterer's hammer [...]. [...] I could hear him knocking out the nails and wrenching up the lids; [...]. Just as I closed the staircase door behind me, I heard a rumbling noise from above; then all was still. [...] I ran up the stairs, and there, on the landing, I saw John lying huddled up in a heap at the foot of the top flight. There was a wound at the side of his forehead from which a little blood was tickling. The case-opener lay on the floor close by him and there was blood on the axe-blade.‹« (*Osiris*: 236-237)

Bellingham died by the tool with which he was about to reveal the Egyptian objects brought to Britain along with Sebek-hotep's mummy, which makes these objects the indirect cause of his death. Tellingly, the tool he was using to open the packing-cases, a hammer-like instrument evocative of a tool used in archaeological excavation, became the deathly instrument instead of fulfilling its purpose. Thus, the whole chain of events that finally leads to Bellingham's death is an inversion of the initial perpetrator-victim relation between Bellingham and the mummy of Sebek-hotep, in that Bellingham finally indirectly becomes the victim of the Egyptian objects, killed by the tool that symbolises *his own* archaeological penetration into the mummy's tomb.[8] This emphasises again the instability of notions of identity and the performative acts attributed to them. On a more abstract level, the fact that Bellingham and the mummy's identities are confused points towards the inversion of the subject-object relation in that Bellingham assumes the position of the passive mummy and the mummy that of the active agent. This is an interesting aspect which reappears in several mummy texts, albeit with slight modifications:[9]

8 This again is reminiscent of the *Frankenstein Discourse*, where Frankenstein also suffers from the consequences of his initial act of creating his creature.

9 In general, this subject-object inversion regarding archaeologist and mummy is typical of texts in which mummies are resuscitated and the mummy takes over the role of the agent, whereas the archaeologist often assumes the role of the passive victim. Another variation of this theme is double identities/alter egos that sometimes even coexist, as for example in George Griffith's *The Mummy*

»These fictions of the mummy's curse threaten and often entail the inversion of traditional subject-object relations between those who steal, purchase, gaze at, or study mummies and the embalmed bodies themselves: possessors are possessed, violators are violated, consumers are consumed.« (Bridge 2008: 138)

In most of these texts a character is either in danger of becoming a mummy herself/himself or assumes a mummy-like position or state.[10] What appears

---

and Miss Nitocris (1906), which »identifies itself as a romance of the fourth dimension to open up the possibility of being alive and dead at the same time, a possibility embodied in the resemblance between Nitocris Marmion, the daughter of the physicist, and a mummy« (Seed 1998: 192).

10 One of these texts is Grant Allen's »My New Year's Eve among the Mummies« (1878), in which the protagonist agrees to become a mummy to be able to spend his ›life‹ with an ancient Egyptian princess he has fallen in love with: »He at once acceded to my wishes, and briefly explained the mode in which they usually treated the corpse. That word suddenly aroused me. ›The corpse!‹ I cried; ›but I am alive. You can't embalm me living‹, ›We can‹, replied the priest, ›under chloroform.‹« (*New Year's Eve*: 43) In another short story by C. J. Cutliffe Hyne, »The Mummy of Thompson-Pratt« (1904), Gargrave, Egyptologist at Cambridge, claims that his colleague Thompson-Pratt is the direct descendant of an ancient Egyptian. Gargrave wants to prove this by means of an experiment involving the mummy and Thompson-Pratt. The description of Thompson-Pratt and the mummy is revealing in this context: »Thompson-Pratt and the man who had predeceased him some 3000 years were lying side by side on the hearthrug [...]. [O]nly the dancing, uncanny firelight illuminated the faces; and as I gazed a little closer, a curious thrill went through me. It was Thompson-Pratt's which seemed to be the dead face now. Behind the shriveled mask of the mummy there was surely some flicker of life.« (*Thompson-Pratt*: 227) Three other texts in which the characters assume the position of a mummy are Sax Rohmer's »The Mysterious Mummy« (1903), Humus Fergus's *The Green Mummy* (1908), and Haggard's »Smith and the Pharaohs« (1913). Although the circumstances differ in each of these cases, in all three texts a character assumes the position of a mummy by lying down in a sarcophagus or a mummy case. While in Rohmerd and Haggard's stories the characters eventually are able to leave the sarcophagus again, in Fergus's novel the character in question, a criminal Professor of Egyptology, is shot inside the mummy case where he was trying to hide. Two

to underlie these different representations of a subject-object inversion is the secret fear of turning into the passive object, that is, the mummy, gazed at, evoking the self as the familiar in the other as the strange. This goes in accordance with Freud's assumption that the recognition of the familiar in the unfamiliar is one of the reasons for why we perceive things as uncanny.

Furthermore, however, the dissolution of boundaries and identities as suggested by the inversion and confusion of the subject-object relation can be read as the basic theme of mummy fiction in general. In all of the texts analysed above, the threat for modern society arises through the failure of classification and categorisation, be it through the confusion of identities, the blurring of reality and fantasy, or the resurrection of the past in the present. Underlying all these aspects is the recognition of the familiar in the unfamiliar in connection with the awareness that beyond contemporary society the ›other‹ can no longer be classified and identified. This confirms the function of the archaeological space as a blank space in which, due to the lack of structure, identity becomes arbitrary and thus interchangeable, reflecting the *Jekyll and Hyde* and the *Dracular Discourse*, where identities are essentially instable and floating. In modern society, stability is only possible momentarily if at all, while continuity is revealed as illusionary.

---

other texts in which the characters assume a mummy-like state are *The Jewel of Sven Stars*, where Trelawny sleeps the sleep of a mummy, and George Griffith's *The Mummy and Miss Nitocris* (1906), where the identities of Nitocris Marmion, daughter of a famous Egyptologist, and that of the mummy of the ancient Egyptian princess Nitocris are essentially blurred.

6.4 PHAROS THE ALMIGHTY: THE SUBVERSION OF VICTORIAN GENDER ROLES

> And yet, how monstrous it was! how unheard of! how entirely beyond all bounds of human experience.
> (*LOT NO. 249*: 70)

In the previous chapter I analysed on the inversion of identities in regard to the self and the ›other‹. The present chapter focuses on the subversion of gender identities as characteristic for the archaeological space and prominently presented in Guy Boothby's *Pharos*, where Pharos, an ancient omnipotent Egyptian, appears in modern London to take revenge for the mistreatment of his own mummy by the British. I suggest that in the omnipotence and immortality of Pharos's character, which stands in stark contrast to that of his victim, Forrester, a British bachelor who inherited Pharos's mummy from his father, central discourses of masculinity as associated with rationality and authority are renegotiated, thereby subverting traditional gender identities. This subversion of traditional gender identities marks an increasing loss of the control characteristically represented in mummy fiction, which prominently reflects collective anxieties and destabilisations towards the end of the nineteenth century.

In *Pharos*, Forrester, upon getting involved with Pharos, finds the love of his life in Valerie, ward of Pharos and mysterious violinist, with whom he accompanies Pharos on the latter's journey to Egypt. Although Forrester's departure from Britain marks his ultimate entrance into the archaeological space as the temporal and spatial other, his strong physical reaction upon his first encounter with Pharos already implies his susceptibility to Pharos's influence. Before meeting Pharos consciously for the first time at Burlington House in London, Forrester *feels* him:[1]

1 Forrester and Pharos's first encounter takes place at Cleopatra's Needle on the Victoria Embankment in London in the dark. At this point, however, Forrester still does not know who Pharos is (cf. *Pharos*: 16).

»I had not been alone more than a few seconds, however, before I became conscious of a *curious sensation*. It was accompanied by a *lowering of the pulse* that was quite perceptible, followed by an extraordinary *feeling of nausea*. I *battled against it in vain*. The room and its occupants began *to swim before me*. I *tottered, and at length, being unable any longer to support myself, sat down on the seat behind me*. [...] Approaching me from the crowd, leaning upon his stick, just as I remembered him on the previous occasion [...] was the old man whose personality had given me such a shock at the foot of Cleopatra's Needle.« (*Pharos*: 35-36, emphasis mine)

As Forrester's strong bodily reactions previous to Pharos's appearance show, Forrester is under Pharos's control even before having made his acquaintance. On the one hand Forrester's strong physical reaction in connection with Pharos's arrival can be seen as foreshadowing the development of the strong bond between Forrester and Pharos and eventually also Pharos's infecting Forrester with the lethal plague. On the other hand, the symptoms and feelings described by Forrester suggest a leaning towards male hysteria, which is interesting if we consider his friend William Betford's letter to George Trevelyan at the beginning of the novel:

»Indeed, to my mind [Forrester] had but one fault, a not uncommon one in these latter days of the nineteenth century. I refer to his somewhat *morbid temperament* and the consequent *leaning towards the supernatural* it produced in him. [...] While saying, however, that he had a *weakness for the supernatural*, I am by no means admitting that he was what is vulgarly termed a spiritualist. I do not believe for an instant that he ever *declared himself so openly*. His mind was *too evenly balanced*, and at the same time *too healthy*, *to permit such an enthusiastic declaration* of his interest.« (Ibid.: 10, emphasis mine)

Even though Betford clearly intends to plead for Forrester's sanity and rationality in his letter in order to justify Forrester's extraordinary experiences, the way in which he characterises Forrester in this passage seems to suggest the opposite. Considering Forrester's strong reaction upon Pharos's arrival, the »morbid temperament« (ibid.) and »the leaning towards the supernatural« (ibid.) point towards an instable and nervous character prone to ›unmanly‹ behaviour such as hysteria rather than an »evenly balanced« (ibid.) and »healthy« (ibid.) character. What supports this assumption is that Bedford further writes that Forrester was »too healthy, *to permit* such an *enthusiastic declaration«* (ibid.). This phrase is significant since, although meant to confirm Forrester's mental health, through its use of the verb »permit« (ibid.) it leaves open, if not even calls into question, whether Forrester really is as »evenly balanced« (ibid.) as he pretends to be. It rather suggests that the opposite is the case, as is implied by the fact that he never »*declared* himself so *openly*« (ibid.). All this reveals Forrester as a character with a certain predisposition to the supernatural and the romantic. That archaeology seems to be connected to these feelings is emphasised by Betford's reference to Egypt as the »ancient country and its mystic past« (ibid.) as well as the »history of the ruined cities of Mashona-land« (ibid. 10-11) in connection with his account of Forrester's character.

In addition to his »morbid temperament« (ibid.: 10) and the penchant for the supernatural ascribed to him by his friend Bedford, as a painter Forrester lived the ideal of decadence, as he himself points out: »›I had lived *for* my art and *in* my art.‹« (Ibid.: 68, emphasis mine) Against the background of the Victorian idea of masculinity, all these attributes and dispositions ascribed to Forrester have to be seen as particularly ›unmanly‹[2] in that they violate contemporary concepts of rationality, power, and self-control. Interestingly, while Forrester seems to be forced more and more towards his dispositions of irrationality and morbidity through the events in the novel, Pharos, in contrast, stands out as the embodiment of almightiness and omnipotence. Even though Forrester tries time and again to regain control over both himself and Pharos, he ultimately is unable to do so: »[F]or such a man as Pharos to exist in this prosaic nineteenth century, and stranger still, for me, Cyril Forrester, who had always prided myself on my clearness of head, to believe in him was absurd.« (Ibid.: 170) Forrester's rational unwillingness to accept the existence of figures like Pharos in the modern world is very much reminiscent of what Van Helsing in *Dracula* refers to as »possible impossibilities« (*Dracula*: 193), that is, apparitions and phenomena which according to the Victorian discourse of science and

---

2 Note that in particular male hysteria and morbidity are read by modern critics as indications of homosexuality, i.e., the most unmanly behaviour conceivable at that time: »However, the ›decadent‹ literati of the late 19th and 20th centuries, particularly in France, staged a kind of protest against conservative masculinity; the terms of this protest linked male hysteria, homosexuality and artistic sensibility in a nexus of interdependency.« (Vicari 2012: 54-55)

rationality cannot exist.[3] And yet, both in *Pharos* and *Dracula* the (male) characters finally come to the painful realisation that there are things beyond the explicable of modern rationality. Tragically, however, it is also obvious that both Pharos and Dracula live on the people who believe in them. Both Pharos and Dracula actually verbalise this at the beginning of each novel: »›You are required for a certain work, and for that reason alone I dare not offend you or excite your anger‹« (*Pharos*: 57)[4] and »›[h]ow dare you touch him, any of you? […] This man belongs to me! […] I promise you that when I am done with him, you shall kiss him at your will‹« (*Dracula*: 39). This suggests that as much as Dracula can be seen as stirring sexual passion in Lucy rather than creating it, the reason why Forrester and Harker are victimised and instrumentalised by Pharos and Dracula, respectively, lies in the character of the two men. Symbolically speaking, both Pharos and Dracula can only exist and control Forrester and Harker because the latter believe in them, that is, because men like Forrester and

3 Note that both Pharos and the Count are undead and thus relics of bygone times – they are the animate past in the present. Although both are immortal, they are finally not invincible, as their respective destruction at the end of the novels demonstrates. Pharos and Dracula are themselves victims of their past lives in that they are doomed to a living death due to a flaw or debt incurred during their ›lifetime‹ in a distant past. Furthermore, they are inhuman or superhuman in various senses in that they possess both physical and mental powers that are inexplicable by means of rationality. Additionally, they are not only widely learned and polyglot but also omniscient as far as the movements and thoughts of their victims are concerned. Moreover, their treatment of their victims is characterised by cruelty and sadism, and both Pharos and Dracula show a close affinity to animals in that they are able to control them or are themselves attributed with certain animalistic traits. All these are features that render them independent of any contemporary social or discursive formations in that both Pharos and Dracula exist beyond any restrictions and limitations dictated by Victorian society. Finally, neither Pharos nor Dracula ever eat or drink but live on »some white powder« (*Pharos*: 124) and blood, respectively.

4 Later in the novel Pharos becomes even more explicit when he states: »›[O]ur comedy is at an end, and for the future you are my property, to do with as I please. You will have no will but my pleasure, no thought but to act as I shall tell you.‹« (*Pharos*: 179)

Harker, who conceive of themselves as rational, are actually irrational and prone to »womanly superstition« (*Pharos*: 136). Thus, both Forrester and Harker are emasculated through Pharos and Dracula's influence. This emasculation, however, is only possible beyond the sphere of the ordinary, in the space created around creatures such as Pharos and Dracula, and is due to the hollowness and morbidity of modern masculinity as such, as formulated by Glover in regard to Dracula:

»By vividly dramatizing the horrors of degeneration and atavism, the figure of the Count underscores the sexualised threat that lay at their core, the assumption of ›a sexual ›instinct‹ capable of turning to such perverse or precocious forms as ›homosexuality‹ or ›hysteria.‹« (Glover 1996: 69)

In both *Dracula* and *Pharos*, this threat is emphasised by the various references to male hysteria[5] and illness (cf. Glover 1992: 995). While Forrester has his hysteric fits right at the beginning of the novel and later when he is locked in the pyramid in Egypt, in *Dracula* there are several instances in which men break down in tears and seek comfort in the bosoms of both men and women. Both Forrester and Harker fall ill after their traumatising experiences, Forrester at the Temple of Ammon, where he gets infected with the plague (cf. Chapter 7.4), and Harker at the Count's castle, where he is kept imprisoned. This is significant in that it can be read as reflecting the weakness and instability of contemporary patriarchal society as such,

5 »Hysteria in modern times has also been a *male* malady. The full historical story of the male nervous maladies, however, has yet to be told, and the reasons for this cultivated silence are themselves highly instructive. Since ancient times, physicians, and natural scientists closely observed and extravagantly theorized female weakness, emotionality, and madness. What this long procession of male experts signally failed to see, to acknowledge, and to ponder was the existence of masculine nervousness and mental illness among all social classes and in diverse guises. They did so despite rampant counter-evidence in the clinic and the laboratory, on the streets and the battlefields. The information we do have about this subject comes to us almost exclusively from letters, diaries, memoirs, novels, and autobiographies, which is to say from nonmedical sources.« (Micale 2008: 6) Cf. also Janet Oppenheim's (1991) *»Shattered Nerves«: Doctors, Patients and Depression in Victorian England.*

which, when confronted with these subversive forces, threatens to collapse.[6] This is emphasised by the fact that for a long time it remains unclear whether these experiences are real or only imagined/dreamt, which points towards the mental instability of the characters.

It is also instructive in this context that both Pharos and Dracula see their victims as their property, which they feel free to use for their purposes and to control as they wish. In other words, Forrester and Harker are kept as their prisoners. Though both Pharos and Dracula seemingly offer their victims a chance to leave them, due to some irrational force[7] neither Forrester nor Harker is able to escape from the realm of his ›abductor‹. Significantly, however, it has to be noted that in both cases Forrester and Harker are the ones that, though unknowingly at first, actively seek contact with their future enemy by entering the realms of the ›other‹, be it the Count's castle in Transylvania or Pharos's yacht bound for Egypt. In other words, Forrester and Harker allow themselves to be carried away by Pharos and Dracula, respectively.[8] Against this background, what Forrester and Harker

6 Note that both Pharos and Dracula use the life force of their victims to nourish themselves, Dracula by drinking Lucy and Mina's blood, Pharos by bereaving Forrester of his energy. This is revealed when Forrester perceives a change in Pharos after the former has been infected by the latter with the plague: »It struck me that [Pharos's] step was more active than I had yet seen it.« (*Pharos*: 229)

7 Forrester feels connected to Pharos by an invisible bond (cf. *Pharos*: 57), and although the wolves around Dracula's castle seem to contribute to Harker's inability to escape from the castle, the latter seems very much determined by an *irrational feeling* of impotence.

8 Interestingly, as far as the method of spreading their ›disease‹ is concerned, Dracula and Pharos employ a similar strategy in adulterating modern Britain in that in a sense they both use a host to insert their ›diseases‹ into society. While in *Pharos* this role is appointed to Forrester, whom, once infected with the plague, Pharos leads to the various places of London's social life in order to contaminate the masses, in *Dracula* the Count uses first Lucy and later also attempts to use Mina as a host for his virulent vampire ›disease‹. Since in both novels the ultimate aim of the invader is to colonise Britain, i.e., to bring it under his control, the Count's choice to attack the ›weaker sex‹ to ultimately destroy the whole of patriarchal society seems very clever and at the same time renders the danger coming from the Count even more threatening. If we con-

experience in the further course of events can be read as the fulfilment of their unconscious longings, which are excluded from the Victorian discourse but become possible in the archaeological space. These longings can be essentially characterised as comprising all experiences, practices, and emotions which were generally proscribed and forbidden as ›unmanly‹. In both *Pharos* and *Dracula* men lose control and are dominated by and become addicted to irrationality – allegedly because they are confronted with larger-than-life creatures in whose presence they are ›allowed‹ to perform in ›unmanly‹ ways.[9]

---

sider now that Dracula lures women by appealing to their own latent sexuality, the very fact that Forrester falls prey to Pharos is illuminating, since as shown above, Forrester is attracted to Pharos because he seems to fathom his own suppressed feelings and emotions in him.

9 Mummy fiction in general seems to be particularly suitable for figures like Pharos, who dominate and control male characters. Another prominent example is Loudon's scientific romance *The Mummy! A Tale of the Twenty-Second Century*, which, though originally published already in 1827, became popular again in the second half of the nineteenth century and was both reprinted as a two-shilling yellow-back edition by Frederick Warne in 1872 (cf. Frost 2008: 4) and reissued in serialised form in the popular weekly *Dicks' English Library of Standard Works* in 1887. In this novel, twenty-second century London is terrorised by the omnipotent mummy of Cheops, who has been resuscitated from his tomb by two Englishmen and escaped from Egypt to invade Britain in a motorised balloon. In the face of Cheops's omnipotence, Father Morris, portrayed in the novel as an extremely rational, cold-blooded, and sarcastic priest character, is terrified and, in much the same way as Forrester, is bereaved of his masculinity in contact with Pharos, losing both power and control, which are essentially considered male attributes: »›Even now he crossed my path and bade me beware, or I should become *his slave*[;] […] *even I trembled* […] when I gazed upon the countenance of that *tremendous being*, and read there the *traces of fierce and ungoverned passions, wild and destructive* in their course *as the raging whirlwind. Even I dreaded the influence he might exert upon our destinies*, and *shuddered at the thought of such a creature's being released from the fetters of the tomb*, and sent back as a destroying spirit upon the earth. The eternal gloom which hangs upon his brow, seems to bespeak a fallen angel, for such is the deadly hate that must have animated the rebellious spirits when expelled

These examples, in which an ›emasculation‹ of a character is caused by his contact with a mummy or another creature from the past like Pharos and Dracula, show how the latter are instrumentalised to function as a subterfuge for the male characters to make possible and justify experiences which under normal circumstances were unheard of.[10] In mummy texts, the archaeological sphere thus offers the space necessary for the deconstruction of the traditional gender roles that dominate heteronormative Victorian and Edwardian society:

»The heterosexualization of desire requires and institutes the production of discrete and asymmetrical oppositions between ›feminine‹ and ›masculine‹, where these are understood as expressive attributes of ›male‹ and ›female‹. The cultural matrix through which gender identity has become intelligible requires that certain kinds of ›identities‹ cannot ›exist‹ – that is, those which gender does not follow from sex and those in which practices of desire do not ›follow‹ from either sex or gender.« (Butler 2007: 24)

---

from heaven. *His look is terrific*; and *my blood froze in my veins at his horrid laugh*, which seemed to ring in my ears like the mockery of fiends when they have involved a human being inextricably in their toils.‹« (*Mummy* Vol. II: 23-24, emphasis mine) Another example of the ›emasculation‹ of a character is E. and H. Heron's »The Story of Baelbrow« (1898), in which a mansion is haunted by a vampire mummy. Here the owner of the mansion, Mr Swaffan, poses the following question to the detective assigned to investigate the mummy attacks at his house: »›Do you imagine there is any danger – I mean *to men like ourselves*? *Hysterical women cannot be taken into serious account.* [...] [A]s a mere matter of *common sense*, I must express my opinion plainly. I believe the whole thing to be *the result of excited imagination*, and I am about to prove it.‹« (*Baelbrow*: 88-89, emphasis mine) It is only a little later that Swaffan himself is attacked by the mummy and losing consciousness displays the ›hysterical‹ behaviour he shortly before scornfully ascribed to women. His behaviour thus resembles that of Forrester and Harker's above-mentioned hysteria in contact with Pharos and the Count, respectively.

10 Other examples are the ancient Egyptian priest Savak in *Iras* and the mummy in *Nemesis of Fire*.

Only in the presence of Pharos and Dracula's all-embracing power are men like Forrester and Harker able to abandon conventional patterns of behaviour and slip into the role of the subjugated, which in Victorian society was designated for women. In Pharos and Dracula's spheres, however, which are beyond everyday reality, ›womanly‹ behaviour for men seems possible and to some extent even acceptable, since Pharos and Dracula take over the ›male‹ role. This is emphasised by the following passage, in which Forrester's perception of Pharos has a strong sexual connotation:

»As [Pharos] said this his diminutive form seemed to *grow larger* and *more terrible*, until *it appeared to have attained twice its ordinary size*. His eyes shone in his head like living coals, and seemed to burn into my brain. I saw Valeric rise from the place where she had hitherto been crouching, and snatch an Oriental *dagger* from a table. Then, swift as a panther, she sprang upon him, only to be *hurled back against the wall as if struck by an invisible hand*. Then, *obedient as a little child, I closed my eyes and slept.*« (*Pharos*: 358, emphasis mine)

The immense growth of Pharos's form to »twice its ordinary size« (ibid.) likens him to a phallus, which is further underscored by the fact that Forrester experiences him as even more terrible than before. The very fact that Valeric's attack with the dagger, which can be read in this context as symbolising ›conventional‹ male sexual power, is warded off without any difficulty by Pharos, emphasises his extraordinary virile power. In the face of such an overpowering display of masculinity, the only means of escape for Forrester is to close his eyes and sleep like a little child, and thus enter a state which is equal to a ›womanly‹ faint.

That Pharos's behaviour, very much like that of Dracula, is homosexually charged[11] is further underlined when Pharos tells Forrester how he managed to infect him with the plague: »›In *obedience to my desire* you followed me to Italy and accompanied me thence to Egypt. It was I who

11 Cf. »This should remind us that the novel's opening anxiety, its first articulation of the vampiric threat, derives from Dracula's hovering interest in Jonathan Harker: the sexual threat that this novel first evokes, manipulates, sustains, but never finally represents is that Dracula will seduce, penetrate, drain another male.« (Craft 1984: 109-110) Cf. also Talia Schaffer (1994) »›A Wilde Desire Took Me‹. The Homoerotic History of Dracula«.

drew you to the Pyramid and *decreed* that you should *lose your way inside*, in order that when *fear had deprived you of your senses I might inoculate you with the plague.*‹« (Ibid.: 357, emphasis mine) It is significant that Pharos lures Forrester into the pyramid, where he waits for him to lose his senses to then molest him as an aboulic and helpless man by infecting him with the plague. Both the vocabulary used in this passage and the very fact that Forrester only later learns that he has been contaminated, that is, inseminated, with the plague by Pharos, liken the act of infection to the rape of a swooning female or male.[12] The association of Pharos's infecting Forrester through a rape-like act is crucial in this context if we consider the other symbolic rape in the novel: the one committed by Forrester's father in the act of violating the ancient tombs of Pharos's *mother* country (cf. Chapter 7.4). In the light of Forrester's ›hereditary guilt‹ for the ›rape‹ committed by his father, Pharos's ›rape‹ of Forrester at the ancient site can be read as an act of retribution for what Pharos and his country have suffered through modern men. This is further emphasised when Pharos reveals to Forrester that the latter is actually the host of the lethal plague and thus responsible for the contamination of Europe: »›In Constantinople, Vienna, Prague, Berlin, Hamburg, wherever you went you left the fatal germs of the disease *as a legacy behind you.*‹« (Ibid., emphasis mine) That Pharos refers to the act of spreading the plague as a »legacy« (ibid.) Forrester leaves behind him is significant, since it once more points to the fact that the plague is used by Pharos as a vengeance on Forrester, who has inherited the ›hereditary guilt‹ in place of Victorian society as a legacy from his father. The legacy Forrester has received in the form of the plague from his father is now being spread by him and thus shared with those others responsible for the violation of Pharos's (ancient) Egypt.

»At once almost tangible and yet separated from the present by the gulf of centuries, these ancient figures were visibly in perpetual exile from their own lives. And the fascination of the audience of that time was, inevitably, shot through with morbid,

12 Cf. the scene in *Dracula* where the female vampires (try to) rape Harker and are interrupted by the Count, who rebukes them: »›How dare you touch him, any of you? How dare you cast eyes on him when I had forbidden it! Back, I tell you all! *This man belongs to me*!‹« (*Dracula*: 38, emphasis mine)

erotic pleasure, granted an aura of respectability by the scientific and archaeological discourses in which the exhibition took place.« (Pearce 2002: 58)

A further aspect that is important to Pharos's act of infecting Forrester in the pyramid is the place itself and the condition in which Pharos abuses Forrester. Both seem to be necessary preconditions for the ›rape‹. First of all, for Pharos it is essential that Forrester loses both his manliness and his senses before he can lay hand on him:

»At last, *thoroughly beside myself with terror*, I began *to run aimlessly about the room in the dark beating myself against the walls and all the time shouting at the top of my voice for assistance.* Only when I had no longer strength to move, or voice to continue my appeals, did I cease, and *falling upon the ground rocked myself to and fro in silent agony*. [...] [T]he horrible silence, the death-like atmosphere, the flapping of the bats in the darkness, and *the thought of the history and age of the place in which I was imprisoned*, must *have affected my brain*, and for *a space I believe I went mad.* [...] Then my *senses left me, and I became unconscious.*« (Ibid.: 177-178, emphasis mine)

This extraordinary situation of being lost alone in the darkness of the pyramid again excites hysterical behaviour in Forrester, which results in him losing consciousness. The fact that Forrester refers to himself as being »beside [himself] with terror« (ibid.: 177) suggests that he is not *himself* but rather someone or something else. This inkling is further supported by his ›womanly‹ behaviour, characterised by fits of hysteria, madness, and swoon, which makes the whole passage reminiscent of a stereotypical scene in a Victorian mental asylum for women. Thus, Forrester's initial comment of being »beside himself« (ibid.) can also be read as an excuse for and justification of his subsequent ›unmanly‹ behaviour.

Interestingly, Forrester mentions »the history and age of the place« (ibid.: 178) as being particularly responsible for the maddening effect the whole situation has on him. This is illuminating if we further consider the following thoughts of Forrester on the event of his breakdown in the pyramid: »For all I knew to the contrary, I might have wandered into *some subterranean chamber never visited by the Bedouins or tourists*, whence my feeble cries for help would not be heard.« (Ibid.: 177, emphasis mine)[13] That Forrester perceives the pyramid chamber as a place that might never have been visited before by Bedouins or tourists, that is, by people of the present time, suggests that he at least speculates on finding himself in a place beyond the present time and space.[14] In this context, Forrester's descent into the subterranean chamber of the pyramid becomes a spatiotemporal trespassing into another sphere, where he is confronted not only with the terrors of the past in the form of Pharos, but also with the terrors of his own present self in the form of suppressed emotions and fantasies. Thus, at an historic site, which until this point has not been exploited by modern contemporaries and therefore constitutes an unknown sphere, both the past and the present turn into a threat. This is underscored by both Forrester's later infection with the plague within the pyramid and by his fits of hysteria, madness, and swoon. Consequently, both Forrester's masculinity and sanity are threatened from within himself once he has entered the archaeological space. That this loss in the »total darkness« (ibid.: 175) of one's own self might be final is underscored by Forrester's strong feelings in the pyramid: »Fighting down the terror that had risen in my heart and *threatened to annihilate* me, I once more commenced my circuit of the walls, but again without success.« (Ibid.: 177, emphasis mine)

While so far the focus has been on Pharos's ›male‹ qualities and their effeminating effect on Forrester, it is crucial to note that there is also a more ›unmanly‹ side to Pharos's character. This side is characterised by spiritual and instinctive knowledge, as referred to by Forrester's thoughts about Pharos upon their nocturnal arrival in the English countryside: »To this day I have no notion whether he was acquainted with the country, or whether he was merely trusting to chance and his own peculiar instinct to

13 This is further underscored by Forrester's thoughts upon returning to the Temple of Ammon a week later: »In a vague fashion I wondered how it was that these rooms had never been discovered by the hundreds of Egyptologists who, since the time of Napoleon, had explored the temple. […] I had studied my guidebooks carefully on our voyage up the river, and was quite convinced that *no mention of such places had been made in any one of them.*« (*Pharos*: 215, emphasis mine)

14 Eventually this is confirmed when Forrester shortly afterwards regains consciousness and finds himself in the past (cf. ibid.: 181-183).

bring him out at a railway station.« (Ibid.: 311) The »peculiar instinct« (ibid.) of Pharos mentioned here is particularly interesting if we consider that frequently in the novel Pharos's appearance and behaviour are likened to those of animals, such as his »cat-like spring« (ibid.: 69) or his »wolfish teeth« (ibid.: 70). Pharos's affinity with animality, however, becomes most explicit through his pet monkey Pehtes, with its »evil-looking little face« (ibid.: 125). Significantly, Pharos himself introduces the monkey to Forrester as follows: »›Let me make you acquainted with *my second self*‹, he said, and then, turning to the monkey, continued, ›Pehtes, make your salutation.‹« (Ibid.:, emphasis mine) Pharos's affinity with the monkey as well as his instinctual knowledge and his beastlike behaviour reveal a very strong animalistic trait in his character, which in more general terms can be read as representing the stereotypical images of both the Orient and the past, functioning as a projection screen for Western dreams and fantasies, as paradigmatically pointed out by Said: »The Orient was almost a European invention, and had been since antiquity a place of romance, exotic beings, haunting memories and landscapes, remarkable experiences.« (Said 1978: 1) This is further emphasised by the drugs Pharos offers Forrester on several occasions: »Little by little the subtle intoxication of the weed was permeating my whole being; a gentle languor was stealing over me, and as a result my brain had never before seemed so bright or my capacity of enjoyment so keen as it did then.« (*Pharos*: 127) Significantly, it is also by means of an unknown drug which Pharos forces Forrester to drink that Forrester is carried back through the centuries into the time of Ramesses II:

»The opiate, or whatever it was, must have been a powerful one, for I had scarcely swallowed it before an attack of giddiness seized me. The outline of the Sphinx and the black bulk of the Great Pyramid beyond was merged in the general darkness. I could hear the wind of the desert singing in my ears, and the voice of Pharos muttering something in an unknown tongue beside *me*. After that I sank down on the sand, and presently became oblivious of everything.« (Ibid.: 180-181)

When Forrester wakes up, he finds himself in ancient Egypt, where he witnesses the vilification of Pharos's ancient double, Ptahmes.

In regard to Pharos's character, both his affinity with animals and the aforementioned passages are revealing in that they show another side to Pharos's character apart from his prominent superhuman virility, in the form of physical and mental strength. On the one hand this other side of Pharos can be read as the embodiment of the sensuality and lure commonly associated with Oriental femininity, but also in particular with ancient Egypt as such. Pharos stimulates and appeals to Forrester's senses by means of the drugs, which are not only mind-expanding and sensualising but also capable of transporting Forrester into a different time. In other words, the drugs work in much the same way as the luring female in that they beguile Forrester and draw him into their power. On the other hand, Pharos's animalistic and brute behaviour can also be seen as pointing towards the common contemporary perception of the Orient and the past as uncivilised and primitive in comparison with modern Western civilisation.

Taking this into consideration, it becomes clear that Pharos embodies not only a superhuman stereotypical masculinity but also a strong stereotypical femininity in that his effect on Forrester is reminiscent of that of a luring Oriental female.[15] Furthermore, Pharos's character also shows a beastlike inhumanness, suggested both by his cruel behaviour and by his affinity with animals, which can be read as representing the alleged primitive state of the Middle East as opposed to the West. In this context, Valeric's remark in regard to Pharos is telling: »›I often wonder, Mr. Forrester, whether he can be human.‹« (Ibid.: 158) After all we have learned about Pharos by this point, he is certainly not human according to the contemporary Victorian definition of humanness. His character is an accumulation of *extra*ordinary abilities and features that all essentially deviate from Victorian norms of humanness, gender, and morality. »Could it be possible that such power was permitted to human beings? No, no – a thousand times no! If he had that influence he must be an agent of the Evil One, whose mission it was to draw to perdition the souls of helpless men.« (Ibid.: 68) The very

15 Moreover, quite similar to what Rauch pointed out in regard to Cheops in *The Mummy!*, Pharos also displays motherly qualities when he nurses Forrester after having infected him with the plague: »Thus, in spite of his gender, the Mummy acts as a kind of mother figure throughout the novel. [...]. Unlike ›mommy‹ though he may be, the Mummy's moral stance, ›to assist the good and punish the malevolent‹, has everything to do with nineteenth-century conceptions of parental responsibility.« (Rauch 1994: xxv) The combination of Pharos's erotic and motherly qualities once more underlines the ambivalence of the archaeological space.

fact that Forrester poses the question as to whether it is »*possible* that such power was *permitted* to *human beings*« (ibid.), in combination with his strong negation of this question, is revealing in this context. Significantly, Forrester, rather than feeling threatened by Pharos's power, seems to be more disturbed by the fact that such power is *permitted* to a human being. Unable to accept that this might be the case, Forrester displaces this discomforting idea by bereaving Pharos of any humanness by means of religious classification schemes that brand Pharos as an agent of the devil. As much as Satan's realm is a parallel universe to the world of mankind, Pharos's deviation from the prevalent norms also means that his existence can be placed beyond contemporary discourses. As a relic of the past, he can be perceived by people in the present. However, apart from the fact that he is conceived as evil, he cannot be classified and categorised and remains essentially amorphous. In analogy to the *Jekyll and Hyde Discourse*, where good and evil are essentially gradual and part of the same thing, Pharos can thus be seen as subverting binary categories such as heteronormative gender constructions.[16] It is this elusiveness of Pharos which renders him both so powerful and threatening to the contemporary Victorian discourse.

16 Again this is reminiscent of the Count in *Dracula*, who changes his form several times in the novel and thus can also be conceived as amorphous. Thus, both Pharos and Dracula underscore what MacFarlane has pointed out in regard to the figure of the mummy: »The ways in which the mummy can undermine the certainty of modern scientific texts are played out in the contrast between the *indeterminate figure* of the mummy and comprehensive knowledge.« (MacFarlane 2010: 20, emphasis mine)

6.5 MARGARET TRELAWNY: THE ›OTHER‹ WOMAN

›Can't you see for yourself I am a re-incarnation?‹
(*MIRACULOUS EXPLORER*: 102)

While in the previous chapter I discussed the subversion of traditional gender concepts with regard to Forrester and Pharos's relationship in *Pharos*, the present chapter analyses Stoker's *Jewel* in this respect. Focusing on Margaret Trelawny, I will show how her character can be read as tresspassing contemporary gender notions, oscillating between conventional and unconventional conceptions of femininity. Similar to Douglas's *Iras*, where Iras is introduced to the novel in her absence, the first encounter with Margaret is an indirect one through Malcolm Ross's dream. The reader is confronted with Ross's dream *in medias res*, which although told by him in retrospect – as indicated by the past tense – has a strong sense of immediacy. The dream recapitulates an earlier encounter between Ross and Margaret on a solitary boat trip including Ross's perception and appraisal of Margaret. Central to Ross's perception of Margaret is her »narrow up-bringing« (*Jewel*: 7), her isolation, and her distant relationship to her father:

»Again, in that blissful solitude the young girl lost the convention of her prim, narrow up-bringing and told me in a natural, dreamy way of the loneliness of her new life. With an undertone of sadness she made me feel how in that *spacious home each one of the household was isolated by the personal magnificence of her father* and herself; that there *confidence had no altar*, or *sympathy no shrine*; and *that there even her father's face was as distant as the old country life seemed now.*« (Ibid., emphasis mine)

Significantly, Margaret is said to recount her new life in a »natural, dreamy way« (ibid.), which seems to be somehow contradictory in itself, as »natural« (ibid.) evokes the notion of reality while »dreamy« (ibid.) points towards the realm of fantasy. Again, it is this ambivalence between reality and fantasy, one of the main themes in the novel, that is also prominent in

Margaret's character and closely links the text to the *Dracula Discourse*, where boundaries between reality and fantasy are continuously blurred. As depicted in Ross's dream, Margaret, who has only recently come to live with her father, suffers from loneliness and the dominance of her father, whom she does not really know. Accordingly, it only seems natural that Margaret contacts Ross, who seems to be her only confidant, though she barely knows him, when her father is mysteriously attacked in his bedroom in the middle of the night: »›You said you would like to help me if I needed it; and I believe you meant what you said. The time has come sooner than I expected. [...] Come at once, if you are able to [...]. I suppose I shall realise later what I have done in asking such a favour; but at present I cannot think. Come! Come at once!‹« (Ibid.: 9) When Ross, who immediately sets out for the Trelawny house, meets Margaret in the hall, we get a first impression of Margaret which points towards her ambivalent character, foreshadowed in Ross's dream. On the one hand we learn that »[s]he was not in any way shy [and that] [s]he seemed to rule all around her with a sort of high-bred dominance« (ibid.: 11), while on the other hand »she was greatly agitated and pale as snow« (ibid.), »impulsively« clasping Ross's hand (ibid.). Furthermore, Ross notes that »Miss Trelawny's hand somehow became lost in [his] own« (ibid.), which he perceives as an »unconscious self-surrender« (ibid.). Ross's perception and depiction of Margaret upon this occasion are significant, since they point towards Margaret's *dual existence* as gradually evolving in the novel, anticipating the *Jekyll and Hyde Discourse*, which is characteristic of many mummy texts of the time.[1] As Dr Jekyll turns more and more into Mr Hyde, the characters in these stories tend to (gradually) lose their modern identities to become increasingly like their former selves prior to their reincarnation in the present. However, though usually the characters, in analogy to the *Jekyll and Hyde Discourse*, experience the recognition of their (ancient) alter ego as liber-

1 This is popularly reflected by the reincarnation motif prominent in Grant Allen's »The Miraculous Explorer«, Guy Boothby's »A Professor of Egyptology« and *Pharos, the Egyptian*, Bram Stoker *Jewel of the Seven Stars*, George Griffith's »The Lost Elixir« and *The Mummy and Miss Nitocris*, C. J. Cutliffe Hyne's »The Mummy of Thompson-Pratt«, Lilian Bagnall's »In the Tombs of the Kings«, Rider Haggard's »Smith and the Pharaohs«, and Sax Rohmer's »The Cat«.

ating, it commonly has a disturbing effect on their surroundings, as pointed out by Deane:

»The modern British characters' discovery that they are really ancient Egyptians, like the projection of imperial power onto immortal queens, threatens to dissolve many of the oppositions that typically structure British identity: science/magic, Christianity/paganism, rationality/superstition, modernity/antiquity, colonizer/colonized, and at times, masculinity/femininity.« (Deane 2008: 402)

Significantly, however, it is the tension of the not yet dissolved opposition that is experienced as disturbing and underlines the ambivalence characteristic of the archaeological space. The anticipated inversion of identities of the present and ancient self is also very strong and thus uncanny in Margaret's character.

Characteristically, Margaret »seem[s] to rule all around her with a sort of high-bred dominance« (ibid.), which reveals the character of the queen she eventually turns out to be in part as the double of the Egyptian Queen Tera, whose mummy is in possession of Margaret's father.[2] In spite of her intrinsic queenly charisma, Margaret's *behaviour* towards Ross suggests helplessness and compliance, which culminates in Ross's interpretation of her behaviour as an »unconscious self-surrender« (ibid.). This is significant insofar as it suggests not only that there are two sides of Margaret's character but also that they are essentially different modes of *re*action towards her surroundings. These modes of reaction are dependent on how Margaret is judged by her counterpart and on the situation she is confronted with.

2 Margaret's alter ego emerging in the form of Queen Tera's powerful character in Victorian society stands in stark contrast to the submissive ›angel of the house‹ female represented by Iras, which Daly refers to as an »anti-New Woman« (Daly 1994: 41): »›To have one belonging to me, depending on me, how sweet the possession.‹« (*Iras*: 90) Lavenham's tenure is confirmed by the passionate exclamation of Iras, the recovered female: »I am not dreaming! I am awake at last; and it is you – *my lord*« (ibid.: 78, emphasis mine) and later on »[o]h *my master*, is it true I am awake at last?« (ibid.: 89). However, that this type of femininity is considered outdated is suggested by Iras's premature death at the end of the novel, which is anticipated *inter alia* by Lavenham's perception of Iras as having »*the air of a woman in a decline*« (ibid.: 154, emphasis mine).

Furthermore, while the strong, queenly, and independent part of Margaret's character is more subtle, her weaker, submissive, and helpless part is more openly expressed and frequently emphasised through Ross's corresponding interpretation of her verbal and non-verbal behaviour.

Although the identification of Margaret as Queen Tera's alter ego is acknowledged by most critics (cf. Glover 1995: 7; Byron 2007: 60), the interpretation of its function in regard to Margaret's character varies.[3] Hence, as Margaret's alter ego of the past, Queen Tera embodies a time-transcending and powerful femininity like that ascribed to the New Woman towards the end of the century (cf. Glover 1996: 90).[4] Unlike Margaret's traditional and submissive femininity, and in analogy to Jekyll's alter ego Hyde, however, Tera's femininity has no place in contemporary society. As Glover points out, Tera's portrayal in *Jewel* »directs us towards Stoker's own ambivalence regarding the changing status of women in an era of growing feminist agitation. Instructively, Queen Tera gives us the triumph of womanhood in a world of men« (Glover 1995: 7-8). Margaret's two parts and the behaviour associated with each part can thus be read as revealing the cultural context as determining in regard to her performance rather than making Margaret's ›nature‹ responsible for it.[5] This once more

3 The danger associated with Tera is emphasised by Byron: »Finally, as many critics have noted, Margaret Trelawny contains two anagrams of Tera, most notably the last four letters of her first name reverse the name Tera. A little more speculative play with names yields some potentially significant results. Tera has its roots in the Greek *teras* or monster, and the name Margaret read backwards becomes a teragram. As ›gram‹ placed at the end of a word is a prepositional compound denoting a thing written or recorded, ›Margaret‹ could then be nothing more than the site for the recording of the monstrous, the site, once again, of both terror and Tera.« (Byron 2007: 61)

4 Cf. Glover (1995). Commenting on Stoker's *Dracula*, Hopkins (1998) states that »[c]ritical response to *Dracula* in particular has identified two possible motivations for the deep-seated ambivalence in Stoker's portrayal of women: his attitude to his own wife and mother, and his uneasiness at the figure of the ›New Woman‹, about whom Mina jokes more than once during the course of *Dracula*« (Hopkins 1998: 146). Cf., in this context, Senf (1990).

5 This corresponds to Butler and Storey, who emphasise the performative character of identity, as opposed to considering it as ›naturally‹ given.

reflects the potential of the archaeological space for deconstructing conventional identities and alludes to the underlying (male) fear »that women cannot be properly objectified: that they are not as they appear to be« (Smith 2004: 86).

One of the few instances in which the double nature of Margaret's character surfaces is in Ross's flashback of their first encounter at a ball in Belgrave Square:

»A queenly figure! Tall and slim, bending, swaying, undulating as the lily or the lotus. Clad in a flowing gown of some filmy black material show with gold. For ornament in her hair she wore an old Egyptian jewel, a tiny crystal disc, set between rising plumes craved in lapis lazuli. [...] For all her gracious bearing towards me, when our hostess introduced me, I was then afraid of her. It was only later at the picnic on the river, I had come to realise her sweet and gentle nature, that my awe changed to something else.« (*Jewel*: 64)

Tellingly, here the »queenly figure« (ibid.) and appearance of Margaret frightens Ross in spite of her »gracious bearing« (ibid.) towards him, whereas once alone with her on the river he discovers her »sweet and gentle nature« (ibid.). Although it is left open as to what Ross's »awe change[s] to« (ibid.) upon his second encounter with Margaret, the interesting point is that Ross's perception of Margaret changes once they have left a certain social circle. As long as Margaret is in society, she remains distant and aloof, embodying with her graceful bearing Queen Tera, while when in close and intimate contact with Ross she turns into the amiable Margaret. With regard to these two different types of femininity, represented by Queen Tera and Margaret, respectively, this can be read as pointing towards the fact that the realisation of the strong and independent female image embodied by Tera is only possible in an isolated distant past, as part of the archaeological space, while in order to exist in modern Victorian society it is necessary for Tera to turn into the »sweet and gentle« (ibid.) Margaret. This is underscored by the fact that Ross perceives Margaret as the »queenly« (ibid.) Tera as long as he watches her from a distance, while he believes to have discovered her true and more feminine character once he is in close contact with her and she thus becomes approachable to him. If one reads Margaret's change in behaviour against the background of the contemporary patriarchal and heteronormative discourse of Victorian society with its fixed binary oppositions of sex and gender roles, in contact with Ross Margaret cannot but assume a traditional ›female‹ identity, since the only other possible identity, the ›male‹, is already occupied by Ross. In other words, in order to make an *inter*action with her nineteenth-century surroundings possible, Margaret has to comply with certain behavioural standards and norms, which she is independent of as long as she is her alter ego, which is associated with the archaeological space.[6]

Thus, to get into contact with her fellow human beings, and in particular with her fellow *men*, Margaret has to abdicate her »queenly« (ibid.: 64) and »high-bred dominance« (ibid.: 11) and abandon herself to »unconscious self-surrender« (ibid.):

»I went over to the sofa and looked down at the senseless man. [...] As I looked and wondered, there began to steal over me again that phase of wandering thought which had last night heralded the approach of sleep. I resisted it, and *held myself sternly to the present. This was easier to do when Miss Trelawny came close to me, and leaning her forehead against my shoulder, began to cry silently.* Then *all the manhood in me woke*, and to present purpose.« (Ibid.: 47, emphasis mine)

This passage underlines the negative power of the past and sleep as the two most dangerous threats to the characters' physical and psychological well-being in the novel. It is crucial that Ross feels stronger and manlier once he knows Margaret *close* to him crying on his shoulder, since it underscores the importance of Margaret's behaviour for sustaining the present *for* Ross, and thus protecting him against the evil forces of the past. Thus, as mentioned above, Margaret's closeness means safety to Ross: first, because she then displays a more helpless and dependent femininity which restores Ross's feeling of masculinity, and second, because she is more approachable and thus easier to grasp and control for Ross.

Nonetheless, the hidden danger which is immanent in Margaret's character repeatedly surfaces and grows more prominent as the novel proceeds,

6 This is further underlined by other powerful women inhabiting the archaeological space, e.g., Queen Ma-Mee in »Smith and the Pharaohs«; Queen Claudia in *The Mummy!*; Queen Ayesha in *She*; Queen of *N'Shabé* in *A Romance of N'Shabé*; Queen Nazir in *The City of Gold*; the legendary Queen Sheba referred to in many texts dealing with Zimbabwean archaeology.

as suggested when Ross states that »[t]here was something in her voice so strange to me that I looked quickly into her eyes. They were bright as ever, but veiled to my seeing the inward thought behind them as are the eyes of a caged lion« (ibid.: 215). This underlying ›other‹ behind the ordinary self points towards the *Jekyll and Hyde Discourse*, where Hyde gains more and more control over Jekyll. That Ross only looks »quickly« (ibid.) into Margaret's eyes forebodes his fear and insecurity upon realising this »strange« (ibid.) »something« (ibid.) he cannot grasp in Margaret's voice. As the simile of the »eyes of a caged lion« (ibid.) indicates, caution is indeed required, since these eyes suggest Margaret's unpredictable character, which underneath its domesticated surface is wild and powerful. The fact that Margaret's »inward thought« (ibid.) remains »veiled« (ibid.) to Ross points towards the impossibility of grasping and controlling Margaret's hidden thoughts and feelings.[7]

7 Another text which presents the character of a woman oscillating between two selves is Sax Rohmer's short story »The Cat« (1914). In this story, the Edwardian protagonist Inez Durward, whose father is a famous professor of Egyptology, is about to decide whether or not to spend her life with Sir Marcus Auckland and marry him: »[Sir Marcus] touched [Inez's] hand as it rested upon the piano. She did not withdraw it, and he clasped it in his own. Still she sat looking up at him with her big, mysterious eyes. In them he could read no answer, no invitation, yet no actual rebuff. He had thought to find her in a different mood, but tonight the mystic phase of her nature asserted itself – the phase he did not understand, had never understood.« (*Cat*: 79) As Ross does in *Jewel*, Sir Marcus here perceives the strange and unfamiliar side of the woman which he does not understand because it is different from what he expects from her as a woman. This is emphasised by Inez's reaction: »›[P]lease do not think there is anyone else, nor imagine that I am trifling with you; but I wish – *I suppose I am odd and queer* – Not a great thing, or perhaps you may think it *is* a great thing.‹« (Ibid.: 80, emphasis mine) What Inez finally asks of Marcus is to give her two more days to find out for sure whether she is ready to become his wife. While Sir Marcus at first feels »ashamed [and] unworthy, as he always [does] when that strange, heatless flame burned in her eyes« (ibid.: 81), he recovers once Inez laughs and touches his arm: »Instantly she was become the Inez he knew, the Inez he understood and *dreamed of possessing.*« (Ibid., emphasis mine) This is significant, since like Ross in *Jewel* Sir Marcus is attracted to one side of Inez

In *Jewel*, apart from the double identity of Margaret finding expression in her alter ego Tera, Margaret's conflicting parts are symbolised by her cat Silvio on the one hand and the cat mummy in Mr Trelawny's study on the other.[8] While the first is described by Ross as »a magnificent animal[,] [a] chinchilla grey Persian with long silky hair; a really lordly animal with a haughty bearing, despite his gentleness; and with a great paw which spread out as he placed them on the ground« (*Jewel*: 31), the true nature of the

– the one he wants to »possess« (ibid.) – while he is repelled by her mysterious and inexplicable side. Similar to what happens in *Jewel,* distance and closeness seem to influence which part gains the dominance in Inez's character, as is emphasised by the fact that she becomes the familiar Inez once she has physical contact with Sir Marcus, while she literally remains distant as long as they do not touch each other. During the time following this interview, the reader repeatedly witnesses what Inez and Sir Marcus have already been feeling for quite a while, namely »›that [Inez is] ruled by two distinct intelligences‹« (ibid.: 80-81). When Inez has to inspect a newly arrived mummy for her father at a befriended antiquarian dealer's, she learns from the latter the mummy is that »of Sekhar, a priestess of Bast or of Sekhet[,] […] [who] symbolised passion« (ibid.: 87). When the sarcophagus is opened a black cat leaps out, which follows Inez home and turns out to be particularly attached to the woman, while it shows a strong aversion against Sir Marcus and even attacks him. Eventually, it is through the cat that Inez finds out about Sir Marcus's fraudulent intent, as he only wanted to marry Inez for financial reasons. Thus, similar to Tera in *Jewel,* the mystical part of Inez, which turns out to be that of the priestess of Sekhet, functions as the emblem of an independent and powerful femininity. The existence of this femininity, which is rejected by the male characters in both texts and finally even turns out to be a threat to their life concepts, is only possible in the archaeological space.

8 The power and strength associated with cats in both *Jewel* and »The Cat« was probably adopted from the Egyptian worship of the goddess Bastet the Goddess of Cats, Lower Egypt, the sun and the moon (alias Bast, Baast, Ubasti, and Baset): »Amongst the ›sacred animals‹ that were mummified and buried in vast quantities by the ancient Egyptian during the last century of their long history, cats held a special place and were accorded a special respect.« (Zivi/Lichtenberg 2005: 106)

latter is only revealed close to the end of the novel, when Mr Trelawny unwraps the cat mummy:

»As the last wrappings came away, we saw the animal seated before us. He was hunkered up; his hair and teeth and claws were complete. [...] The whiskers had been pressed down on the side of the face by the bandaging; but when the pressure was taken away they stood out, just as they would have done in life. He was a magnificent creature, a tiger-cat of great size. But as we looked at him, our first glance of admiration changed to one of fear, and a shudder ran through each one of us; for here was a confirmation of the fears which we had endured. His mouth and claws were smeared with the dry, red stains of recent blood.« (Ibid.: 228-229)

As much as Margaret can be read as the »sweet and gentle« (ibid.: 64) alter ego of the powerful Queen Tera, Margaret's cat Silvio seems to function as the domesticated alter ego of the wild and dangerous Egyptian »tiger-cat of great size« (ibid.), which, through the »red stains of recent blood« (ibid.) on its mouth and claws, can eventually be identified as the nocturnal attacker of Mr Trelawny. Significantly, Silvio shows a strong aversion to the cat mummy and, when confronted with it, displays aggressive behaviour: »Whilst [Margaret] was fondling [Silvio], he suddenly gave wriggle like an eel and slipped out of her arms. He ran across the room and stood opposite a low table on which stood the mummy of an animal, and began to mew and snarl.« (Ibid.: 31) If we read Silvio and the cat mummy as the representatives of the two women Margaret and Tera, respectively, the aggression displayed by the domesticated cat Silvio against the wild Egyptian cat mummy becomes representative of the two opposing types of femininity embodied by Margaret and Tera. As much as Silvio seems to hate the presence of the wild tiger-cat of the past *in* the present, Margaret seems to try to suppress, though much less aggressively, the »caged lion« (ibid.) representing Tera within her own self in the manner of Jekyll trying to suppress Hyde.[9]

In this regard, it is also crucial to consider the fact that in the novel Margaret is the only major female character surrounded by at least six

9 To some extent, women like Margaret and Inez are the domesticated version of the past, which is also true of Iras, whom, although she is shyer by nature, Lavenham is anxious to assimilate into modern British society.

prominent male figures.[10] Furthermore, apart from being a male-dominated party, the team at Trelawny's house, with Ross as a lawyer, Winchester as a doctor, and Dolan and Daw as policeman and detective, respectively, mirrors the contemporary Victorian discourse of rationality and science, in which women, due to the prevailing power structures, were essentially absent (cf. Byron 2011: 56-57).[11] Considering this, Margaret's »domesticated« (ibid.) behaviour, in particular towards Ross, with whom she has fallen in love, becomes plausible, since it seems to accommodate itself to the male discourse at the Trelawnys' house, in which women – as in contemporary society – have to submit to the male idea of femininity. The only means Margaret has of expressing herself more freely is in talking about and through Tera, which becomes more and more explicit towards the Great Experiment at the end of the novel:

»I was beginning to doubt Margaret! [...] It was not her love, or her honour, or her truth, or her kindness, or her zeal. [...] It was herself! Margaret was changing! At times during the past few days I had hardly known her as the same girl whom I had met at the picnic, and whose vigils I had shared in the sick-room of her father. [...] Now she was generally *distrait*, and at times in a sort of negative condition as though her mind – her very being – was not present. [...] At such times, though she would say to me sweet and pleasant things which she had often said before, she would seem most unlike herself. It was almost as if she were speaking parrot-like or at dictation of one who could read words or acts, but not thought. After one or two experiences of this kind, my own doubting began to make a barrier; for I could not speak with the ease and freedom which were usual to me. And so hour by hour we drifted further apart.« (Ibid.: 204-205)

Characteristically, Ross experiences Margaret's behaviour changed to the negative, and he now, in contrast to the time of their vigils at Mr Trelawny's sickbed, when he felt her very close to himself, perceives Margaret as if she »was not present« (ibid.). In spite of the »sweet and pleasant

10 The other two minor female characters are the two nurses, while the six major male characters are Malcolm Ross, Abel Trelawny, Doctor Winchester, Superintendent Dolan, Sergeant Daw, and Mr Corbeck.

11 This again is reminiscent of the *Dracula Discourse*, in which a team of rational men tries to eliminate Dracula.

things« (ibid.) Margaret says to him, Ross conceives her as »most unlike herself«, since she seems to speak the words of an*other*, »parrot-like or at dictation« (ibid.).[12] Although we do not know what these words spoken by Margaret are, the fact that Ross describes them as »words or acts, but not thoughts« (ibid.) suggests that he cannot identify them as a genuine expression of Margaret's character and thereby more or less disqualifies them by implying that they are empty phrases without an idiosyncratic meaning endowed by Margaret. Furthermore, Ross says right from the beginning that it is not her attitude towards him, with »her love, or her honour, or her truth, or her kindness, or her zeal« (ibid.), that has changed but rather she »herself« (ibid.), suggesting a change in her very character (cf. *Jekyll and Hyde*). In other words, Ross can feel the core of Margaret's individuality change, while her behaviour seems to remain the same, which suggests that Margaret is a *changing woman* in spite of herself. What makes this development so hard to grasp for Ross is the fact that he rather feels than sees it and that there is no visible evidence he could name. Unlike the Margaret of the present he believes to know, the woman Ross notices in Margaret at these moments is the very distant and remote woman he first met at the ball and was frightened of. It is the distant alter ego of Margaret in the form of Queen Tera of the past which Ross has no access to and thus cannot control. This is underscored by Ross's comment that he and Margaret gradually »drifted further apart« (ibid.). Thus, the more Margaret changes, the more she becomes like Tera, a distant woman of the past, in whom, however, Ross has no interest: »I had looked for Margaret to ask her to come with me; but when I found her she was in one of her apathetic moods, and the charm of her presence seemed lost to me.« (Ibid.: 205)

Nevertheless, it appears to be not only the missing »charm of [Margaret's] presence« (ibid.) which makes her seem less attractive to Ross but also the subtle fear and mystery Ross repeatedly feels coming from her. However, it is not until shortly before the Great Experiment, that is, the resuscitation of Tera, that the danger and threat associated with Margaret momentarily surfaces, when her cat Silvio attacks the cat mummy brought into the cave at Cornwall by Mr Trelawny and the other men:

12 Cf. *Jekyll and Hyde Discourse*. Dr Jekyll's friends and acquaintances also experience him as increasingly controlled by an unknown force even before they know about the existence of Mr Hyde.

»*[Margaret's] eyes blazed, and her mouth took a hard, cruel tension which was new to me. Instinctively* she stepped towards Silvio as if to interfere in the attack. But I too had stepped forward; and as she caught my eye a *strange spasm came upon her, and she stopped.* Its intensity made me hold my breath; and I put up my hand to clear my eyes. When I had done this, she had on the instant recovered the calm, old grace and sweetness she swept over and lifted Silvio, just as she had done on former occasions, and held him in her arms, petting him and treating him as though he were a little child who had erred. As I looked a strange fear came over me.« (Ibid.: 201, emphasis mine)

On the basis of the assumption that Silvio, as the domesticated alter ego of the cat mummy, functions as Margaret's familiar in analogy to the relationship of Queen Tera and her familiar, the cat-mummy, Silvio's attack on the wild but at this moment defenceless cat mummy once more points towards Margaret's inner conflict in regard to her own female identity. This is further underlined by Margaret's wild reaction towards Silvio's attack on the cat mummy. Thus, Silvio's action and Margaret's reaction in this passage emphasise the two conflicting parts within Margaret, of which one is attracted to the wild and strong side represented by Tera, while the other is trying to fight this ›other‹, yet dormant side, as emphasised by Silvio's attack on the cat mummy.

That the domesticated part of Margaret represented by Silvio eventually also represents the Victorian idea of femininity, that of a gentle, soft, and submissive woman, is suggested by the fact that Mr Trelawny finally destroys the cat mummy: »›Fire will take away all danger for the future; even an astral body cannot materialise from ashes!‹ [Mr Trelawny] signed to follow him. Margaret turned away with a sob. [...] ›Oh! it seems like murder! The poor Queen's pet ...!‹ The tears were dropping from under the fingers that covered the eyes.« (Ibid.: 229) By burning the mummy of the tiger cat, which, as the familiar of the powerful Queen Tera, was responsible for the bloody attacks on Mr Trelawny, Mr Trelawny does not only eliminate Tera's most potent weapon and reliable guardian in the present but also bereaves her of that part of her female identity which is most threatening to his own male identity. This is underlined by MacFarlane: »It becomes the role of the hero, in these cases, to restore a more natural order, (represented as a stable, unified body of knowledge) by destroying the ancient texts, the mummy, or both.« (MacFarlane 2010: 21) Margaret can only but cry upon the destruction of what used to be part of her strong alter ego embodied by Tera.[13]

The brutality performed at this moment against a passive and helpless mummy is reminiscent of the destruction of the Count and his vampire victims in *Dracula,* who have to be staked and beheaded in order to eliminate them forever:

»But most of these Egyptian revenge fantasies re-establish safety, order, and reason by shooing their ill-behaved mummies back into their coffins and crypts: the undead disintegrate into dust, return to their shrunken, desiccated states, or simply fall into a state of de-animation just as inexplicably as they were revivified; otherwise they are hacked up, stabbed, shot, burned, and stuffed unceremoniously back into their sarcophagi by the heroic forces of the living.« (Bridges 2008: 141)

In all of these cases, the archaic violence brutally performed on these bodies expresses the immense threat posed to modern society by these strange intruders from the past. As they have powers that surpass those of

13 Two other contemporary texts in which mummies are burnt in order to protect modern British society against their evil power are Doyle's »Lot No. 249« (1892) and Blackwood's »The Story of Baelbrow« (1908). In both of these stories a mummy has been identified as the cause of evil and thus a threat to society, which has to be eliminated. However, before being burnt, as in *Jewel* the mummies in question are first brutally attacked, as the following passage illustrates: »For a moment Swaffan stood over the thing; then with a curse he raised his revolver and shot into the grinning face again and again with a deliberate vindictiveness. Finally he rammed the thing down into the box, and clubbing the weapon, smashed the head into fragments with a vicious energy that coloured the whole horrible scene with a suggestion of murder alone.« (*Baelbrow*: 93) A similar scene takes place in »Lot No. 249«, where the owner of the mummy is forced to dismember it before throwing it into the fire: »In frantic haste he caught up the knife and hacked at the figure of the mummy, ever glancing round to see the eye and the weapon of his terrible visitor bent upon him. The creature cracked and snapped under every stab of the keen blade. A thick, yellow dust rose up from it. Spices and dried essences rained down upon the floor. Suddenly, with a rending crack, its backbone snapped asunder, and it fell, a brown heap of sprawling limbs, upon the floor.« (*Lot No. 249*: 79)

(modern) humans, the means of destroying them have to be particularly effective, hence first the mutilation of the body and then the burning in the case of the mummies, and first the staking and then beheading in the case of the Count and his kind, respectively.

That in *Jewel* the burning of the cat mummy is associated with the destruction of a dangerous form of femininity is emphasised by the fact that while Margaret is deeply upset by Mr Trelawny's procedure, comparing it to murder, Ross and the other men seem greatly relieved after the cat mummy has been destroyed: »A few minutes of fierce conflagration; and then we breathed freely. Queen Tera's Familiar was no more!« (*Jewel*: 230) Thus, to eliminate the wild familiar of the ›old‹ New Woman Tera, essentially means to protect modern masculinity not only for the present time but also for the time to come, as »›fire will take away all danger for the future‹« (ibid.).[14] Against this background, Mr Trelawny's plan to resurrect the great Queen Tera appears careless, unless we assume that he intends to resurrect Tera as a different woman from the one she originally was.

The original power and potency of Queen Tera is confirmed by Corbeck's accounts of her life and death in ancient Egypt:

»›Perhaps the most remarkable statement in the records, both on the Stele and in the mural writings, was that Queen Tera had power to compel the Gods. [...] In [a] statement it was plainly set forth that the hatred of the priest was, she knew, stored up for her, and that they would after her death try to suppress her name. This was a terrible revenge, I may tell you, in Egyptian mythology; for without a name no one can after death be introduced to the Gods, or have prayers said for him. Therefore, she had intended her resurrection to be after a long time and in a more northern land.‹« (Ibid.: 129-130)

Here Tera is portrayed as a powerful woman discriminated against by male priests, and though she appeared to be immune against their hatred during her lifetime, she feared their wrath after her death. Significantly, Corbeck describes Tera's fear as being of the kind that her name might be suppressed by the priests once she is dead. Apart from the fact that this is a

14 The same is true of Dracula, who poses a central threat to modern patriarchal society.

»terrible revenge […] in Egyptian mythology« (ibid.), it would also mean the exclusion and thus elimination of Tera from historical records, *damnatio memoriae* (cf. Assmann 2006: 105), and thus from the cultural memory as such. At the same time, the suppression, exclusion, or deletion of a female name from communicative and collective memory – as already mentioned in regard to Iras's fading signature in *Iras* – mirrors women's situation in society and their position in historiography. For a powerful and independent woman like Tera, used to fame and tribute, the idea of having her name suppressed posthumously by men must be even more threatening than to a woman of the nineteenth century, used to discrimination and suppression: »The mummies of desirable women in these stories are almost all connected with lost and powerful knowledge: they are priestesses, sorceresses, and witch-queens. […] In this sense, what the eminent Victorian Egyptologists fear and desire is not the reanimated mummy herself, but their relation to the gaps that she reveals in their own systems of knowledge.« (MacFarlane 2010: 6) Although this might sound grim, it points towards the central conflict of the resurrection allegedly planned by Tera for herself, since, as in Iras's case, it seems extremely ironic that a woman fleeing from male suppression in ancient Egypt would hope to find a better life in nineteenth-century Britain, as implied by Margaret's statement:[15]

»›From first to last the cryptic writing and symbolism of that wondrous tomb of that wondrous woman is full of guiding light; and the key of the many mysteries lies in that most wondrous Jewel which she held in her dead hand over her dead heart, *which she hoped and believed would beat again in a newer and nobler world*!‹« (*Jewel*: 181, emphasis mine)

15 »In her lecture ›The Social and Political Position of Women in Ancient Egypt‹ (c. 1890), Amelia Edwards […] liked to place particular emphasis upon the fact that thousands of years ago Egyptian marriages were subject to a ›rule of contact‹ that gave women far more economic and legal freedom than their modern British sisters.« (Glover 1996: 90) In *Pharaohs, Fellahs, and Explorers* (1891), Edwards dedicates a whole chapter to the ancient Egyptian Queen Hatasu and her use of power (cf. Edwards 1891: 261-300).

It does not seem convincing that a woman like Tera, who is repeatedly referred to as a provisional and capable woman »›of extraordinarily subtle mind‹« (ibid.: 221), would plan her resurrection at a time of female discrimination and impotence. In this regard it becomes plausible that the Great Experiment would fail, since Tera's alleged plans for her own resurrection very much depend on Mr Trelawny's interpretation and are not concordant with Tera's actual plans – if she had had any in this respect at all.

Interim Findings: The Mummy as the *less familiar strangeness*

In the present chapter I have shown how the reception of Egyptian archaeology as the *less familiar strangeness* in mummy fiction reflects a growing destabilisation and unease of contemporary society. Unlike Greek archaeology, which as the *most familiar strangeness* served as a means of (re)constructing and stabilising identity, the popular negotiation of Egyptian archaeology was characterised by a strong ambivalence. Similar to Greek archaoelogy, Egyptian archaeology was seen as a means of reconstructing the past in order to stabilise the present, which I have demonstrated in my reading of *Iras* and *Osiris*. In both novels archaeological reconstruction is used to legitimise and confirm identity in the present. In *Iras* both Lavenham and Iras's identity depends on Lavenham's archaeological search for evidence in the more recent and remote past. In *Osiris* the detective investigation following the murder of Bellingham turns into an archaeology quest with the aim to destinguish Bellingham's modern identity from that of an Egyptian mummy. As I have shown in regard to these novels, however, persisting stability is endangered as in both novels modern identity is threatened by the gradual shift and dissolution of boundaries. This threat eventually culminates in the deconstruction of binary categories, as I have demonstrated in regard to the subversion of gender concepts in my reading of *Pharos* and *Jewel*. Common to all of the novels is the ›otherness‹ associated with the past, represented by the archaeologist, the mummy, or the (ancient) alter ego of a protagonist, revealing the Hyde-part of the *Jekyll and Hyde Discourse*. Significantly, this ›otherness‹ surfacing in the inversion of the object-subject relation and gender roles is experienced in most cases as disturbing, since it fundamentally threatens and destabilises the binary categories of the Western patriarchal order, as proclaimed by Butler in regard to gender norms: »The loss of gender norms would have the effect of proliferating gender configurations, destabilizing substantive identity, and depriving the naturalizing narratives of compulsory heterosexuality of their central protagonists: ›man‹ and ›woman‹.« (Butler 2007: 200) Thus, the ambivalence and arbitrariness of identity concepts in the archaeological space reveal the constructed character of these concepts, while at the same time exposing the double standards of Victorian and Edwardian society: »Just as bodily surfaces are enacted *as* the natural, so these surfaces can become the site of a dissonant and denaturalized performance that reveals the performative status of the natural itself.« (Ibid., emphasis mine) As I have shown in my reading of the four mummy novels, the reception of Egyptian archaeology was characterised by a growing destabilisation and insecurity in regard to the characters' identity, anticipating fears of reverse invasion and annihilation, which I will discuss in the following chapter.

7. The Mummy and Great Zimbabwe as the *most unfamiliar strangeness*

›Today the last portion of that vengeance which has been decreed commences, and when all is finished I got to that rest in ancient Thebes which has been denied me those long three thousand years.‹
(*PHAROS*: 220)

So far I have shown how the archaeological space as an ambivalent and indefinite space served as a counter-discourse to Victorian and Edwardian society, in which collective and individual identity was created, renegotiated, subverted, and inverted. The present chapter looks at the different ways in which contemporaries dealt with the archaeological space as the space of the ›other‹. On the basis of what I have termed the *Dracula Discourse*, the temporal and spatial ›other‹ will be analysed in terms of a hostile force which, in analogy to the figure of Count Dracula, infiltrates modern Britain in an act of reverse invasion. As in previous chapters, the *Frankenstein* and the *Jekyll and Hyde Discourse* will also be included in the analysis. Unlike my reading of Greek and Egyptian archaeology as the *most* and *less familiar strangeness*, respectively, in the present chapter I present Egyptian archaeology and Zimbabwean archaeology as collectively perceived as the *most unfamiliar strangeness*. Focusing on Guy Boothby's *Pharos* and Theo Douglas's *Iras*, the first two chapters elucidate how the archaeological space was experienced as an alterity and consequently perceived as threatening to modern society. In Chapter 7.3, I will illustrate how the threat of the past culmitates in the inexplicable evil of Pharos's character. In Chapter 7.4, I will demonstrate in regard to *Pharos* how the initial threat perceived as coming from the archaeological space in an act of identification wanes, allowing for underlying feelings such as guilt and repentance on the part of the British in regard to the treatment of the past. In addition to the *Dracula Discourse*, as part of which the Count might be seen not only as a victimiser but also as a victim,[1] this is reminiscent of the *Frankenstein Discourse*, in which the atrocities committed by Frankenstein's creature are a consequence of the creature's isolation and loneliness, for which Frankenstein through the initial act of creation is responsible. The last two chapters focus on the reception of the Zimbabwe ruins, which in a process of negation and assimilation were used by the British as a means of creating an intentional history. Essential for the creation of this intentional history was an identification of the British colonists with the ruins, which, however, eventually resulted in an over-identification with the past, causing fears of degeneration and doom in regard to the British Empire. On a symbolical level, this over-identification with the archaeological ruins of once powerful civilisations also marks a moment of anagnorisis in regard to the incompleteness in the Lacanian sense: Wholeness as experienced in the state of the Real, represented by the past, can never be regained.

1 Cf. Margaret L. Carter (1997) and Carol A. Senf (1979).

7.1 »THE DISEASE TRAVELS FAST«:[1] THE INVISIBLE THREAT

I am of opinion that the soul is not *spirit* but *matter*; matter of such infinitely minute particles as to be perfectly invisible to even the most powerful microscope yet made.
(*DR. NOSIDY*: 31)

In *Pharos*, the diffuse power of the past over the present associated with the amorphous character of Pharos materialises in the plague[2] instrumentalised by Pharos as a collective revenge on the modern world. When Forrester wakes up in his bedroom at a hotel in Cairo after spending the night with Pharos at a pyramid, he has been infected by Pharos with the plague without knowing it. As after a vaccination, Forrester feels feeble and dizzy and also discovers the spot of the injection on his arm, which he inspects carefully:

»What was stranger, as soon as I stood upon my feet, I was conscious of a sharp pricking sensation on my left arm, a little above the elbow; indeed, so sharp was it, that it could be felt, not only on the tips of my fingers, but for some distance down my side. To examine the place was the work of a moment. On the fleshy part of the arm, three inches or so above the elbow, *was a small spot, such as might have been made by some sharp-pointed instrument*[3] – a hypodermic syringe, for instance – and which was fast changing from a pale pink to a purple hue. My wonderment was increased when I discovered that the spot itself, and the flesh surrounding it for a distance of more than an inch, was to all intents and purposes incapable of sensation. I puzzled my brains in vain to account for its presence there.« (*Pharos*: 184-185, emphasis mine)

---

1 *Pharos*: 300.

2 Similar to the Egyptian smell in *Jewel* (cf. Chapter 4.2), the plague is invisible, impalpable, and thus uncontrollable.

3 Note the similarity to the vampire bite here (cf. *Dracula Discourse*).

Pharos has obviously used a hypodermic syringe to inject the pathogen of the plague germ into Forrester's arm, linking the archaic past, in the form of the plague, to the modern present, in the form of a technically advanced medical instrument. The two spheres of past and present are further contrasted by the time gap between Forrester's experiences at the pyramids and his awakening at the hotel: »When Pharos bade me drink the stuff he had poured out, we were standing before the Sphinx at Gizeh; now, when I opened my eyes, I was back once more in my bedroom at the hotel in Cairo.« (Ibid.: 183) While the historical site at the pyramids of Gizeh at night both represents and evokes the past,[4] the bedroom at the hotel in Cairo with the »[b]rilliant sunshine [...] streaming in through the jalousies« (ibid.) clearly stands for the present. The very fact that Forrester is infected in the pyramid turns the plague itself into a relic of the past which is inserted into the modern world by Pharos, who uses Forrester as a host for its germs. Thus, the plague is turned into a symbolic threat of the past to the present, and, in a more general sense, can be read along the lines of the *Dracula Discourse* as a reverse invasion.

Generally, the idea of an invasion of the present by the past is realised in many mummy texts in the form of the mummy curse.[5] The mummy

4 Note that it is from the pyramid that Forrester is transported into the past.

5 This is illustrated in Clive Pemberton's short story »The Bulb« (1906), where the Allinghams, a married couple, find an ancient bulb in a mummy case they have purchased at an auction. Curious as to what flower it will grow into, Mr Allingham plants the bulb, before he learns from a befriended Egyptologist, who translates the papyrus accompanying the mummy for him, that the bulb is supposed »to work the vengeance of the [g]reat [Tanes Iscarus]« (*Bulb*: 125) on anyone disturbing the eternal peace of his daughter, whose mummy was purchased by the Allinghams. Although Allingham and the Professor, after learning of the bulb's purpose, immediately return to the Allingham house to look for Allingham's wife, they arrive only to find her lifeless body on top of the mummy: »It was true! The flower – that wonderful flower that had grown from the bulb in the Egyptian sarcophagus – was one moment a living emblem of perfection perfected; the next it was shrivelled into brown tinder as by a flash of lightening! The Professor sprang forward and looked sharply at Allingham's wife. He looked at the face of a dead woman!« (Ibid.: 128) While in »The Bulb« the mummy curse is materialised in the bulb and thus becomes palpable, in

curse essentially functions in the same way as Pharos's lethal plague and Dracula's infectious bites in that it infects the ›body‹ of modern society in an act of revese invasion with its venomous, often lethal ›germs‹.[6] Significantly, as with a bacterial or viral infection, it is only through its *effect* on a befallen individual/organism that a mummy curse becomes visible.[7] Similar to *Dracula*, in *Pharos* the danger and potency coming from the plague is foreshadowed immediately after Forrester's infection, as underlined by the above-quoted scene at the hotel room. Although Forrester suffers from the symptoms of infection and detects the respective marks on his body, – the

---

many other texts the mummy curse remains much more diffuse and invisible. A prominent twentieth-century example is the mummy curse that allegedly befell Howard Carter, Lord Carnarvon, and other members of the excavation team of Tutankhamun's tomb: »On November 26, 1922, archaeologist Howard Carter, his immensely rich benefactor Lord Carnarvon, and daughter Lady Evelyn unsealed the royal tomb of Tutankhamun in the Valley of the Kings, revealing an extraordinary find, the only such unravaged site to be discovered in the modem [sic] era. The initial worldwide publicity was quickly followed by a tidal wave when Lord Carnarvon died five months later of complications after a mosquito bit him at the site. A not very specific inscription on the tomb was now read by the press as an unmistakable curse. This resuscitation of the old taboo against grave robbing was intensified when two other members of the expedition also died suddenly. Such sensational elements begged for exploitation by Hollywood – and so they were. The tomb of King Tut, as the press called him, had just been relieved of the last of its treasures when THE MUMMY appeared.« (Larsen 2000: 1)

6 Similar to Pharos's plague and Dracula's bites, the mummy curse eventually is also the ›vengeance‹ for a (bodily) sin/misdeed committed by modern men/ society (cf. below).

7 Cf. *Jewel of Seven Stars*, »The Story of Baelbrow«, *Nemesis of Fire*, *Iras: A Mystery*, »The Ring of Thoth«, and »The Cat«. The analogy of the mummy curse and germs is illustrated by the popular reception of the deaths among the excavators of Tutankhamun's tomb in 1922. Speculation as to the cause of death of the people involved in the excavation included theories of fungal spores as well as an alleged mummy curse befalling those who disturb the tomb of the Pharaoh. Cf. also Laurence A. Rickels (1992) »Mummy's Curse« and Roger Luckhurst (2006) »The Mummy's Curse: A Genealogy«.

puncture, swelling, and discolouring of his arm – he is unable to interpret these signs accordingly and is only able to do so when, back in Britain, Pharos confronts him with the truth. This again shows very strong parallels to *Dracula*, where it also takes a considerable time before Van Helsing is finally able to interpret and identify the Count's bite marks on Lucy's throat for what they really are and take up the required actions together with the other men. Unlike *Dracula*, where Van Helsing is able to offer an explanation for the seemingly inexplicable and to take up the necessary precautions, in *Pharos* Forrester's anagnorisis comes (too) late:

»It was as if a bandage had suddenly been removed from my eyes, enabling me to see everything plainly and in its proper light. […] The dreadful secret was out. In less than five minutes the mystery of two months had been solved. Now I knew the meaning of the spot I had discovered upon my arm on the morning following my terrible adventure in the Pyramid; now I could understand my illness in the desert, and the sudden death of the poor Arab who had nursed me. In the light of this terrible truth, everything was clear as daylight.« (*Pharos*: 350-351)

However, in spite of this anagnorisis, there is nothing Forrester can do at this point. The plague has already spread rapidly and is by now out of control, and Forrester is completely isolated with his knowledge, lacking an expert like Van Helsing who would support him. Consequently, Forrester's attempt to report to the authorities fails[8] and the only possibility to save

8 The reason why Forrester fails to inform the authorities about Pharos is that it would not only risk his own reliability but also question his sanity. That Forrester is well aware of this is supported by the various instances in the novel where he refers to his fear of not being taken seriously: »I am oppressed by the fear that my narrative may seem too extraordinary to carry with it any semblance of reality.« (*Pharos*: 215) In fact, when upon returning from Egypt to England Forrester decides to report to the authorities and confides all he has experienced to the Home Secretary, who is an acquaintance of his, his fear is confirmed: »›I can see that you do not believe what I have told you‹, I cried with great bitterness. ›Sir Edward, I implore you to do so. I assure you, on my honour as a gentleman, I will swear, by any oath you care to name, that what I say is true in every particular. Pharos is still in London, in Park Lane, and if you are quick you can capture him. But there is not a moment to lose. For God's sake

Britain from the plague is to leave the country: »›Hush‹, [Valeric] said, ›you must not talk! Let it suffice that I have saved you, and now we are away from England, and at sea together. Pharos is dead, and the past is only a bitter memory.‹« (Ibid.: 376)[9]

What makes the infectious character of Dracula and Pharos as well as the mummy curse in general so dangerous is, *inter alia*, the fact that the ›disease‹ they spread is invisible and can only be diagnosed for certain once it is already well advanced. In regard to coeval science, this can be read as a reference to the ›discovery‹ of certain hitherto invisible bacteria which recent inventions and research had made visible for the first time (cf. also Chapter 2.2):

»The first photographs of bacteria, now viewed as one of the most significant early advancement in the history of scientific and medical imaging, were made on the continent in the 1870s [...]. At a time when Victorian scientists and others regarded photographs as consequential for proving matters of fact, the camera seemed to offer objective access to a world of microbial wonders. But photographs of bacteria were neither transparent in meaning nor determined by a single interpretation outside the contexts where visual techniques and viewing practices informed how Victorians

believe me before it is too late!‹ ›I have listened to all you have said, my dear Cyril‹, he answered soothingly, ›and I can quite understand that you believe it to be true. You have been ill, and it is plain your always excitable imagination has not yet recovered its equilibrium. Go home, as I say, and rest. Trust me, you will soon be yourself once more.‹« (Ibid.: 369) This scene is reminiscent of Frankenstein's confession to the authorities in that in both cases the motivation for Forrester and Frankenstein's confession lies in the immediate threat Pharos and the creature pose to society. Significantly, although in both cases Forrester and Frankenstein are confronted with benevolence and indulgence, neither of them is taken seriously. The stories they have to tell are dismissed as fantasies resulting from a period of severe illness.

9 That Forrester's departure is a final one is emphasised in his letter to his friends preceding Forrester's report: »›Matthews Simpford, in the Strand, is keeping two pictures for me. [...] I ask you each to accept one, and when you look at them, try to think as kindly as possible of the friend who is gone from you forever.‹« (*Pharos*: 18)

saw the war against germs, including magic lantern shows, atlases, medical laboratories, and the illustrated popular press.« (Tucker 1997: 389-391)

These new insights gave rise to all kinds of popular speculations and fantasies as to what other invisible threats the air around us might contain. Furthermore, since the second half of the nineteenth century public vaccinations had been carried out on a grand scale.[10] In this context, the threat coming from an invisible disease that subtly infects the human body and rapidly takes on an epidemic character can be read as a fear of these unknown bacteria but also as distrust in the consequences of novel scientific methods, such as inoculations, which became prominent in the anti-vaccination movement in Britain. The impotence of modern medicine is emphasised in the novel when Valeric contracts the plague during her and Forrester's escape from Pharos, and Pharos finally turns out to be the only one able to help her (cf. *Pharos*: 269, 357).

What is interesting in this context is that while Van Helsing and his supporters are able to chase the Count away from British soil across Europe to his home country, where they finally succeed in destroying him, in *Pharos* it is Forrester and Valeric who have to leave Britain after Pharos has committed suicide. Furthermore, it is significant that before Pharos, Forrester, and Valeric arrive in Britain, Forrester and Valeric attempt to escape from Pharos by secretly leaving Prague by train for Germany, where Pharos finally catches up with and regains control over them in Hamburg. Thus, instead of pursuing Pharos and chasing him out of Britain, as the characters in *Dracula* do with the Count, Forrester and Valeric are chased across the continent and finally have to leave their home country in order to save Britain.

Characteristically, time and space also play a prominent role not only during Forrester and Valeric's escape from Pharos but also before and during their journey with Pharos:

»In travelling either with Pharos or in search of him, it was necessary to accustom one's self to rapid movement. I was in London on June 7th, and had found him in Naples three days later; had reached Cairo in his company on the 8th, of the same

10 Cf. Diana Brunton (2008) *The Politics of Vaccination: Practice and Policy in England, Wales, Ireland, and Scotland, 1800-1874.*

month, and was four hundred and fifty miles up the Nile by the 27th. [...] From this itinerary it will be seen that grass was not allowed to grow under our feet. [...] On land the trains could not travel fast enough; on board the yacht [Pharos's] one cry was, ›Push on, push on!‹« (Ibid.: 233)

The affinity with speed emphasised here in regard to Pharos is very much reminiscent of the rapid movement and action required for the successful persecution and destruction of the Count in *Dracula*. However, while in *Dracula* the characters avail themselves of time and speed against the Count, in *Pharos* Forrester, Valeric, and finally also Europe fall prey to these forces once the plague has started to spread, as explicated by Pharos: »›[B]ut you forget that the disease travels fast – faster than you do when you run away from me, my dear Forrester.‹« (ibid.: 300) In other words, while in *Dracula* the characters are in control of time and use it against the intruder, in *Pharos* the characters are controlled by time used by the intruder against them.[11] Considering the growing significance attributed to time as such during the nineteenth century, the very fact that in *Pharos* the characters lose control over time can be read as a serious threat to society. This is in particular the case against the background of Victorian capitalist progress and technologisation, which essentially aimed at the control of time. In Victorian society, not being in charge of time literally meant being ›out of time‹.[12]

---

11 In *Dracula*, Quincey Morris points out: »›As there is no time to lose, I vote we have a look at his house right now. Time is everything with him; and swift action on our part may save another victim.‹« (*Dracula*: 242)

12 The terrifying fear of being caught up with by the past is most vividly demonstrated in Doyle's short story »Lot No. 249« (1892), in which Smith, a student at Oxford, is chased by a mummy, reanimated by his fellow student Bellingham, through the night: »It moved in the shadows of the hedge, silently and furtively, a dark, crouching figure, dimly visible against the black background. Even as he gazed back at it, it had lessened its distance by twenty paces, and was fast closing upon him. Out of the darkness he had a glimpse of a scraggy neck, and of two eyes that will ever haunt him in his dreams. He turned, and with a cry of terror he ran for his life up the avenue. There were the red lights, the signal of safety, almost within a stone's-throw of him. He was a famous runner, but never had he run as he ran that night. [...] As he rushed madly and wildly through the

In *Pharos*, the importance and danger of rapidity are further emphasised by the fact that both Pharos and the plague speed into the present from the past, spreading their invisible pathogenic germs. Thus, in a more general sense this could also be read as a collective fear of economic and political failure towards the end of the nineteenth century after a period of progress and prosperity. At such a time fantasies of invasion were common, as the popularity of invasion literature in the last decades of the century and at the beginning of the twentieth century highlights (cf. Clarke 1997). The very fact that in *Pharos* the past, in the form of Pharos and the plague, literally seem to catch up with the present underscores the deep contemporary fear of degeneration and decline.

What is crucial in this context is the reason why in *Pharos* the characters fail to avert the invasion of Britain while in *Dracula* it is successfully thwarted. Apart from the fact that the characters in *Dracula* are in charge of time and use it for their means, both their professions[13] and their actions are characterised by rationality, and even Mina, the only woman in the anti-vampire campaign, is endowed with a clear mind and common sense that would have been regarded as ›atypical‹ for a woman at the time. In *Pharos*, however, the character of Forrester, a decadent artist who »live[s] for [his] art« (*Pharos*: 68), embodies a strong contrast to the rationality and practicality associated with the doctors, the lawyers, and the military men in *Dracula*. Accordingly, Forrester's attempts to fight Pharos fail, since they essentially lack rational and systematic thoughts and actions. Against the background of contemporary fears of decline and degeneration, the failure of Forrester in the person of a decadent artist can thus be read as criticism and a warning against what might happen if society gets too engaged with decadence and consequently neglects fundamental principles of rationality, objectivity, and morality.

---

night, he could hear a swift, dry patter behind him, and could see, as he threw back a glance, that this horror was bounding like a tiger at his heels, with blazing eyes and one stringy arm out-thrown.« (*Lot No. 249*: 72-73) Though Smith is eventually able to save himself from the mummy by entering his friend's house, it is only due to his extraordinary speed as a »famous runner« (ibid.) that he escapes from his pursuer from the past.

13 Note that Seward, Von Helsing, Harker, and Quincey Morris are either doctors, lawyers, or soldiers.

Once Forrester has been infected with the plague, though yet unaware of it, he desperately wishes to return to Britain and to leave behind all the threats he has encountered during his travels abroad:

»How I longed to be in England, no one can have an idea. The events of the last few months, if they had done nothing else, had at least deprived me of my taste of travel; and as for the land of Egypt, the liking I had once entertained for that country had given place to a hatred that was as vigorous as I had deemed the other sincere.« (Ibid.: 289)

Ironically, Forrester thinks himself and Valeric to be safe from the plague once they have disembarked on British soil. This, however, soon turns out to be an illusion, as much as Forrester's belief that a British doctor could diagnose and cure the mysterious disease he has contracted in the pyramid: »I had half made up my mind to set off for Luxor on my own account, in the hope of being able to discover an English doctor, from whom I could obtain some medicine and find out the nature of the ailment from which I was suffering.« (Ibid.: 229) In fact, however, as the rapid spread of the plague on the continent and in Britain suggests, the »›meddlesome European doctors‹« (ibid.: 230) are able neither to offer a proper diagnosis of the symptoms nor to come up with any medicine to fight it. In other words, medicine and science, like the authorities, are completely helpless in the face of the foreign epidemic, as becomes clear when Pharos informs Forrester of the following: »›A very philanthropic decision on your part‹, he answered sarcastically. ›Unfortunately, however, I am in a position to be able to inform you that your charity is not required. Though the authorities are not aware of it, the plague has already broken out in England.‹« (Ibid.: 300) In *Pharos* the plague functions as the materialisation of an invisible, amorphous, and hostile threat associated with the past. In an act of reverse invasion it infiltrates modern Britain, leaving its contemporaries as helpless and anxious as the characters in Stoker's *Dracula*. As in the *Dracula Discourse*, the infested and weakened human body thus becomes the emblem of the vulnerability of British society.

7.2 »The advancing Shadow«[1] of the Past: The Consummation of the Present

How could such a thing as this stride about the streets at Oxford, even at night, without being seen?
(*Lot No. 249*: 75)

In Douglas's *Iras,* as in many other mummy stories of the time, the mummy of the ancient Egyptian Iras is afflicted with a curse which turns out to be caused by Savac, an immortal lovesick Egyptian priest from the past.[2] Already at the beginning of the novel, the reader is informed of strange and disturbing happenings accompanying the transport of Iras's mummy from Egypt to Britain. These incidents, which in an Orientalist manner are initially dismissed as the products of the Arabs' superstitious and childish fears, eventually turn into a threat to Victorian middle-class society, foreshadowed by Lavenham's thoughts upon having seen Savak for the first time at Mrs Payne's party: »I felt as a man might who is drunk with the inhalation of a narcotic. What spell had bewitched me in that *ordinary middle-class house*, with its *good-natured hostess* and *crowd of common-place guests* [...]?« (*Iras*: 22, emphasis mine) Although Savak's hatred seems to be directed solely against Iras and Lavenham, there is more at stake than their own lives, as Savak has no qualms to appear in the midst of London's »ordinary« (ibid.) and »common-place« (ibid.) »middle-class« (ibid.) society. Again, like Forrester in *Pharos*, Lavenham is the one who imports the threat from the foreign past into the present (cf. *Dracula Discourse*) in the form of the mummy, naïvely assuming that »the man who was [Iras's] enemy has been dead and dust for centuries [and that] [i]t is beyond his power to harm [Iras] or come between [them], either in Egypt or [England]« (ibid.: 83). This links Lavenham, who is (indirectly) responsible

1 *Iras*: 154.

2 Other mummy texts in which people are affected by mummy curses are: »The Ring of Thoth«, *Jewel of Seven Stars*, »The Story of Baelbrow«, »The Bulb«, and *Nemesis of Fire*.

for Iras awakening in the present, to Frankenstein, by whose hand the creature comes to life and becomes a threat to contemporary society.

It is only shortly afterwards that Lavenham has to realise that, in analogy to *Dracula*, where all efforts to prevent the Count from invading peoples' private spheres fail, he can neither protect Iras's and his own life against the attacker from the past nor can they escape from his powerful influence. This is symbolised by the seven pendants Iras wears on a chain around her neck, which become the emblem of her and Lavenham's life in the present. This present is gradually consumed by the past in the form of Savak: »›These seven drops are seven spaces of time – whether long or short I know not; the time I have granted me to spend with you. So long as I can keep them I live in the life of this world, were it a hundred years; but when they go I die and am a spirit. While they are safe I am yours.‹« (Ibid.: 85) Yet the price Lavenham and Iras have to pay in order to protect the seven pendants against Savak's hand is a high one, since instead of settling down after their marriage they become fugitives in their own country: »I think throughout the journey we both had the feeling that we were fugitives, and were glad when stately stonebuilt Edinburgh closed about us once more.« (Ibid.: 128) What under different circumstances could have turned into a traditional middle-class marriage, as suggested by both Iras's character and the longings Lavenham projects on her (cf. Chapter 6.1.), becomes a relationship on the run, characterised by the transitoriness of the present.

Likewise, happiness becomes confined to the moment, as is underscored by Lavenham's notes: »It is said that the nation is happy which has no history; and for the first seven days following our marriage my history is only that we were happy.« (Ibid.: 118) Characteristically, Lavenham indirectly compares his happiness to that of a nation without a history, which can be read as referring not only to the lack of history of Lavenham and Iras's relationship but also to Lavenham's lack of personal and family history (cf. Chapter 6.1).[3] As suggested by the transitoriness of Lavenham and Iras's happiness, however, lasting happiness in the form of a stable relationship and identity is only possible if there is a (family) history, repre-

3 Since Lavenham and Iras's relationship is based on love at first sight – they decide to stay together forever upon seeing each other for the first time –, it essentially lacks a development.

senting a collective memory and a social frame of reference, which provides both a foundation and continuity. Since Lavenham lacks both, his identity, like that of Forrester in *Pharos*, is essentially threatened, as his premature death at the end of the novel shows.

At the same time, Lavenham's instability can be read as mirroring the self-perception of a whole society towards the end of the nineteenth century, which is underscored by Lavenham's comment on Iras's multilingualism: »This girl, who had lived her waking life when the savage dwellers in our islands of the sea were grubbing in fellowship with the cave-bear and gabbling some dialect of Norse, had spoken to me in perfect English.« (Ibid.: 78) Here, the very fact that Lavenham refers to his British ancestors as »savage dwellers [...] grubbing in fellowship with the cave-bear and grabbling some dialect of Norse« (ibid.), while Iras simultaneously spends her »waking life« (ibid.)[4] in the advanced civilisation of ancient Egypt, can be read as hinting at a growing unease in regard to the instability of British national identity caused by economic, social, and cultural changes towards the end of Queen Victoria's reign.

Instead of settling down in a comfortable middle-class home and starting a family, Iras and Lavenham have to be on the move, as quick movement is the only thing which seems to protect them at least temporarily from Savak's attacks:

»The resolve I renewed in the watches of that night was to keep moving onward, and to sojourn nowhere for a longer period than twenty-four hours; during that time, so far as experience had borne out, we might consider ourselves secure. It would be a weary life of perpetual journeying, and might tax her diminishing strength; still, if it lengthened the span of her continuance with me, that was motive for all.« (Ibid.: 165)

On their journey through Northern England and Scotland, Lavenham and Iras travel mainly by rail (cf. Chapter 6.1). In spite of their rapid movement, however, time and again Savak catches up with them and diminishes Iras's life force by taking away one by one the pendants from her necklace. It is significant that Lavenham and Iras have no chance of escaping from the powers of the past although they make use of the British railway system,

4 In this context, »waking life« (*Iras*: 78) is equivalent to an enlightened life.

one of the most modern and progressive technical and infrastructural achievements of the nineteenth century.[5] This once more is reminiscent of the *Dracula Discourse*, where modern transport becomes a means of chasing Dracula, as well as in *Pharos*, where Forrester and Valeric try to escape from Pharos by train. That modern technology, as in *Dracula*[6] and *Pharos*, has no real chance against the powers of the past is further emphasised when Lavenham has to realise despite his initial confidence that even a Chubb lock is not strong enough to protect Iras's necklace against Savak's hands: »›I have faith in a Chubb lock […] [your] trinket is safe this time, my darling; you need not trouble about it any more.‹« (Ibid.: 139) On the following day, however, Lavenham has to learn that in spite of the high-security lock, one more pendant has »perished by some agency beyond [his] power to define« (ibid.: 140) inside the box.

In the light of contemporary *fin-de-siècle* anxieties, this once more can be read as the subconscious fear of an industrial and a cultural regression, a reverse development to nineteenth-century industrialisation, symbolising a consummation of the present through the past. This reverse consummation of the present through the past rather than the other way around, symbolised in the novel by the successive disappearance of the pendants on Iras's necklace, is further hinted at in one of Iras's dreams: »I was wearing this Englishwoman's dress that you have given me, which belongs to the present, not the past. […] And there before me among the pillars, unchanged, implacable, commanding, *he* was looking at me as the snake looks at the bird.« (Ibid.: 145) Here, the »Englishwoman's dress« (ibid.) Iras is wearing is explicitly linked to the present, while both her surroundings and Savak, only referred to as »*he*« (ibid.), are ascribed to the past. The fact that Savak, as much as Pharos, as the figuration of the past, is »looking at [Iras]« (ibid.), who symbolises the present in her Englishwoman's dress, »as the snake looks at the bird« (ibid.) suggests an imminent consummation, that is, a consummation of the present by the past.

5 At some point they also travel aboard a steamer, likewise symbolising modern technology (cf. *Iras*: 146).

6 Dracula can only be kept away with garlic cloves and the crucifix. Moreover, his final elimination is only possible by very archaic means (staking and beheading).

In this context, as in *Pharos* and *Dracula*, time and speed also play a prominent role in the novel, essentially mirroring the contemporary obsession with the economy of time as one of the determining factors of nineteenth-century industrialisation. Consequently, in *Iras* time means both security and danger, freedom and confinement, always depending on who is in control of it at the very moment. Accordingly, Lavenham and Iras eventually fail because they run out of time, that is, lose control over it, due to Lavenham's wrong decision to travel further into the countryside:

»It was a piece of singular unwisdom, I reflected after, which had prompted me to embark on that cross-country journey, where the means of conveyance forward might often become difficult [...]. I calculated we could reach Inverardoch possibly in another day, and Calander by the following evening; and from there the railway would be available for Edinburgh and the South.« (Ibid.: 165)

While earlier in the novel Lavenham fled from the bigger cities to protect himself and Iras from social preconceptions and denunciations, he now realises that if there is one thing that can save them from the past as personified by Savak it is the technology and infrastructure of the modern world to be found *in* the cities.[7] However, Lavenham's insight comes too late, and his last-ditch attempt to save Iras by carrying her through the snow to a mansion nearby culminates in his delirious breakdown and Iras's death: »I would wade through the snow as the man had waded in the morning, and carry Iras to the manse. [...] I was strong enough to carry her, light as she was, in my arms; and at the manse they could not refuse us shelter – the church would afford that for the night, even if the house were full.« (Ibid.: 177-178)

Lavenham's reference to the church here as one of the last places of refuge is significant, since it points towards a religious theme which, though subliminal, is strong in the novel. At the beginning of his relationship with Iras, Lavenham thinks of his new life in terms of the biblical creation: »›And the evening and the morning were the first day.‹ That

7 As mentioned above, in many archaeological texts the characters are most threatened by the past as soon as they are isolated from modern society, e.g., at night or in the countryside. This is emphasised in *Jewel*, where the resurrection of Tera is carried out at night in a secluded cottage in Cornwall.

fragment of the Mosaic account of creation floated through my mind not inappropriately; for had I not come upon new heavens and a new earth, and was not all my life created afresh?‹« (Ibid.: 90) This Christian imagery reappears more strongly in one of the feverish dreams Lavenham has at the foot of the mountain Ben Cruachan during their flight:

»One night, when I was at the worst, it seemed to me that I awoke and lay looking at the white mountain with some soft light about it such as the stars or moon, when, between me and it, uplifted into space, I saw a cross. Not the window-bars, for they were gone, but such a form or wood as that with which painters have made us familiar in ghastly representation of the Crucifixion. To this instrument of torture my wife was bound, in the white dress she wore in the coffin [...] and the green gleam of the necklace about her throat; while I saw plainly the four remaining lotus-buds where they hung below.« (Ibid.: 150-151)

In Lavenham's fantasy Iras's body becomes that of Christ, who has to endure earthly pain before achieving final salvation and resurrection in another world. As suggested by the »four remaining lotus-buds« (ibid.), this time is about half way there. To Lavenham this is nearly unbearable, since although the four pendants mean four »spaces of time« (ibid.: 84) that Iras will be able to spend with him, this time, as enunciated by the crucifixion, also means torture to Iras, as it is to be a time spent at Savak's mercy. However, it is through Iras, who refers to her approaching end with faith and confidence, that Lavenham finally finds comfort and assurance in the Christian faith:

»›Since I have grown weaker, it is as if something had widened in my mind – as if a power were given me to look forward, and to see dimly that what seems the end may be only a stage of our beginning – not for me alone, in the great change, but for me and you together. [...] O my love! I am glad you are gray-haired – I am glad you are not a young man – so that the time may not be long.‹« (Ibid.: 176)

Thus, this Christian idea of a reunion after death and eternal peace promoted here by Iras as a source of hope and comfort after her demise is echoed by Lavenham, who also seems to be assured that there will be an ultimate reunion with Iras after his own death: »[T]he power of the enemy had spent its permitted force against our union on earth, and could touch us no more: the Hereafter – that Hereafter of which, I a doubter, had received such tremendous evidence – alone would decide with whom the victory lay.« (Ibid.: 247) The very fact that the only way to ultimately escape from the power of the past in the person of Savak is through the entrance into the Christian afterlife, while the modern world cannot offer any permanent safety, seems to suggest both the need for and the reliability of the traditional Christian belief system. The only thing which can stand up against the pagan threat of the past is the Christian church.[8] This is underlined earlier in the novel when Lavenham, feeling threatened by Savak, seeks refuge in a church: »The preacher was very impressive [...]. I listened in a species of warm drowsiness which had succeeded my wakeful night and was perhaps further induced by the notion of security under that roof greater than I could feel under any other.« (Ibid.: 32-33)[9] Against the background of contemporary Britain, the return to Christianity, which itself had lost influence and meaning due to new insights in fields like biology and geology and due to the crisis of the Anglican Church in the second half of the nineteenth century, can be read as both the attempt and need to retrieve stability and comfort in a time characterised by decline and regression.[10] Lavenham and Iras's fate seems to show what happens once progress and prosperity fail and the uncivilised and pagan past regains control over British middle-class lives, making impossible a settled and enduring family life. That Iras refers to the pendants on her necklace as »›seven drops [symbolising] [...] seven spaces of time – whether long or short [she does not] know‹« (ibid.: 85) hints at the unpredictability of the approaching doom, which mirrors the general apocalyptic atmosphere towards the end of the

8 This is also supported by the indignation of Mrs Mappinbeck, Lavenham's landlady, upon the arrival of the mummy case at Lavenham's apartment in London, as alluded to by Lavenham: »She was a lone woman and a widow, and had taken me in as a gentleman; and now I was taking her in by turning her respectable lodgings into a charnel and a dead-house for disreputable *heathen* corpses.« (*Iras*: 62, emphasis mine)

9 Although Lavenham also discovers Savak in the crowd at the church, the Egyptian does not threaten him this time (cf. *Iras*: 33).

10 The disappearing of the seven pendants from Iras's necklace could also be read as a reversed biblical genesis, i.e., a process of degeneration as opposed to a process of creation.

century. The collective fear of the coming end naturally resulted in the re-strengthening of traditional middle-class values and belief systems such as Christianity.

7.3 THE INEXPLICABLE EVIL

> The eternal gloom which hangs upon his brow, seems to bespeak a fallen angel, for such is the deadly hate that must have animated the rebellious spirits when expelled from heaven. His look is terrific; and my blood froze in my veins at his horrid laugh, which seems to ring in my ears like the mockery of fiends when they have involved a human being inextricably in their toils.
> (*MUMMY* VOL. I: 24)

Whereas the evil of the past and its threat to the present discussed so far are still causally connected, in that this evil can be seen as a reaction to a wrong suffered in the past, the past is also often associated with causeless evil. This is particularly the case for the seemingly inexplicable evil of Pharos's character, which calls to mind the *Jekyll and Hyde Discourse*. Although at first glance it could be argued that the figure of Pharos is thoroughly evil and thus is essentially lacking the ›Jekyll‹ part, the fact that Pharos, in spite of his Hyde-like appearance,[1] is an accepted and esteemed member of (British) society[2] shows that he possesses qualities equal to that of the respected Dr Jekyll. However, as in *Dr Jekyll and Mr Hyde* it is the ›Hyde‹ part of Pharos that is most instructive in regard to Pharos's relevance for the novel and coeval society.

1 Similar to Hyde, Pharos is repeatedly referred to as being of a small stature and having an ugly and evil countenance (cf. *Pharos*: 28, 36, 142, 315). Furthermore, he walks on a stick, which is also reminiscent of the adnate and crippled impression Hyde's figure evokes (cf. *Jekyll & Hyde*: 34).

2 This is emphasised by Forrester's following comment: »How Pharos became a member of [the Antiquarian Club], I do not know; but there, as at Pompeii, Cairo, Luxor, Prague, and, indeed every other European city, he seemed to be not only quite at home, but equally well known.« (*Pharos*: 226)

An instance in the novel which is most prominent in regard to the inhuman wickedness of Pharos's character, and essentially likens him to Hyde, is Pharos's reaction upon witnessing a suicide at the Embankment late at night. Pondering on the past in front of Cleopatra's Needle,[3] Forrester is also close to the scene and becomes aware of the incident, when he hears the suicide victim call out for help from the water. When Forrester hastens towards the river in order to offer help, he perceives another man close by, who later turns out to be Pharos, watching the drowning man without interfering:

»Thirty seconds must have elapsed since we had heard the cry for assistance, and now, as I looked, the drowning man was washed in at the foot of the steps upon which we stood. It would have needed but the least movement on the part of the man below me to have caught him as he swept by and to have saved him from a watery death. To my amazement, however – and even now, after this lapse of time, my gorge rises at the very thought of it – the other did not offer to help, but drew himself back. Before I could return my eyes, the wretched suicide had passed out of sight and had vanished into the darkness again. As he did so a pronounced chuckle of enjoyment reached me from the man below – a burst of merriment so out of place and so detestable that I could scarcely believe I heard aright.« (*Pharos*: 26)

The course of events in this scene and the inhuman cruelty displayed by Pharos bears strong parallels to *Dr Jekyll and Mr Hyde*, where Enfield's first encounter with Hyde is also characterised by the brutal behaviour towards another person, in this case a little girl, who is run down by Hyde in a deserted street late at night.[4] Though Enfield, in contrast to Forrester, is able to hold the delinquent to account, his shock and hatred in the face of such unheard-of behaviour of a human being is nearly identical to that of Forrester.[5] Worse than the cruelty of the respective deed, however, seems to

---

3 In this context, Cleopatra's Needle also functions as a site of memory that both within Forrester (his thoughts about the past) and around him (Pharos) evoke the past.

4 In fact, the two scenes are so similar in regard to the actions, reactions, and happenings that Boothby might have used Stevenson's novella as a model.

5 Forrester and Enfield are equally struck by the extraordinary appearance and bearing of the offender, with a particular emphasis on the inhuman and cruel

be the causeless character of the crime as well as the (obvious) enjoyment the offenders gain from it.[6] This is central to the frightening effect Pharos and Hyde have on their ›opponents‹ and eventually also on society. They both personify evil for the sake of evil and thus are not only immoral but amoral in their deeds. Even though we learn eventually that an explanation can actually be found for Pharos's inhuman behaviour, it remains in the background for quite a long time and then is only revealed to Forrester and Valeric. For society in general, however, the motivation for Pharos and Hyde's deeds remain undisclosed, and thus their evil remains inexplicable.[7] The combination of an amoral and inexplicable evil within a society characterised by moral values and rationality is perceived as horrifying and shocking:

»One thing was quite plain, however profoundly I may have been affected by my proximity to this singular being, I was not the only one who came within the sphere of his influence. Indeed, it was strange to notice the manner in which the polite crowd drew away from him, and the different expressions upon their face as they stepped aside in order to give him room to pass. Had he been a snake, they could scarcely have shown a more unanimous desire to withdraw from his neighbourhood.« (Ibid.: 36)[8]

---

eyes and gaze of the delinquent (cf. *Pharos*: 27-28 and *Jekyll & Hyde*: 33). Likewise, both Forrester and Enfield feel a strong desire to murder the offender (cf. *Pharos*: 27 and *Jekyll & Hyde*: 32).

6 Unlike Pharos, who openly enjoys the drowning of the suicide (cf. above), on this occasion Hyde does not show any traces of enjoyment. However, we later learn from Jekyll's notes that in the role of Hyde he too indulges in his evil deeds.

7 In Hyde's case, this is also true of the murder he commits on Sir Danvers Carew.

8 The collective repulsion Pharos stirs is further made explicit in the following observation by Forrester: »Many and curious were the glances thrown at my companion by the people about us as we entered, and I noticed, with a satisfaction for which I could scarcely account at the time, the expression of fear and undisguised abhorrence his unique personality inspired in them.« (*Pharos*: 332)

In the light of Victorian society, Pharos's subversive form of inhumanness can be read as another expression of the fear of moral decline and degeneration that eventually leads to the prevalence of evil. Characteristically, it is this inhumanness that both Forrester and Enfield claim to perceive in the appearance of Pharos and Hyde: »›There is something wrong with [Hyde's] appearance; something displeasing, something downright detestable. [...] He must be deformed somewhere; he gives a strong feeling of deformity, although I couldn't specify the point‹« (*Jekyll & Hyde*: 34). Much as Hyde's physical deformity mirrors his inner moral corruption, Pharos's appearance points towards his evil character, as illustrated by Forrester's description: »It was his eyes, the shape of his face, the multitudinous wrinkles that lined it, and, above all, the extraordinary colour of his skin, that rendered his appearance so repulsive. [...] His eyes were small, deeply sunken, and in repose apparently devoid of light and even of life.« (*Pharos*: 28) Although the motivation for Pharos and Hyde's behaviour is unknown to Forrester and Enfield at this point, they account for it by using physical characteristics, which bear strong racial, anthropological, and medical connotations.[9]

9 This is again reminiscent of the discrimination against Frankenstein's creature. Cf. Howard L. Malchow (1993). A similar pattern in regard to these categorisations can be observed in Loudon's *The Mummy!*, where Cheops's appearance is introduced in terms of ›otherness‹ from the beginning: »The dried distorted features of the Mummy looked yet more hideous than before, when animated by human passion, and his deep hollow voice, speaking in a language [Edric] did not understand, fell heavily upon his ear, like the groans of fiends.« (*Mummy* Vol. II: 254) Significantly, this ›otherness‹ is interpreted by all the characters in the text as the expression of pure evil, as illustrated by Father Morris's statement: »›I have again seen the Mummy! that fearful spectre from the tomb. [...] I am not timid, but my very soul recoiled from the hideous aspect of that awful being; the cerecloths of the grave are still wrapped round him, his fearful eyes glare with unearthly lustre, and his deep sepulchral voice thrills through every nerve.‹« (Ibid.: 23) Here the mummy's »hideous aspect« (ibid.) is enforced by the fact that he is a »spectre from the tomb« (ibid.), which connotates both death and the past. This underscores how the ›otherness‹ of the mummy as perceived by the characters is enforced by associations with death, decay, and the past, which all seems to legitimise and explain the »unearthly lustre« (ibid.) in the

In this respect, Enfield's concluding remarks get to the point when he sums up Hyde's appearance as follows: »›He's an *extra*ordinary looking man.‹« (*Jekyll & Hyde*: 34, emphasis mine) Pharos and Hyde's physical features make them different from the ordinary, revealing both men as abnormal.[10] The marked ›otherness‹ of Pharos and the mummy in fiction in general also points towards the common idea of the similarity of allegedly ›primitive people‹ to less developed human civilisations of the past and, associated with that, the fear of modern society of being infiltrated by these less developed, degenerate civilisations (cf. below).

The categorisation and classification of people according to their ›normal‹ and ›abnormal‹ appearance as well as its assumed significance for a person's character immediately calls to mind the Victorian obsession with physiognomy and phrenology.[11] Since these pseudosciences were based on the idea that a sound state of society could be ensured by identifying certain

mummy's eyes. In Doyle's »Lot No. 249«, the hideousness of the mummy is even transferred to its ›owner‹, Edward Bellingham, as described by his fellow student Jephro Hastie: »›There's *something damnable* about him – *something reptilian. My gorge always rises at him.* I should put him down as a man with secret vices – an *evil liver*.‹« (*Lot No. 249*: 50, emphasis mine) In place of his mummy, which at this point has not made its appearance yet, Bellingham is stigmatised as the ›other‹, as emphasised by Hastie associating him with a reptile, implying both danger and exoticism. Interestingly, similar to Forrester's reaction upon meeting Pharos for the first time, Hastie feels *physically* sick in contact with Bellingham, which suggests an allegedly *natural* and thus *true* reaction of the body against any *unnatural* and thus dangerous elements invading a (sound) community. Before actually being able to see and rationally grasp the intruder, the characters can *feel* him, which once more renders him similar to an unknown disease befalling the human body, which at first, though felt by the person concerned, is not yet ›known‹, i.e., identified, and consequently a potential danger to the whole community. This is also reminiscent of the smell in *Jewel* associated with the past, as discussed in Chapter 1.2.

10 Cf. also: »Evil […] had left on that body an imprint of deformity and decay.« (*Jekyll & Hyde*: 84)

11 Cf. John Van Wyh (2004), Sharrona Pear (2010). On the relevance of physiognomy in Stoker, cf. Glover (1992) and (1996).

individuals on grounds of their alleged ›otherness‹, it seems interesting to question what makes Pharos appear to be different.

Apart from his misshapen body, Pharos's face is also referred to by Forrester as horrifying upon their first encounter: »To understand what I mean you must think first of old ivory, and then endeavor to realize what the complexion of a corpse would be like after lying in an hermetically sealed tomb.« (*Pharos*: 28) The image »of a corpse« (ibid.) lying »in an hermetically sealed tomb« (ibid.) evokes the mummified body of a long-dead but well-preserved individual, an impression which is further emphasised by the old age Forrester ascribes to Pharos (cf. ibid.: 28).[12] Thus, though Forrester is yet unable to interpret what he perceives in Pharos's face, his description of it reveals Pharos's genuine identity and origin.[13]

The fact that it is Pharos's mummy-like face »that render[s] his appearance so repulsive« (ibid.) to Forrester is instructive, since it reveals not only an underlying fear of death but also a threat associated with the past, that is, a dread of the past to the present. Yet unlike the noble past of ancient Egypt propagated by Pharos later in the novel, the association here is clearly that of old age and decline: »[Pharos] walked feebly, supporting himself with a stick, upon which his thin yellow fist was clutched till the knuckles stood out and shone like billiard balls in the moonlight.« (Ibid.) Pharos amounts to the remains of a high culture which has vanished completely, and like the ancient ruins in modern Egypt he is only a shadow of his former self, as his old and fragile frame suggests. That Pharos finally fails to prevail in the modern age is made explicit when at the end of the novel his process of physical degeneration is completed shortly before he destroys himself:

»In a large armchair before a roaring fire, though it was the middle of the summer, sat Pharos, but so changed that I hardly knew him. He looked half his usual size; his skin hung loose about his face, as if the bones had shrunken underneath it; his eyes, always so deep-set in his head, were now so much sunken that they could scarcely

12 The corpse-like appearance of Pharos is also very similar to that of Cheops, as described by Father Morris.

13 Pharos turns out to be the alter ego, if not the living embodiment, of Ptahmes's mummy (representing Pharos's true identity) from the time of the ancient Egyptians (standing for Pharos's origin).

be seen, while his hands were shrivelled until they resembled those of a mummy more than a man.« (Ibid.: 372)

Apart from Pharos's resemblance to a mummy, his shrivelled and shrunken body also evokes the association with earlier stages of human development, which is further underscored by Pharos's permanent companion, the monkey Pehtes.[14] Against the background of contemporary society, one main aspect of the threat people perceive as coming from Pharos is the fact that he embodies, and in the course of the novel undergoes, the degradation of a once powerful man to an ape-like creature, a sharp and disconcerting parallel to Darwin's evolutionary theory, conceived by many as a ubiquitous dread.[15] That degenerated creatures like Pharos and Hyde are unnatural and bound to perish is emphasised by the suicide both Pharos and Hyde finally commit. Pure evil, which finds its expression in the degenerate forms of Pharos and Hyde, ultimately leads to self-destruction.

In spite of the abhorrence Pharos's character and appearance evokes, there is also something to it which deeply fascinates and attracts Forrester. This is underlined by Forrester's description of Pharos's face when Forrester watches him secretly in his sleep: »Anything more fiendish could scarcely be imagined. It was not the face of a human being, but that of a ghoul, *so repulsive and yet so fascinating was it.*« (Ibid.: 151, emphasis mine)[16] This scene bears a close resemblance to the scene in *Jekyll and Hyde* in which Jekyll for the first time sees himself in the small and deformed body of Edward Hyde in the mirror of his laboratory: »›And yet when I looked upon that ugly idol in the glass, I was convinced of no repugnance, rather of a leap of welcome.‹« (*Jekyll & Hyde*: 84) In the further course of his letter to his friend, Jekyll goes on to elaborate on the »provinces of good and ill which divide and compound man's dual nature« (ibid.: 81). Jekyll, however, as much as any of his contemporaries, was never allowed to live this »profound duplicity of life« (ibid.), since he had to

14 This bears strong parallels to Ayesha's devolution at the end of Haggard's *She*.

15 This parallels the reverse creation mentioned in Chapter 7.2 in regard to the disappearing pendants on Iras's necklace.

16 The sensationalist pleasure Forrester experiences upon looking at Pharos's face bears similarities to the pleasure people got out of freak shows or mummy unwrappings.

›hyde‹ his ›negative‹ side behind a »grave countenance before the public« (ibid.). The very fact that Forrester is both attracted and repelled by Pharos as the embodiment of evil can be seen in a similar light. In a society which claims morality and severity as its core values, there is no place for anything that deviates from these paradigms. As a consequence, the sides of human nature that are considered negative by the collective are suppressed, as Jekyll depicts in his letter. As in the case of Jekyll, this may ultimately lead to the radical eruption of evil and bring forth creatures like Hyde: »Edward Hyde, alone, in the ranks of mankind, was pure evil.« (Ibid.: 85) The extreme embodiment of evil in figures such as Hyde is therefore the result of the repressive circumstances and double standard of a society that requires rigid and restrictive behavioural patterns of its members.

Forrester's attitude towards Pharos is characterised by the ambivalence of fascination and repulsion, since on the one hand Pharos embodies what is banned from the contemporary Victorian discourse, although it is essentially part of »man's dual nature« (*Jekyll & Hyde*: 81). On the other hand, Forrester also knows that Pharos personifies the forbidden fruit which it would be sin to taste. In other words, Forrester's strong bond to Pharos is not only due to the fact that in contact with and through Pharos Forrester is allowed to display a certain ›unmanly‹ behaviour (cf. Chapter 6.4), but also because Forrester recognises himself to some extent in Pharos.

That Forrester does not stand alone in this respect can be read in society's reaction towards Pharos, as observed by Forrester at a ball: »Much to my astonishment, [Pharos] seemed to have no lack of acquaintances, and I noticed also that everyone with whom he talked, though they paid a most servile attention to his remarks while he was with them, invariably heaved a sigh of relief when he took his departure.« (*Pharos*: 334) The fact that Pharos is likewise known and dreaded by everyone present can be read symbolically in the sense that people recognise in Pharos their own evil nature, which, however, they are anxious to suppress. Consequently, they are relieved once Pharos has left them again without revealing this secretive and hidden part of them in public. Another instance that points towards a connection between Pharos's evil character and contemporary British society is Forrester's reference to Pharos's face as not being that of »a human being, but that of a *ghoul*« (ibid.: 151, emphasis mine). If we take the word »ghoul« (ibid.) to refer to »[a]n evil spirit supposed (in Muslim countries) *to rob graves and prey on human corpses*« (*OED*, emphasis mine) and further consider that in the eyes of Pharos the English are »people who rifle the tombs of the dead kings and queens of Egypt« (*Pharos*: 331), what Forrester sees in Pharos's face can be read as a subliminal recognition of his own fiendish self in place of Victorian society as a whole. This, then, is part of the reason why Forrester feels both fascination and repulsion while watching Pharos's face.

7.4 The Legacy of Bygone Times: The Power of the Past and the Corruption of the Present

> Bad smells indicate the presence of corruption. Whenever corruption is present, there is danger to health. IZAL removes bad smells and destroys the germs which are the cause of infection.
> (*Illustrated London News*, May 4, 1907: 705)

Although the exploitation of the British colonies in the realm of the Empire had found broad support throughout the century, there were also more critical voices that increased towards the end of the nineteenth century. This is underlined by Haggard's article »Trade in the Dead« (1904),[1] in which Haggard openly criticises and condemns the British violation of ancient Egyptian tombs and their mummies: »Should not we English shudder if some seer told us that within a given number of years, say 3,000 […], those who rest in Westminster Abbey were destined to be treated in just this fashion, to satisfy the curiosity of men unborn?« (*Trade in the Dead*: 145) The parallel Haggard draws between his modern British contemporaries and the ancient Egyptians, on the basis of a hypothetical similarity in the treatment of their dead, suggests an underlying feeling of guilt as the consequence of an identification with the past.[2] In the present chapter, I will

---

1 »The Trade in the Dead« was published in the *London Daily Mail* on June 4, 1904.

2 This identification process of the present with the past is also underlined by Margaret Trelawny's reaction to her father's intention to unroll the mummy of the Egyptian Queen Tera in Stoker's *Jewel of Seven Stars*: »›Father, you are not going to unswathe her! All you men …! And in the glare of light! […] Just think, Father, a woman! All alone! In such a way! In such a place! Oh! it's cruel, cruel!‹ […] However, [her father] began to appeal to her reason: ›Not a woman, dear; a mummy! She has been dead nearly five thousand years!‹ ›What does that

use *Pharos* as an example to show in what way British Imperialism, as it materialises in the archaeological discourse, can be seen as characterised by a suppressed guilt with its treatment of the temporal and spatial ›other‹.

When Pharos appears in the middle of the night at Forrester's studio, Forrester, although he has met him only thrice,[3] is already under Pharos's spell: »Lightly as I proposed to myself to treat it, [Pharos's] extraordinary individuality must have produced a deeper effect upon me than I imagined, for, as in the afternoon, I soon discovered that, try to divert my thoughts from it how I would, I could not dispel his sinister image from my mind.« (*Pharos*: 48) It is in this state of mind that Forrester returns to his studio, which is »decorated in the Egyptian fashion« (ibid.: 49), late that night. Walking through his room and looking at the various ancient Egyptian decorations he has seen many times before, Forrester suddenly becomes aware of a change in them: »It was as if the quaint images of the gods, which decorated the walls, were watching me with almost human interest, and even the gilded face upon the mummy case, in the alcove at the farther end, wore an expression that I had never noticed on it before.« (Ibid.) While his previous encounters with Pharos have quickened Forrester's interest in ancient Egypt, occupying his thoughts accordingly, the change in his perception alluded to at this point represents an important instance in regard to his further development in the novel. It demonstrates how Forrester's inner mental state comes to influence his perception of his surroundings, which he now sees in a changed light and which marks the beginning of an identi-

matter? Sex is not a matter of years! A woman is a woman, if she had been dead for five thousand centuries!‹« (*Jewel*: 230-231), as well as by a statement of Monny, who together with her friends penetrates an ancient Egyptian royal tomb in Williamsons's *It Happened in Egypt*: »›[W]hy, after all, where's the great difference between opening the coffin of a woman dead thousands of years ago, or a few months? Supposing people wanted to dig up Queen Elizabeth, to see what had been buried with her? Or Napoleon? What an outcry there'd be all over the world. This poor queen is defenceless, because her civilization is dead, too. Could *you* force open the lid of her coffin [...] and take the jewels off her neck?‹« (*It Happened in Egypt*: 510)

3 »Monsieur Pharos, the man I had met at the foot of Cleopatra's Needle some weeks before, at the Academy that afternoon, and at Medenham House only a couple hours since.« (*Pharos*: 51)

fication process with the ancient Egyptian past. Motivated by the impression Pharos has left on him,[4] Forrester projects his own strange and ominous feelings onto the Egyptian images and artefacts in his studio, thus alienating them from their modern context and empowering them at the same time. Not only do the things seem to have changed their meaning, they also seem to have come alive, turning from inert objects into anthropomorphised things, »*watching* [him] with an almost *human interest*« (ibid.), as noted by Seed in regard to Stoker's texts: »Less luridly these narratives generally describe a fantastic process where, as it were, the commodity ceases to be an inert object and in some senses come alive.« (Seed 1998: 189) Instead of being watched by Forrester, the things now watch *him*, thus inverting the subject-object relation by turning him into the object,[5] which is being watched with »almost human interest« (ibid.):[6]

»Thus, the centripetal force exuded by the objects of antiquity further threatens to supplant interpersonal relations with bonds between people and things – bonds that imply the reversal of the traditional power structure governing the objects's subordination to the subject.« (Bridges 2008: 144)[7]

In the novel this is further emphasised by the fact that Forrester thinks of the golden face on the mummy case as having changed its expression (cf. *Pharos*: 49) and further even imagines the mummy addressing him: »It

4 This is emphasised by Forrester's address of Pharos: »›For some reason or other I had never read the translation until tonight. I suppose it must have been my meeting with you that put the idea into my head.‹« (*Pharos*: 55)

5 Cf. also Chapter 6.3.

6 This impression is emphasised if we consider Pharos's accusation of modern excavators later in the novel, where he refers to his ancestors as being displayed at museums to be looked at: »Look about you. Where are those mighty ones now? Scattered to the uttermost parts of the earth, stolen from their resting-places *to adorn glass cases in European museums*, and to be sold by auction by Jew salesmen at so much per head, according to their dates and state of preservation.« (*Pharos*: 206-207, my emphasis)

7 Cf. also »Drawn into the history of the object he seeks to collect, the would-be collector is likely to become a collectible object himself rather than enhancing his status as a subject.« (Daly 1994: 41)

might have been saying, ›Ah, my nineteenth-century friend, your father stole me from the land of my birth, and from the resting-place the gods decreed for me; but beware for retribution is pursuing you, and is even close upon your heels.‹« (Ibid.: 49) Here, the implicit menace coming from the anthropomorphised Egyptian artefacts is made explicit through a verbalised threat, which is enforced by the announced temporal closeness. However, this threat is not simply pronounced as an inexplicable threat but as a call for retribution for a misdeed committed in the past. The very fact that the mummy has been stolen from its tomb by Forrester's father represents the collective abuse of Egyptian mummies by the Western world (cf. ibid.), as made explicit by Pearce:

»The mummies were torn from their tombs [...] by early nineteenth-century explorers such as Giovanni Battista Belzoni and Henry Salt, and [...] deprived of their substance by examiners like Thomas Pettigrew and Giovanni D'Athenasi. The people they once were, were robbed of their own life stories to be turned by exhibition into a carnival for London audiences where the fragments of their own stories written on their coffins and bandages were used as historical titillation, rather than history. The audience, however, like all audiences, had their own stories of exile.« (Pearce 2002: 55)

Thus, Forrester, with the death of his father, inherited not only the mummy but also the guilt attached to it, which turns the original guilt into a hereditary debt. This feeling of guilt, which becomes the underlying controlling principle of the novel, is significant for Forrester's further actions and refers to the *Frankenstein Discourse*, since Frankenstein is also possessed by the feeling of guilt for the atrocities committed by the creature he has formed: »The mummy's curse sequence of resurrection-for-revenge entails a return of the repressed – or rather, a return of the *oppressed* and (temporally) *regressed*.« (Bridges 2008: 139, emphasis mine) Even though Forrester has not committed the initial sin himself, as will become clearer later, he stands for the collective crime the Western colonisers have committed against the spatial and temporal ›other‹. Against this background, the above passage is important since it illustrates how – triggered by his first few encounters with Pharos – Forrester seems to have become susceptible to a mysterious influence that draws him into the past. It is through that past that his own family history is interlinked with that of ancient Egypt, which essentially marks the beginning of identification with the spatial and temporal ›other‹. Under the influence of Pharos, Forrester, confronted with the familiar antiquities in his studio, for the first time seems to become aware of their very presence and meaning. This recognition takes him away from the present time and space as he takes up his father's notebook to look for further information on the mummy:

»Entries there were in hundreds; records of distances travelled, of measurements taken, evidence as to the supposed whereabouts of tombs, translations of hieroglyphics, paintings, and inscriptions [...] but for some time no trace of the information I required. At last, however, it struck me to look in the pocket contained in the cover of the book. My diligence was immediately rewarded [...]. According to the statement here set forth, the coffin contained the mortal remains of a certain Ptahmes, Chief of the King's Magicians – an individual who flourished during the reign of Merenptah (Amenepthes of the Greeks, but better known to the nineteenth century as the Pharaoh of the Exodus).« (*Pharos*: 50)

The Egyptian antiquities, and in particular the mummy case, function as a doorway to another world and time beyond everyday life in London. As has been shown above, this retreat from the present to the past metaphorically also implies the individual's orientation away from an outer world towards an inner world, that is, a preoccupation with one's feelings and memories. In Forrester's case this is emphasised by the fact that he reflects on the character and feelings which the mummified body in front of him might have had when alive:

»Once more I stood looking at the stolid representation of a face before me, wondering what the life's history of the man within could have been, whether his success in life had equalled his ambition, or was commensurate with his merits, and whether in that age, so long since dead, his heart had ever been thrilled by thoughts of love.« (Ibid.: 51)

The confrontation with the past in the form of the mummy causes Forrester to project individual human characteristics and feelings such as love onto the dead body, thereby subjectifying the past by endowing it with alleged human emotions. In doing so, Forrester betrays his growing identification with the mummy.

In Forrester's case the mummy as well as the other Egyptian relics inherited by his father thus function both as remains of an anonymous general past *and* the personal past of the Forrester family and thus represent both a collective and cultural memory. This is further emphasised by the fact that Forrester inherited not only the Egyptian artefacts from his father but also his father's fascination[8] for ancient Egypt:

»As the world has good reason to remember, his father was perhaps the most eminent Egyptologist our century has seen; a man whose whole mind and being was impregnated with a love for the ancient country and its mystic past. Small wonder, therefore, that the son should have inherited his tastes, and that his life should have been influenced by the same peculiar partiality.« (Ibid.: 10)

In other words, in the mummy and the other Egyptian things Forrester inherited from his father, individual family and collective world history symbolically merge. This is underlined by Pharos's lamentation on the mummy upon his intrusion into Forrester's studio that night: »›Oh, mighty Egypt! hast thou fallen so far from thy high estate that even the bodies of thy kings and priests may no longer rest within their tombs, but are ravished from thee to be gaped at in alien lands.‹« (Ibid.: 56) Although later in the novel we learn that the mummy is actually Pharos's own mummified body, at this point it stands for a whole collective of royal and priestly bodies stolen from their tombs, thus representing a collective ancient Egyptian past. This collective Egyptian past has been disintegrated and desecrated by modern men like Forrester's father, who have removed single mummies and artefacts from their integrated historical context. Disintegrated from their historical and geographical background, both mummies and artefacts are bereaved of their identity and meaning. »›Gaped at in alien lands‹« (ibid.), they are reduced to mere objects which serve as a projection screen for the curious (male) gaze of the modern observer: »›Your father was, I know, an ardent Egyptologist, one of that intrepid band who *penetrated* to every corner of our sacred land, *digging*, *delving*, and *bringing to light* such tombs, temples, and monuments as have for centuries lain hidden from the sight of man.‹« (Ibid.: 57, emphasis mine) In this comment by Pharos, the

8 Cf. also *Osiris*, where the fascination for ancient Egypt runs in the Bellingham family.

verbs »penetrate« (ibid.), »digging« (ibid.), and »delving« (ibid.) clearly have a sexual connotation, while the phrase »bringing to light [...] [that which has] lain hidden from the sight of man« (ibid.) can also be read as a reference to giving birth (cf. Chapter 4.1.). In other words, the archaeologist here is seemingly both the procreator and the bearer of what is finally delivered to the eyes of the modern world. The very fact that in this context the role attributed to Forrester's father can be read as that of a male creator again calls to mind the creation of Frankenstein's creature.

However, that this symbolical creational act through Forrester's father – like that of Frankenstein – is depraved right from the beginning is revealed by the fact that Pharos refers to Egypt as »our sacred land« (ibid.) and thus exposes the ostensive act of creation as the rape of his *mother* country, which renders the things finally produced to the world stained with guilt (cf. Chapter 6.4).[9] This is pointed out by Pearson, who refers to John Barrell quoting »William Prime's 1857 account of his amateur archaeological experiences in Egypt to suggest the potency of Egypt to invoke Western male fantasises. Prime imagines a tomb-robbery and ›rape‹ of a female mummified corpse« (Pearson 2000: 223). It is this guilt which, along with the Egyptian artefacts, is bequeathed by Forrester's father to his son. Forrester, however, is unaware of this guilt, since for him the mummy is essentially part of his own family history, and therefore he refuses to return it to Pharos: »›[Y]onder mummy should pass into my possession.‹ [Pharos said] ›Impossible!‹ [Forrester] answered. ›I could not dream of such a thing! It was one of my poor father's greatest treasures, and for that reason alone no consideration would induce me to part with it.‹« (*Pharos*: 58-59)

Yet even though Forrester fails to recognise the inherited guilt attached to his father's antiquities, he is very much aware of his moral responsibility as regards the two capital offences he suspects Pharos to be involved in – the first being a suicide at the beginning of the novel, the second a murder at a curiosity shop in Forrester's neighbourhood during the night of Pharos's visit to Forrester's studio. The fact that Forrester is on the one hand unable to recognise and accept responsibility for his father's theft of the mummy but on the other hand feels responsible for Pharos's alleged crimes is particularly illuminating if we consider it in regard to the *Fran-*

9 In a broader context this can also be read as a reference to the aggressive colonialisation of the Orient in the realm of the British Empire.

kenstein Discourse. As with *Frankenstein*, where the initial act of making the creature is closely linked to the creature's future behaviour, in *Pharos* the act of excavating the mummy can be seen as being indirectly responsible for Pharos's deeds in the further course of the novel. In both cases, the initial act of creation and excavation involves a certain guilt and moral digression attributed to the creator and excavator, in the former case for disturbing the laws of nature and in the latter for disturbing the peace of the dead. However, both Frankenstein and Forrester fail to accept the responsibility for their guilt, which becomes explicit through the fact that they refuse to provide the creature with a mate and to return the mummy to Pharos, respectively.

As a consequence of their unwillingness to act according to the wishes and needs of the initially victimised creature and Pharos, Frankenstein and Forrester become indirectly responsible for the creature and Pharos's outrages. That Frankenstein and Forrester are eventually to blame for the creature and Pharos's deeds, respectively, is underlined by the fact that neither of them is able to report to the authorities. While Frankenstein seems to fear the impact his confession might have on society, the motivation for Forrester's inability to inform the police of his suspicions concerning Pharos seems less obvious. When the police inquire at his door and inform him of the ghastly murder at a curiosity shop close by, Forrester finds himself unable to tell them of his suspicion concerning Pharos: »An irresistible force was at work within me, compelling me, even against my will, to screen [the policeman], and to tell *the first deliberate lie* to which I had ever given utterance in my life.« (Ibid.: 67, emphasis mine)[10] The fact that Forrester claims »to tell the first deliberate lie« (ibid.) of his life here is crucial, since it marks the first step towards both an alienation from his self and an isolation from his surroundings. By not communicating his know-

10 This is further emphasised by the fact that after the suicide at Cleopatra's Needle, during which both Pharos and Forrester are present, Forrester flees from the scene of the crime once the police arrive without informing the authorities of what he has witnessed: »A police boat was pulling towards the steps, and by the light of the lantern on board I could make out the body of a man. My nerves already strained to breaking pitch, were not capable of standing any further shock. I accordingly turned upon my heel and hurried from the place with all the speed at my command.« (*Pharos*: 29)

ledge to the police, Forrester enters into a sphere which for the time being is only shared by Pharos and himself. As in *Frankenstein*, this is a sphere which is beyond controllable reality, dominated by laws of its own. Isolated from society, Frankenstein and Forrester become obsessed by the urge to find their respective opponent and to confront him with the crimes he has committed in order to salve their own consciousness: »I would seek Pharos out, accuse him not only of the theft, but of the murder, and make him understand, with all the earnestness of which I was master, that justice should be done, and that I would no longer shield him for the consequence of his villainy.« (Ibid.: 72)

From this point on, Forrester is determined to find Pharos, since he will »›know neither peace nor happiness until [he] [has] seen him and spoken to him face to face‹« (ibid.: 81). Furthermore, he is sure that »›[i]f [he] ever wish[es] to be able to look upon [himself] as an honourable man, [he] *must*, do it‹« (ibid.). Again, Forrester's second statement emphasises that he feels a strong moral responsibility for the crime committed by Pharos. However, as with Frankenstein, it is not only the strong suspicion Forrester has as far as Pharos is concerned which makes him feel responsible, but rather that he subconsciously knows that he himself has indirectly induced the murder at the curiosity shop. This is alluded to towards the end of the novel when Pharos finally confesses to the murder of the curiosity shop dealer: »›I stabbed him because he would not give me a certain scarabeus.‹« (Ibid.: 357) Like Forrester, the curiosity shop dealer refuses to return an ancient Egyptian relic once stolen from Egypt to Pharos, who, as a representative of his ancient people, is the rightful owner of the scarabeus. Consequently, Pharos's killing of the curiosity shop dealer can be read as an act of retaliation by the initial victim on the perpetrator. Pharos and the curiosity shop dealer stand respectively for collective revenge and guilt: Pharos kills for the wrongs suffered by his people, the ancient Egyptians, and the curiosity shop dealer dies for the wrongs committed by his people, the modern Europeans. Accordingly, the curiosity shop dealer also dies in place of Forrester, who perpetuates his father's initial guilt of violating ancient Egyptian tombs and sites by refusing to give up the mummy to Pharos. To some extent Forrester seems to be aware of this when he reflects on the mummy's disappearance after Pharos has knocked him out and taken the mummy with him during the night of the murder: »I went to the place where, until a few hours before, the mummy had stood – that *wretched mummy which had been the cause of all the trouble.*« (Ibid.: 71, emphasis mine)

In the further course of the novel, Forrester is bonded to Pharos by an invisible cord, like Frankenstein to his creature. Pharos already alludes to this during his nocturnal visit at Forrester's studio: »›Though you are not aware of it, there is a *bond between us that is stronger than chains of steel.*‹« (Ibid.: 57, emphasis mine) That Pharos is proved correct is shown by the fact that after he has disappeared with the mummy, Forrester starts to stalk Pharos in much the same way as Frankenstein haunts the creature. Inquiring after him at the Egyptian Museum in London and learning that he has left England for Italy, Forrester immediately follows Pharos to Naples. By leaving England for Italy, Forrester makes a further crucial move away from his familiar everyday life towards the unknown archaeological space. The fact that Forrester seems to know Naples (cf. ibid.: 83), as well as its relative proximity to Western culture, suggests that at least at this point Forrester still has the theoretical chance of influencing the further course of events. Also, Forrester's »thoughts no longer [centre] themselves on Pharos« (ibid.: 85) but rather on »the lovely face of his companion« (ibid.), which is significant in regard to Forrester's future decisions. Yet in spite of its closeness to the modern world, Naples also evokes death, decay, and bygone times in the summer with its deserted and dusty places:

»›If there is one place more than another for which I entertain a dislike that is akin to hatred, it is for Naples in the summer time – that wretched period when everyone one knows is absent, all the large houses are closed, the roads are knee-deep in dust, and even the noise of the waves breaking upon the walls of the Castello del'Ovo seem unable to detract from the impression of heat and dryness which pervades everything.‹« (Ibid.: 83)

Thus, Naples can be seen as a liminal space, since through its geographical and cultural situation it marks the border between the western and the eastern hemisphere, which in the novel symbolises on the one hand the modern world of reality and on the other the ancient world of adventure and fantasy.

However, it is not at Naples that Forrester is eventually able to arrange a meeting with his opponent, but at Pompeii. And yet, what other city would be better suited for an encounter with Pharos than Pompeii, the city of the dead? Although the buried and ruined city evokes multiple associations, it can be connected with Pharos for two reasons: First, as the gravesite of countless deaths it foreshadows Pharos's ultimate plan, which is a mass murder on modern humanity by means of a lethal plague. And second, Pompeii is a place where the past is most immediately sustained in the present. Like a snapshot in time, the very presence and actuality of a bygone time has been preserved here over centuries in its immediacy by the deathly volcanic ash. In other words, Pompeii can also be seen as symbolising the past *in* the present. This is significant in regard to Pharos, who is himself a modern anachronism. Tellingly, it is at this historical site that Pharos informs Forrester of his family history and the alleged meaning the mummy has for him:

»›You have already been informed by our mutual friend, Sir George Legrath, [...] that I am of Egyptian descent. Perhaps you do not understand that, while the ancient families of your country are proud of being able to trace their pedigrees back to the time of the Norman Conquest, a beggarly eight hundred years, or thereabouts, I, Pharos, can trace mine, with scarcely a break, back to the nineteenth dynasty of Egyptian history, a period, as your Baedeker will tell you, of over three thousand years. It was that very Ptahmes, the man whose mummy your father stole from its ancient resting-place, who was the founder of our house. For some strange reason, what I cannot tell, I have always entertained the belief that my existence upon this earth, and such success as I shall meet with hereon, depended upon my finding that mummy and returning it to the tomb from which sacrilegious hands had taken it.‹« (Ibid.: 104)

By contrasting Egyptian with English history, Pharos points out and scorns the relativeness of »a beggarly eight hundred years« (ibid.) of English history, which, in comparison to Egyptian history, seems ridiculously short. This is revealing in regard to the contemporary *fin-de-siècle* discourse with its fears of decay and degeneration, since in this context Pharos's reference to the shortness of English history can also be read as the lack of continuity and identity of the English race as opposed to that of ancient Egypt, as pointed out by Edwards in regard to *Dracula's* Count and *Jewel's* Tera:

»The Count himself and Queen Tera in the latter novel, are living, authentic embodiments from the past, surviving, by virtue of wherever occult forces into the present. [...] They have their own genealogies which are rooted in ancient times. Their existence in the Victorian present is deeply disturbing for types of history that seek retrospectively to explain and quantify, whether through narrative guile or methodological rigour.« (Edwards 1998: 99)

A people's history in the form of cultural memory, with its myths, legends, and tradition, is part and parcel of a people's collective identity, and the further backwards it can be traced in time the more stable and safe it is perceived to be (cf. Assmann 2011: 35-36; Gehrke 2004: 26). As pointed out above, the fear of moral and cultural decline after an age of progress and prosperity was further stirred up by contemporary theories of evolution and degeneration:

»Late-Victorian popular fiction charts a paradoxical anxiety about Great Britain's imminent and inevitable decline during the decades when the empire was in fact at its height, solidifying its worldwide dominance. Such literature responded to and perhaps aggravated widespread concerns about late nineteenth-century British deficiencies, real and perceived: Britain's loss of global economic ascendancy, its often fraught relationship with its colonies, its internal weakness due to social and cultural degeneration and decadence.« (Hurely 2008: 193)

Against the background of these fears, the lack of tradition and continuity, which are essential identity-forming components, attested by Pharos for the British in comparison with the Egyptians, can be read as a deep-rooted collective fear on the part of the British of not being able to perpetuate their hold on modern civilisation.

Pharos continues to scorn the British by introducing the Baedeker, the emblem of superficial and shallow pseudo-knowledge, employed by masses of British tourists to appropriate and domesticate foreign countries. The shallowness of the Baedeker knowledge, however, can also be seen as an allusion to the ignorance of modern Britain compared to the wisdom of ancient Egypt. The only thing Britain, as a nation lacking substance and continuity both in the past and the present, is capable of doing is to gorge itself on the rich history of another nation by abducting their ancestors and stealing their antiquities: »The mummy's forced removal from the ›slow symbolic order‹ of burial and history into the modern order of scientific inquiry effectively challenges any easy opposition between western science and the ›barbarism‹, criminality and superstition of past civilizations.« (Macfarlane 2010: 18) In the eyes of Pharos, the British are no more than barbaric tomb raiders that return from their conquests to gape at their loot: »›And since you are such a stickler for what is equitable, perhaps you will show me [your father's] justification for carrying away the body from the country in which it had been laid to rest, and conveying it to England to be stared at in the light of a curiosity.« (*Pharos*: 105) This is significant, since in spite of Forrester's weak attempt to bring in moral issues of his own by informing Pharos »that [the mummy] is not [Pharos's] property« (ibid.), Pharos ironically reveals the British to already be what they most deeply fear to become: a race of immoral and uncivilised men, which was, in fact, how they conceived others: »The Victorians tended to regard present-day non-Europeans contemptuously or condescendingly, as uncivilized barbarians, whereas the ancient Egyptians were respected as a scientifically and culturally advanced people.« (Hurley 2008: 183-184) On the one hand, Pharos's accusation towards Forrester can be read as a criticism of the Empire and British Imperialism, while on the other it can also be seen as a reference to the Victorian double standard in general. In much the same way as the mummy, Forrester only seemingly belongs to the Forrester family. Britain seems to be adorning itself with borrowed plumes, as underlined by Wieczorkiewicz: »Egypt, the land, the culture, and the history – are taken into protection and ›adopted‹ by the West.« (Wieczorkiewicz 2005: 58) As with Pharos's mummy, what is created is a sham continuity, and with it a sham national identity in terms of Gehrke's intentional history.

Forrester's inner admission of defeat following his discussion with Pharos – »I did not know what to answer« (*Pharos*: 106) – only emphasises that Pharos is essentially right. However, Forrester, instead of openly acknowledging his and his father's guilt, representative of the British nation, changes the subject and instead accuses Pharos of the murder at the curiosity shop: »›What you have to do now is to clear yourself of the more serious suspicion that exits against you. I refer to the murder of the curiosity dealer.‹« (Ibid.) Even though Pharos is able to give an explanation and convince Forrester of his innocence for the time being, the essential function of Forrester's accusation, to shift the focus from his own guilt to that of Pharos, is fulfilled. Thus, once again, instead of reflecting on the moral implications brought forward by Pharos in regard to his father and his nation, Forrester acquits himself and Britain of any guilt by accusing Pharos of murder. This is particularly illuminating when one considers that Pharos's action, like that of Frankenstein's creature, really is the *re*-action to unjust treatment in the past.

Forrester's encounter with Pharos, however, also exhibits other parallels to Frankenstein's meeting with the creature, since in both cases these confrontations mark a crucial stage in regard to the further course of events. For both Forrester and Frankenstein, the confrontation with their opponent is what they have been craving for a considerable time, allegedly to confront their antagonist with their guilt and re-establish moral order. Yet instead of yielding to Forrester and Frankenstein's accusations, Pharos and the creature point out the motives for their actions, which eventually inverts the victim-perpetrator relation in that Forrester and Frankenstein are now revealed to be guilty:

»Pharos's accusations strategically evoke anxieties about the exhumation and the study of the ancient dead. By conflating the titillation of public display with the crimes of sacrilege and grave robbing, Pharos here articulates the complexity of the threat that is posed by the disinterred mummy in mummy stories. He insists that the mummy is ›a body‹ stolen from a grave, not an artefact to be collected and studied.« (MacFarlane 2011: 18)

As a consequence, both Forrester and Frankenstein are ready to accede to the requests of their opponents by providing the mummy and creating a kindred mate, respectively. This suggests that Forrester and Frankenstein assume responsibility for their actions to some extent. At the same time, this bargain and alleged consent between Forrester and Pharos and Frankenstein and the creature also quickens and intensifies the relation between the two couples. In Forrester's case this is marked by his accepting Pharos's invitation for the evening in spite of himself: »If there had been any way of getting out of it, I think I should have taken advantage of it; but as I could not discover one, I was perforce compelled to accept his invitation.« (*Pharos*: 111)

Ironically, what Forrester finally learns when he accompanies Pharos to return the mummy to its secret tomb at the Temple of Ammon in the middle of the night is that the mummy is not that of Pharos's ancestors but Pharos *himself*, that is, that Pharos and Ptahmes, the alleged ancestor, are one and the same person:[11]

»I drew nearer and looked upon the mummified remains. Dried up and brown as they were, the face was still distinctly recognisable, and as I gazed I sprang back with a cry of horror and astonishment. Believe it or not as you please, but what I saw was none other than the face of Pharos. The likeness was unmistakable. [...] And yet it seemed so utterly impossible, so unheard of, that the man stretched out before me could be he whom I had first seen at the foot of Cleopatra's Needle, at the Academy, in Lady Medenham's drawing room, and with whom I had dined at Naples after our interview at Pompeii.« (Ibid.: 226)

That Pharos *is* in fact Ptahmes is significant, since it also means that the mummy, which has hitherto functioned as proof of Pharos's descent, suddenly becomes meaningless in this regard. Pharos's descent can no longer be proved by means of concrete samples, and Pharos's family history as related by himself turns out to be no more than *his*tory, and his identity thus essentially constructed. Instead of representing family continuity and lineage, the mummy now underscores Pharos's isolation and loneliness, which is further fortified if we consider Ptahmes's life history as related by a stranger Forrester meets during his sojourn in the ancient Egyptian past:

»›Learn, then, that [Ptahmes] hath fallen from his height estate, inasmuch as he made oath before Pharaoh that the firstborn of the King should take not hurt from the spell this Isrealitish sorcerer, Moses, hath cast upon the land. Now the child and all the firstborn of Egypt are dead, and the heart of Pharaoh being hardened against his servant, he hath shamed him, and driven him from before his face.‹« (Ibid.: 182)

As a consequence, Ptahmes/Pharos flees into the mountains, where he finally meets a lonesome death: »Little by little the strain upon his health gave way, he grew gradually weaker, and in the fiftieth year of his life Osiris claimed him for his own.« (Ibid.: 199) And although Ptahmes was

11 »As he finished speaking, the disgraced man withdrew his robe from his face and I realized the astounding fact *that Ptahmes the Magician and Pharos the Egyptian were not ancestor and descendant, but one and the same person*!« (*Pharos*: 182)

said to be »cursed with perpetual life« (ibid.) for lying to the king in the name of the gods and bringing misery upon the Egyptian people, his mummy had been seen buried in the tomb from where it was later stolen by Forrester's father. Thus, Ptahmes's tragic life story is essentially characterised by the two antipodes of fame and downfall, as Forrester realises as he witnesses Ptahmes's funeral in his vision of the past at the Temple of Ammon: »High up on a rocky spur a tomb had been prepared, and to it the body of the man, *once so powerful and now fallen so low*, was being conveyed.« (Ibid.: 223, emphasis mine)

Read against the background of Britain's situation towards the end of the nineteenth century, which, after a period of progress and prosperity, was characterised by collective fears of doom and downfall, Ptahmes's life story is illuminating, since again it can be seen as mirroring coeval anxieties of British society. The fear of digression was on the one hand rooted in a general economic slowdown and the decline of the Empire, but on the other also in new insights of science and changes in regard to gender and religion. Many contemporaries were sure to find in the ruin and decline of the once powerful and glorious ancient civilisations like Egypt, Babylon, or the Roman Empire a prequel of present *fin-de-siècle* Britain. In the novel this is emphasised by Forrester's description of the ancient sites in Egypt:

> »If my drive to the Pyramids, a week before, had been a singular experience, this camel ride among the ruins of ancient Thebes at midnight was much more so. *On every side were the relics of that long-departed age when the city, through the remains of which we were now making our way, had been the centre of the civilized world*!« (Ibid.: 210, emphasis mine)

Thus, among the ruins of ancient Egypt Forrester seems to become aware that grandeur and civilisation eventually are no safeguard against decline, and consequently even the powerful British Empire might some day be prone to perish through these forces. In this respect, the fate of Ptahmes can be seen as representing the ruin of the once highly adept, sagacious, and civilised culture of ancient Egypt. It could further be argued in this context that the person of Ptahmes/Pharos stands not only for »›the splendour and the degradation of the man‹« (ibid.: 224) but also for his isolation and the construction of a feigned family descent. This can be transferred to Britain's self-perception at the end of the century, which was partly responsible for the aforementioned collective fears.

In what way the degeneration and decline of British society is linked in the novel with a collective guilt is illustrated when Pharos, upon returning with Forrester and Valeric to Britain, invites Forrester to »show [him] London as [he] see[s] it in [his] character of Pharos the Egyptian« (ibid.: 325). Forrester agrees and sets out with Pharos on his vespertine tour through London, which opens with a dinner at the Antiquarian Club, where they meet »with as fine a collection of well-born, well-dressed, and well-mannered men as could be found in London« (ibid.: 327). However, while Forrester is immensely impressed by both the architecture of the building and the society present (cf. ibid.), he is soon informed by Pharos that what seems to be noble and long-established is really nothing but an illusory world which is corrupted to the core: »›This is the first side of London life I am desirous of presenting to you[,] […] one side of that luxury and extravagance which is fast drawing this great city to its doom.‹« (Ibid.) Upon this comment, Pharos proceeds by pointing out seemingly dignified individuals in the room, providing information on their *real* character and life, thereby revealing them as immoral, dishonest, and even criminal. In addition, the fact that Pharos makes society's »luxury and extravagance« (ibid.) responsible for the city's doom is a straightforward criticism of decadence common at the time and thus again reflects contemporary discourses. Without allowing Forrester to draw any conclusions at this point, however, Pharos leaves with him for another place of social and cultural significance, the Renaissance Theatre. Here, Pharos points out the ignorance, indecency, and selfishness of »›London Society‹« (cf. ibid.: 330). According to Pharos, men take their families to the theatre without being interested in the plays and instead of following the performance discuss their personal affairs, leaving their shocked daughters to deal alone with the titillating scene performed on stage: »›What think you of the authors of such a play and of the men who bring their wives and daughters to listen to it, and, while admitting its indecency, excuse themselves by saying that the lesson it teaches is a salutary one? I beg you to observe the faces of the young girls.‹« (Ibid.: 331)

The next two stops on Pharos and Forrester's tour lead them to the Occidental Music Hall and to Westminster, where they enter the House of Commons to witness a debate on the current threat by the plague, which has by now arrived in England. »As at the Occidental, as soon as [Pharos] had pointed out to me the sordid side of their lives and characters, and the pettiness of their ambition, the show of party feeling they exhibited even in the face of a grave national peril, he led me away.« (Ibid.: 333) Together with Valeric, Pharos and Forrester now attend a ball at Carlton House Terrace, where again they are surrounded by aristocrats and celebrities whose corruption Pharos continues to reveal to Forrester:

»[D]o what I would, I could not shake it off. The pictures I had seen that evening in Pharos's company rose before my eyes continually; and when later in the evening he came to my side and, with the same look upon his face that I had seen at the Renaissance, pointed out to me the hollowness of this ball also, and showed me the scheming mothers, the false wives and husbands, and the inner lives of many for whom up to that time I had entertained the greatest reverence, my cup was full – *I could bear no more.*« (Ibid.: 334-335, emphasis mine)

Finally, Pharos takes Forrester to a fancy-dress ball at Covent Garden, which Forrester would have enjoyed had his mind not still been dominated by the shocking revelations of Pharos, who goes on to »find[] evil in everything, and hinting always at the doom which was hanging over London« (ibid.: 336). This causes Forrester, who by this time is shocked and nauseated both by Pharos and the people surrounding them, to be reminded of »Dante's Inferno [more] than anything else to which [he] could liken [the ball]« (ibid.). Still, Pharos has not finished with Forrester yet, since there is one more place Pharos wants to take Forrester to: »›We must now descend a grade lower‹« (ibid.), namely to the notoriously deteriorated area of Seven Dials, where they meet all kinds of criminals and shady figures. It is significant that Pharos chooses the neighbourhood of Seven Dials and »other equally sad and disreputable quarters of the city« (ibid.: 340) as the last stop on his London tour, since he thereby links the corruption of the upper classes he has been pointing out at the various social, cultural, and political events directly to the poverty, abjection, and criminal energy of the lower classes. Predicting the coming doom of contemporary society, the scenes at Seven Dials are presented as the end of a chain of development, as a corollary of the corruption of the upper classes. Modern civilised society, comprising all areas and classes, is rotten to the core and there is no way of denying or changing this since the development is too well advanced.[12] The only thing possible at this stage is to finally recognise the evil and rottenness of society and to await atonement for the committed sins, as Forrester's association with Dante's *Inferno* at the fancy-dress ball and Pharos's frequent references to the coming doom suggest.

Thus, Pharos's tour through London functions as an instrument of revelation that shows British society's true character. Underneath the shining surface of progress and development lurk decline and degeneration. What seems to be long established and time-honoured is really hollow and rootless, which renders the depiction of Forrester and Pharos's arrival at the Antiquarian Club deeply ironic: »We *ascended* the *marble staircase* and passed through the ante-room, decorated with the *portraits of many distinguished members past and present*, to the dining-room itself.« (Ibid.: 326, emphasis mine) While the vocabulary used here can be read as symbolising progress, evolution, prosperity, moral values, and tradition, Pharos's subsequent tour through London and his comments on society demonstrate that the opposite is the case, that degeneration, decline, and immorality are the determining forces of modern society. The very fact that the entrance hall of the Antiquarian Club is described here is further telling: As an institution that is supposed to represent tradition and scholarly concern with the past, the revelation of its hollowness and the corruptness of its members is even more drastic, since this eventually questions the pillars of a whole society, including its treatment of the past.

As at Pompeii, however, Pharos not only points out the shocking state of contemporary British society but further contrasts it with that of ancient Egypt by referring to the immoral behaviour of the British excavators and scientists abroad: »›And these are the people who rifle the tombs of the dead kings and queens of Egypt, and write and talk patronizingly about the civilization of the Ancients. But it will not last. A time, however, is coming, and is even now at hand.« (Ibid.: 331) The immorality and evil of society is one thing, but the true sin of modern society is its hypocrisy and its double standards, with which it empowers and legitimises itself while degrading and condemning others both within and beyond national borders. Against this background, Forrester's father's act of stealing Ptahmes/Pharos's mummy from its tomb, as well as his and Forrester's inability to recognise

12 All the places Pharos takes Forrester to are prominent social, cultural, and political spaces, thus representing the pillars of society.

this act as a crime and to admit their own guilt, is representative of the collective crime of modern society and its refusal to accept this guilt and take responsibility for its behaviour. Consequently, since as in *Frankenstein* both society and the creator/excavator are guilty, Pharos uses the plague as a means of collective revenge as much as the creature murders to avenge the wrong it suffered from society in the first place.

Although revenge seems to be Pharos's ultimate aim, the revelatory tour through London shows that he wants Forrester to see and recognise the disastrous state of contemporary British society first. However, before Forrester is ready for this insight and anagnorisis forced upon him by Pharos, he has to leave the contemporary discourse in a spatiotemporal sense to then re-enter society and be able to look at it from a different angle. Along the lines of Forrester's changed perception of the Egyptian artefacts and decorations in his studio at the beginning of the novel, Pharos is a relic of the past that functions as a catalytic force, transporting Forrester beyond the limits of seemingly fixed parameters such as reality, convention, and tradition. In a way, this again shows parallels to *Frankenstein*, if we consider that Frankenstein comes closest to understanding and being sympathetic to the creature when he finally confronts it in the mountains, that is, in the wilderness, which marks the essential antipode to civilisation.

7.5 The Survival of the Whitest

> We had terrible difficulty in traversing this reedy district; but perseverance conquered, and we finally had the satisfaction of standing on terra firma on the further side of the broad but shallow stream, and in full view of a mountain range, in the valleys of which I confidently hoped to find a white race ruled over by a lineal descendant of the greatest of Jewish kings!
> (*N'Shabé*: 148)

While in *Pharos* the underlying guilt of the British colonisers in regard to the past and present of the ›other‹ as the *most unfamiliar strangeness* becomes more tangible in the course of the novel through a process of identification, in Zimbabwe fiction feelings of guilt are largely excluded from the Empire discourse. Instead, these texts are characterised not only by the colonisers' inability to accept their guilt, but by a radical negation of the history of the colonised through their colonisers as a means of dealing with their ultimate ›otherness‹. It is on the basis of this radical negation of the ›other's‹ history that the *most unfamiliar strangeness* is isolated, appropriated, and incorporated into the colonisers' intentional history. In the present chapter I will show how in Zimbabwe fiction archaeology is instrumentalised by the colonisers to create a feigned continuity between past and present on the grounds of alleged racial similarities.

In most of the texts dealing with the Zimbabwe ruins, the white explorers penetrate into the African ›wilderness‹ to explore the ancient ruins usually only known by hearsay. What follows once the ruins have been discovered is an attempt to claim a connection to the ruins, initiating a process of self-identification on the part of the explorer. This process of dealing with the ruins is typically characterised by two main tendencies: first, the attempt to create a *collective* connection to the past on the grounds of racial similarities and biblical legends, and second, once this collective connection has been established, the attempt to build an *individual* connection to this past by means of personal affiliations with that past. Both aspects are necessary steps in the process of the Western colonisers' self-exploration that finally help reconfirm and stabilise individual identity.

In Haggard's *She*, this twofold identification process is illustrated by the collective connection to the past through racial similarities – skin colour – and the individual connection through family history – the Vinceys' lineage. This corresponds to the popular imperialist conception that the Zimbabwean past was essentially white and thus part of a Western history: »The amateur archaeology and ethnography of Bent, Hall and Peters provided a scientific foundation for Haggard's claim that in bringing settlers to Mashonaland Rhodes *was restoring an ancient settlement rather than initiating a new one.*« (Chennells 2007: 15, emphasis mine)

In Haggard's adventure romance, the process of both collective and individual identification is initiated right at the beginning of the novel through the reference to Ayesha, a white queen dwelling in the African ›wilderness‹, and the delineation of Leo Vincey's family history. In regard to Ayesha, Holly and the other men are not only confronted with a female lover and mother figure, as pointed out in Chapter 4.1, but also with an ›unnaturally‹ *white* woman. This we first learn from Mr Vincey's letter, which the men discover in a chest along with the Sherd of Amenartas: »›I learnt also that the people there [in Kôr] speak a dialect of Arabic, and are ruled over by a *beautiful white woman* who is seldom seen by them, but who is reported to have power over all things living and dead.‹« (*She*: 38)[1] The expedition that follows thus not only becomes the attempt to find out whether what is being said in Mr Vincey's letter and outlined on the Sherd of Amenartas is true, but also whether Ayesha's *white* identity, against all likelihood, is true: »›We have been looking for a mystery, and we certainly seem to have found one.‹« (Ibid.: 39) The mystery Leo is referring to here thus also alludes to the improbability of actually encountering a white woman in the ›darkness‹ of the African ›wilderness‹, »in a hitherto unex-

---

1 The Western explorers' search for a white people in the African outback is also a prominent theme in Edward Markwick's *The City of Gold* (1896): »Possibly they might lead me whither my thoughts and aspirations had long turned; to that mysterious region in the very heart of Africa where a strange race of white men was reputed to guardian against contact with the outside world.« (*City of Gold*: 24)

plored region« (ibid.: 38) where »far inland [there] are great mountains, shaped like cups, and caves surrounded by measureless swamps« (ibid.). Ayesha's white identity, together with the Vinceys' lineage on the potsherd, consequently becomes the major trigger and motivation for the men's subsequent exploration. Once they leave for their expedition, the men's quest is to confirm both the originality of the Sherd of Amenartas and Ayesha's identity as a white woman. Through the potsherd outlining Leo's lineage, Ayesha and her skin colour are thus directly connected to Leo's family history, reaching from the British present to a remote Egyptian past. Therefore, the potsherd does not function only as a means of delineating the evolution of civilisation over the centuries (cf. Malley 1996: 179) but also as a means of creating a family continuity on grounds of racial similarities. In this way, the potsherd and the information we get from Vincey's letter create both a genotypical (lineage on the potsherd) and phenotypical (reference to Ayesha's skin colour) continuity between the past and the present.[2]

The creation of continuity on the basis of racial similarities in general and skin colour in particular is characteristic of both fictional and factual texts dealing with the Zimbabwe ruins. In Walmsley's novel *The Ruined Cities of Zulu Land*, the white explorer Captain Weber informs the African king Mosilikatze[3] of the stone houses he and his men are looking for:

> »These houses must have once been the dwellings of the *white man*, and the legends *our fathers* have taught tell us of such *white men*, who came many thousand years since from the regions of the rising sun, landing on these shores. To reach these ruins, to prove that *our fathers* spoke the truth, is our object, and in the *name of our ancestors* we ask thy protection, chief.« (*Ruined Cities* Vol. I: 58, emphasis mine)

Here the link between the past and the present is established by Captain Weber, first by his assumption that the inhabitants of the stone houses were »white men« (ibid.) and second by his reference to the ancestors' legends they want to confirm. Thus, in a way similar to *She,* continuity is created by

2 On the function of the creation of continuity as part of identity formation in Haggard's *She*, cf. also Malley (1996a, 1997), and Coates (2003).

3 Mosilikatze, alias Mzilikazi (c. 1790-1868), was a powerful African king who founded the Matabele kingdom (Matabeleland) in former Rhodesia, present-day Zimbabwe.

an alleged commonness on grounds of genotypical (»our fathers«) and phenotypical (»white men«) characteristics. Significantly, Captain Weber points out that the aim of their exploration is to »prove that [their] fathers spoke the truth« (ibid.). Again this is reminiscent of *She*, where Holly and the other men leave for Africa to confirm what they have learned from Vincey's letter and the Sherd of Amenartas. As pointed out by Lucas, »[a]ffirming the past as belonging, affirms it as ours – as soon as we describe prehistory as a forgotten epoch, it has already become *our* past« (Lucas 1997: 9, emphasis mine). By relating themselves to the African past, the Western explorers thus appropriate it and turn it into their own past. This is further emphasised in *An Imperial Adventure*, where the protagonist Sannie remembers stories she has gathered about ancient ruins in Mashonaland inhabited by a white race:

»She remembered still more mysterious and magnetic stories – of ruins now deserted and desolate as the grave; of cities so ancient that no trace of their founders could be known; of giant walls built up without mortar, so immense of girth and width that a waggon could roll along their coping, yet perforated by the growth of trees which were themselves many centuries old; of great gates with inscriptions above their portals that no man could read, nor even tell of what tongue they were; of stone-lined pits, of huts closed by one enormous slab and joined each to each by underground passages. She remembered, too, faint rumours she had heard from time to time, of Kafir traditions about a *white race* which had visited the Zambesi basin from the east coast and been exterminated at last by a mad upheaval of the barbarous tribes it ruled.« (*Imperial Adventure*: 24-25, emphasis mine)

Again the ancient ruins are connected with a »white race« (ibid.) that was eventually extinguished by »barbarous tribes« (ibid.). Before we are informed of the »white race« (ibid.) as the possible inhabitants of the ruins, however, the ruins as such are described in detail. In particular, their considerable age is pointed out, which is emphasised through the fact that the ruins are »perforated by the growth of trees which [are] themselves many centuries old« (ibid.).[4] In addition, the greatness of the ruins is underscored

4 The information that the ruins are »perforated by the growth of [ancient] trees« (*Imperial Adventure*: 24) is a reference commonly used to emphasise the old age of the Zimbabwe ruins.

by a reference to the »immense [...] girth and width« (ibid.) of the walls and their »great gates« (ibid.) and »enormous slab« (ibid.). The old age and greatness of the ruins is further emphasised through the initial reference to the »mysterious« (ibid.) and »magnetic« (ibid.) character of the stories. Additionally, the fact that »the inscriptions above their portals« (ibid.) cannot be deciphered by modern man (cf. ibid.) underscores the mystery surrounding the ruins. In regard to the »white race« (ibid.) that is assumed to have inhabited the ruins in former times, these references to the old age and greatness of the ruins is telling, since these aspects were often associated with white skin colour and functionalised to prove an alleged white past for Africa.[5] The mystery surrounding the ruins in this context opened up the necessary space for speculation and offered room for projection on the part of the Western beholder. This is further emphasised by the fact that the inscriptions could not be read, let alone understood by any modern man. While the description of the ancient ruins thus evokes stability and power on the one hand, the impossibility to classify them creates uncertainty and leaves room for interpretation on the other.

Once the old age and greatness of the ruins have been established by Sannie's recollections, Sannie also remembers rumours of a »white race« (ibid.) in connection with the ruins. Similar to the link established in *She* between Britain and Ancient Egypt, the reference to the »white race« (ibid.) functions as a means of connecting not only the past to the present but also African to European civilisation on the basis of racial similarities, that is, skin colour. Accordingly, skin colour is here again introduced as a way of proving an alleged affiliation of the white Western explorer of the present with an advanced African civilisation of the past. Sannie's recollection of stories about Mashonaland and its ruins can thus be read in terms of both an individual and collective memory. She not only remembers the stories she has heard, but also what was stored and archived, ›re-membered‹, in these stories. Sannie feels drawn to the country and its ruins because it is at these sites of memory – and in her thoughts about them – that she can connect both with her own self and a collective white origin:

5 In particular, the ruins' old age and their greatness are repeatedly described as fascinating for the Western explorers, which suggests a longing for power and stability on the part of the beholder.

»Identity depends precariously on a register that exists only as representation after the fact. In the second place, and to make matters more precarious, memory unavoidably splits identity between a past self and a present self, which observes that past self and each second reenacts this splitting. Memory therefore becomes both the necessary proof of the continuity of identity and, simultaneously, the proof of its unavoidable discontinuity.« (Franklin 2001: 149)

The blurring between individual and collective memory in connection to the Zimbabwe ruins is quite typical of various texts at the time. In A. A. Anderson and A. Wall's novel *A Romance of N'Shabé*, this is underlined by Montfort's reflections on the city of N'Shabé and its queen: »I reflected that very shortly our party would be quietly but triumphantly moving through the streets of the city of N'Shabé, that we should stand face to face with Solomon's last descendant, and perhaps hear from her lips curious history of her people – a history sure to be brimful of archæological lore.« (*N'Shabé*: 174) The fact that Montfort and his comrades expect to meet »Solomon's last descendant« (ibid.), the city's queen, emphasises their urge to discover traces of a Western, Christian collective identity beyond the boundaries of the hitherto known. Thus, this once more marks the attempt to establish cultural and racial continuity on the grounds of the Bible as well as skin colour over time and space.

The reference to Western, biblical traditional legends such as those of Queen Sheba, King Solomon, and Ophir is frequently used to connect these to actual sites in the present, such as the Zimbabwe ruins, as illustrated by the following statement of Montfort, the narrator of *A Romance of N'Shabé*:

»›I came across gold mines so extensive and so ancient, that I was struck with wonder. But my attention was chiefly enchained by the discovery of a number of forts built of hewn granite, and put together without mortar.‹ ›Unfortunate but unmistakable evidence that the Portuguese have an earlier claim on Mashonaland than you‹, I said, jestingly. ›The size of the trees growing midst the ruins entirely negatives your assertion‹, replied Clavering. ›Some are more than three feet in diameter, whilst others have actually enclosed large blocks of hewn stone with their substance. The Portuguese had established no fortified posts before the sixteenth century, and as these trees are a thousand years old or more, it is evident that the structures were the work of a very ancient race. [...] I was prepared to hear something wonderful, but the traditions, as they were narrated to me, surpassed my expectations. In the far distant past, a white race called ›Abalambas‹ [...] possessed the country, lived in stone dwellings, and worked the gold mines with implements of tsepe, or iron! In the most palmy days of the nation it was governed by a beautiful queen, who believed herself the most powerful ruler on earth; but rumours of the vast dominion, riches, and wisdom of a king far away over the sea so disquieted her and so excited her envy that she determined to see for herself what truth there was in these reports.‹« (*N'Shabé*: 6-7)

In regard to the creation of an alleged white Zimbabwean past, this passage is significant in that it presents several characteristic aspects of texts dealing with the Zimbabwe ruins. The first contact with the ruins is usually an indirect one, through stories that somehow connect the ruins to a white race, as illustrated by the following example from the *Imperial Adventure*:

»›That is it, that is just what I have heard. Those ruins are temples that Solomon built, and there he refined all his gold for his own temple. That is sure proof of the value of the land; and everyone brings the same tale. Oh, I have spoken with scores of those who have been there! Why, they say the very sands over which the rivers run are all of gold, that the Queen of Sheba herself came from thence.‹« (*Imperial Adventure*: 7)

Here Sannie and her uncle randomly learn from a merchant in Cape Town about the ruins and their alleged connection to the Solomonic legend. What usually follows in this formulaic system is a discussion of the old age of the ruins, which most of the time involves the Portuguese settlers and the possibility of them having built the stone houses, thus strengthening the link between the ruins and white people. In most cases, however, this idea is eventually dismissed in favour of the assumption that the ruins must have been constructed by people inhabiting the region *before* the Portuguese.

In the passage from *A Romance of N'Shabé,* as elsewhere, the very old age of the trees growing among the ruins[6] and the fact that the ruins are built without mortar[7] are used as proof for the fact that »the structures [are]

6 Cf. above.

7 The information that the stone ruins were built »without mortar« (*N'Shabé*: 6) was a detail popularly taken as an indication of the ruins' old age (cf. *Elissa*: 59-

the work of a very ancient race« (*N'Shabé*: 6) and thus must have been built prior to the Portuguese settlement in South Africa. Once this has been established, what typically follows is the claim that the builders of the ancient stone houses must have been of *white* origin. Characteristically, this is once more accomplished by reference to well-established myths and legends:

»›In the most palmy days of the nation it was governed by a beautiful queen, who believed herself the most powerful ruler on earth; but rumours of the vast dominion, riches and wisdom of a king far away over the see so disquieted her and so excited her envy that she determined to see for herself what truth there was in these reports.‹« (Ibid.: 7)

Without explicitly mentioning Queen Sheba and King Solomon, this reference is sufficient to trigger in the Western reader the story of the legendary king and queen, which sooner or later is usually openly referred to in the texts:[8]

»›By Jove!‹ I exclaimed, as Clavering paused to take breath, ›I begin to see into the mystery. Ophir, the land whence Solomon obtained his gold, is translated as Sophara in the early Greek versions of the Old Testament, and your Queen of Saba is evidently the Biblical one of Sheba. Sophara sounds very much like the modern Sofala, a district contiguous to Mashonaland.‹« (Ibid.: 8)

Quite typically, as mentioned above, the legend of Queen Sheba and King Solomon is immediately linked to a present reality by the reference to names of existing places that are etymologically explained and identified as modern derivations of the original names mentioned in the Bible and elsewhere. Frequently, different alternative names are mentioned in regard to the ancient ruins at one instance, as illustrated in Glanville's *Fossicker*:

---

60 and *Benita*: 78). In addition to that, the »conical towers« (*Elissa*: 60) associated with the ruins are repeatedly referred to in the novel (cf. *Benita*: 78).

8 This is the case in the following texts: *The Ruined Cities of Zulu Land*, *King Solomon's Mines*, *Elissa*, *The Fossicker*, *An Imperial Adventure*, and *John Prester*.

»›I remember now – your friend Drury told me you were going in search of Mauchberg, or Ophir, or some such place where gold is to be found in bucketsful.‹ ›Ah! he was probably romancing.‹ ›Very likely‹, said Foster, with a dry smile; ›nevertheless, there are treasures of some sort to be found there, left probably by the old Phœnicians.‹« (*Fossicker*: 13)[9]

The British explorer Smedley's use of the names Mauchberg *or* Ophir for the ruins is once more revealing in regard to the connection between past and present, myth and reality. This becomes clear when we consider that the name Mauchberg was introduced to designate the Zimbabwe ruins after Mauch had rediscovered them in 1871, whereas the name Ophir is obviously derived from the Bible. Thus, by mentioning Mauchberg, connotating the present and reality – Mauch had recently (present) explored the ruins (reality) –, and Ophir, connotating the past and legend – as a name in an ancient (past) story (legend) of the Bible – to refer to the same thing, the Zimbabwe ruins, here fact and fiction are almost imperceptibly merged. This emphasises again how in the creation of an intentional history fact and fiction are indiscriminately woven into a history, which in this case serves the purpose of establishing continuity between an alleged *white* past and present.

Additionally, Foster also brings in the »old Phœnicians« (ibid.), which underscores the attribution of a white identity to the former inhabitants of the ruins, implicitly already evoked by the names Mauchberg and Ophir mentioned by Smedley. After inspecting Great Zimbabwe in 1871, Mauch was convinced that he had discovered the remains of a white settlement and eventually identified them as the ruins of Ophir. This is reflected in Bent's account in *The Ruined Cities of Mashonaland*:

»These reports of an Eldorado northwards continued, and produced periodical excitements amongst the young colonists of South Africa. The Boers were everlast-

9 Another example in this context is *King Solomon's Mines*: »›Lad, did you ever hear of the Suliman Mountains up to the north-west of the Mashukulumbwe country?‹ I told him I never had. ›Ah, well‹, he said, ›that was where Solomon really had his mines, his diamond mines, I mean‹. ›How do you know that?‹ I asked. ›Know it; why what is ›Suliman‹ but a corruption of Solomon!‹« (*King Solomon's Mines*: 51-52)

ingly getting up *treks* with a view to reach it; the vague mystery about King Solomon's mines existing there, and the palace of the Queen of Sheba, whetted their appetites when they heard these rumours; but still nothing was definitely done until a German traveller of more than ordinary energy penetrated as far as the Zimbabwe ruins in the year 1871. This man was Karl Mauch: he examined them carefully and wrote an accurate account of them, but, unfortunately, he ventured on a speculation as to their origin which at once cast discredit on his discoveries in the eyes of unbelieving archæologists. He maintained that the fortress on the hill was a copy of King Solomon's temple on Mount Moriah, that the lower ruins were a copy of the palace which the Queen of Sheba inhabited during her stay of several years in Jerusalem, and that the trees in the middle of it were undoubtedly Almug trees.« (Bent 1896: 244)

Although Bent questioned Mauch's hypothesis that the Zimbabwe ruins could be attributed to King Solomon, the many popular stories conceiving Great Zimbabwe to be the remains of the biblical king suggest that Mauch's interpretation corresponded at least to what people wanted to believe:

»What is perhaps of greater interest is the extent to which prehistorians were themselves influenced by the constructions of the novelist […]. This influence was expressed in two ways. First, through the structure of narrative itself. Here the large novels and complex plots, favoured by artists such as Dickens and Trollope, allowed the novelist to situate characters in more complex contemporary and historical contexts. Second, a cultural context was created in which meanings were given (by novelists and poets as well as by prehistorians) to a potentially unintelligible past. *These meanings became popular, almost natural, knowledge and gave expression to cultural presuppositions which were rarely, if ever, subjected to explicit scrutiny.*« (Murray 1993: 179, emphasise mine)

Consequently, similar to Schliemann's claim to have excavated the remains of Homeric Troy, Mauch's claim to have discovered relics from the time of King Solomon was extremely popular. Also similar to the Homeric stories for which Schliemann argued to have found the material proof, the story of King Solomon had the power to fascinate people due to its being a story from the Bible, one of Western society's core texts.

The repeated references to the Bible as well as skin colour in regard to the Zimbabwe ruins thus created a connection between past and present and thereby helped establish a collective (imperial) identity contemporaries could associate with. It is on the grounds of this collective self-identification in terms of racial and cultural affiliation that the ruins eventually also became a site of individual self-exploration and self-recognition for the Western explorer.

This is emphasised by the various instances in which the ruins are associated not only with the retrieval of a collective identity in terms of racial and cultural resemblances, but also with an individual self-discovery, as for example in Gertrude Page's *The Rhodesian*:

> »In this far Rhodesia, how strange that she, the product of the most modern and presumably enlightened age, should linger there amidst these broken walls, and feel strange kinship and fascination about those old people in that remote age; should stretch a hand out to them, as it were, across the centuries, with this feeling that their thoughts had been even as her thoughts, and that the passing of the ages could never eradicate the essential likeness of one people to another in those old eternal questions of whence and why and whereof.« (*Rhodesian*: 117)

This passage underlines how the British woman Meryl, stimulated by the Zimbabwe ruins, feels a personal connection to a collective past. By pointing out the »essential likeness« (ibid.) between her and the people of the past that cannot be eradicated by time (cf. ibid), Meryl establishes a continuity between *their* past and *her* present, which can be read as an instance of self-recognition and an attempt to reconfirm her own identity by linking her individual self to a collective other, as has already been shown for Sannie in *An Imperial Adventure*. This is only possible, however, due to a prior collective racial and cultural identification with the ruins, that is, once certain frames of references have been established.

Furthermore, the melancholy and longing for something indefinite underlying this passage, which is frequently associated with the African landscape and its ruins, is seemingly satisfied by the materiality of the ruins themselves. Considering that feelings of melancholy and longing are often closely related to people's nostalgia for a long-lost past, in particular in nineteenth-century industrial Britain, where many people felt alienated and experienced feelings of individual loss, it seems plausible that the Zimbabwe ruins, popularly associated with a white past, often became sites of both collective and individual self-recognition.[10]

This is also reflected by Major Peter Carew's stream of thoughts at the Zimbabwe ruins upon having learned of King Edward's death in *The Rhodesian*:

»So he climbed on up the winding pathway, enfolded with mystery and romance concerning the feet that trod it in the far-off centuries, and made his way between the mighty natural boulders out to the high platform, where eyes, all those long centuries ago, must have looked out even as his, across the lovely land. […] A man may call England his country because he was born there, and his father before him; but, perhaps, after all, that is a small thing compared to standing upon a high eminence, and looking across a quiet world which is your country because of all you yourself have given to it of hope and faith and steadfast purpose. In some such spirit soothing came to the quiet man on the top of the Acropolis Hill, whispering to him that, after all, this was *his* country, and if the beloved dead did indeed seem so far away in fact, in spirit he was perhaps nearer to his Empire-builders than he had ever been before. He turned his head at last, and his eyes rested upon the circular wall, four hundred feet below, that enclosed the temple ruins. Then for a moment a wave of depression swept over him, blotting out the landscape's loveliness.« (*Rhodesian*: 12)

Carew's thoughts once more point towards various important aspects in regard to the perception of the ruins. First of all, there is again the reference to the old age of the place that »enfold[s] [it] with mystery and romance« (ibid.). In addition, Carew perceives a strange feeling of belonging upon overseeing the landscape and its ruins, and although he is aware of the fact that »England [is] his country because [he] was born there, and his father before him« (ibid.), in the presence of the Great Zimbabwe this becomes »a small thing […] because of all you yourself have given to it of hope and faith and steadfast purpose« (ibid.). It is this question of belonging, including the process of identity formation, of the colonists and explorers that is central to many fictional texts dealing with the Zimbabwe ruins. The departure from the discursive structures of the home country becomes, both in a literal and a figurative sense, the voyage into an unknown space where the

10 Cf. Schliemann's search for Troy and Mycenae.

process of forming a new identity takes place. Africa, and in particular Mashonaland, becomes more than any other place the space of construction and negation of both a collective and individual identity, underlining the function of the archaeological space as a counter-discourse to contemporary Victorian and Edwardian society in general. In order to accomplish the formation of a (new) identity, old conventions and traditions in terms of belonging had to be given up. Even though Britain was their country of birth, for many the African landscape, with both its past and present, held a stronger reality than the home country once it had been assimilated:

»Thus discourses of memory yoked together the individual and the Empire and tied troubling questions about identity – personal and national – to the even more troubling questions raised by evolutionary theory, coming full circle back to what was perhaps the primary impetus behind the nineteenth-century memory crisis.« (Franklin 2001: 144)

That the process of self-recognition is also often closely linked to the retrieval of a personal past is underlined by a number of examples in which missing people are either rediscovered or their true identity is revealed at and in connection with the ruins, respectively. McIver's *Imperial Adventure*, in which the true identity of Sannie/Adrian, the novel's protagonist, is revealed to his/her comrade Jim Sugden at the ruins, is a case in point.[11] After having frantically searched for his missing comrade Adrian, alias Sannie, Sugden eventually finds him/her among the ruins they all have been dreaming of discovering for so long:

»[Sugden] left the apprehensive Abraham and struck up the hill with great strides. He was paying no heed to Adrian's tracks just then; he was intent upon the undecipherable ruin and the many amazing secrets it might conceal. He had ransacked many such another through the length and breadth of that wild land, partly from eager interest in their origin, wrapped in the fables of an unchronicled past, partly because of the gold they sheltered or to which they were supposed to give a clue. He

11 Other texts where the ruins serve as sites of self-discovery are *The Rhodesian*, *The Fossicker*, *A Romance of N'Shabé*, and *The City of Gold*. Texts in which people's identity is discovered or revealed in the process of searching and finding the ruins are *The Fossicker*, *Benita*, and *She* (Leo's identity).

hastened up to the yawning gap in the wall, kicking the intervening stone aside; and there, in confident proximity to a skeleton, lay Adrian, the boy of his love, of his search, – livid, deathly. [...] He tore hastily at the shirt, where it buttoned round the throat, in some alarm. The stillness was ominous, the face so rigid and ghastly. The fastening sprang asunder, the discoloured cotton fell back, and laid bare a bosom the more startling white for the soft sunburn of the neck and the neck and the dark-brown, knotted hand that lay upon it, arrested by surprise. Jim Sugden threw himself back; he snatched his hand away as if that marble scorched him. ›Good heavens!‹ he gasped, trembling from head to foot, the perspiration breaking out in great beads on his forehead. ›Good heavens! So this was – her secret!« (*Imperial Adventure*: 196-197)

Sugden not only hits upon the ruins and the unconscious Adrian, but also finds out about Adrian's ›true‹ identity.[12] In addition, however, this particular anagnorisis can also be read as representative of various experiences and encounters at the Zimbabwe ruins in terms of self-recognition or self-identification in general. In addition to the retrieval of collective identity, where the ruins function as a material confirmation of a legend, myth, or story that is often only confirmed after a long search and subsequent discovery of the ruins, the ruins also fulfil a consolidating function in regard to individual Western identity. The process of mapping the unknown, penetrating into the darkness, discovering the ruins, and creating histories around them thus marks the way to both collective and individual self-recognition: »Collecting, classifying, narrating: these are all ways of putting something in its proper place, of affirming that it belongs – ultimately to us.« (Lucas 1997: 9)

A similar anagnorisis in connection with the ruins takes place in regard to Jim Sugden. Having concealed his identity for years after murdering a man in self-defence, Sugden reveals his identity in front of his recovered brother and comrades: »›There's not a woman living would marry me‹, Jim burst out violently, ›not if she knew; and she knows. You saw her. I'm not a man; I haven't even a name; I'm a memory, a soiled memory; I'm only – a ghost. Eustace knows it all; he's learned in these things.‹« (*Imperial Adventure*: 338) It is shortly after this desperate outcry that Sugden is iden-

12 The cross-dressing of Sannie/Adrian underlines again the archaeological space's potential for the inversion and subversion of (gender) identity (cf. Chapter 6.4).

tified as Eustace's younger brother and is finally able to explain and clarify the events of his past life. Significantly, however, this only seems to be possible *after* his discovery of the ruins and his anagnorisis in regard to Adrian, alias Sannie, at the ruins. Thus, the ruins again have a solidifying and confirmatory function, as is further emphasised by Sugden referring to himself as a man without a name, who is only a soiled and immaterial memory, a ghost. It is through his encounter at the ruins at night[13] that he is finally able not only to recognise Adrian's true identity, but also to retrieve and reveal his own identity. By dismantling his fake identity, Sugden thus finally manages to restore substance to his ghost-like and empty personality.

Another aspect important in this context and frequently introduced in connection with the Zimbabwe ruins is the search for the ruins being related to the search for missing people.[14] In *An Imperial Adventure*, the old hunter Piet Marias talks about his son who disappeared during a hunting trip in Mashonaland: »›I went there twenty years ago; no one cared for gold then. I made a shooting expedition. There was my son Jan who went with me, my son Jan; but he never came back. He went out from our outspan one morning, but he forgot to come back.‹« (Ibid.: 6) Characteristically, it remains unclear what happened to the son and who was responsible for his disappearance. The statement that »he forgot to come back« (ibid.) even suggests that he himself might be responsible for his disappearance, that he

13 As has already been pointed out above in regard to mummy fiction, encounters with the self often take place at nighttime, when people are usually isolated from contemporary modern society.

14 The danger of getting lost/going missing in search of the ruins and the gold associated with them is formulated programmatically in the *City of Gold*: »›Beyond the hills whence cometh the morning sun there dwells a Great White Witch with many people. His kraal is a mountain in size, his cattle are as pebbles in the river for number. Gold he has in abundance, and precious stones, and every other thing that is good. *But they that go to him never return.*‹« (*City of Gold*: 27) Fiction dealing with the Zimbabwe ruins in which people go missing are, for instance, *The Ruined Cities of Zulu Land*, *King Solomon's Mines*, *The Fossicker*, *A Romance of N'Shabé*, *An Imperial Adventure*, and *The Rhodesian*.

might have lost himself in the ›wilderness‹.[15] This is further confirmed later in the novel, when Adrian, alias Sannie, is sure to have discovered the remains of the missing son:

»[Adrian's] hand fell and closed upon a smooth rounded impediment, it touched a fleshless bone; and his eyes, before which the stars were already blotted out, discerned a misty whiteness of outline at his side, the blanched skeleton of a man, that yet was not a skeleton but a recumbent Boer in rough corduroys and a heavy beard, a [sic] in his stiff unyielding fingers. He knew it; it was the body of Jan Marais who lay beside him, of that familiar hunter who had gone out from his father's outspan and had also *forgotten, forgotten to return*. They two looked into each other's eyes and they smiled, for they saw the waste peopled again, and a milder reign of concord and security setting up its tabernacle whose transitory tent-pegs they were fastening.« (Ibid.: 193, emphasis mine)

Here, as in many other texts where people go missing, the site of the ruins becomes the place of anagnorisis, where (a lost) identity is rediscovered and restored. In Markwick's *City of Gold*, Captain Vincent, who is searching for a missing brother, has not yet realised how closely the ancient city they have discovered is linked to his brother. Talking to the city's queen Nazir, he explains that he first has to find his brother before he will be able to appreciate the ancient city and its people: »›I see around me evidence of civilisation far surpassing that even of the land of my birth. Willingly would I sojourn here for the increase of knowledge; but I have left without a comrade who came from afar with me, and for whose life I am answerable. He is my brother, and I cannot desert him.‹« (*City of Gold*: 163) Eventually, it turns out that Captain Vincent will find his lost brother at the City of Gold, which again points towards the ruins' function as a place of anagnorisis and epiphany.

Hence, the ruins in Zimbabwean fiction repeatedly function not only as the material proof and realisation of a collective memory but also as a materialisation of the individual self through self-discovery and self-recognition as well as the restoration of lost identities and the rediscovery of missing people. The close link between individual and collective self-dis-

15 This once more points towards the danger of losing oneself in the archaeological space both in a literal and symbolic, i.e., psychological, sense.

covery is also emphasised in Haggard's *She*. Following his father's instructions and the inscriptions on an ancient potsherd, Leo discovers his true identity as the ancient Kallikrates, Ayesha's lover, at the ruins of Kôr. Leo's individual self-discovery can also be read as a collective self-discovery of Western society on the grounds of racial affiliation, as pointed out by Hinz: »Leo Vincey, lineal descendant of Kallikrates and Amenartas, is literally the epitome of Western civilization.« (Hinz 1972: 419)

As this chapter has shown, Great Zimbabwe is repeatedly used in fictional texts to link the Western present to the African past. I have argued that the origin of the ruins, representing the *most unfamiliar strangeness* within the Empire discourse, had to be negated and reinveted by the colonisers because their true origin was perceived as a confounder to the Empire ideology. In this process of negation, the ruins became part of an intentional history created by the Western colonisers. After mapping and penetrating a hitherto unknown space, the Western explorers in the texts identify with this newly disclosed space, first by means of establishing a collective link to that space on the grounds of traditional legends, like those of King Solomon, Queen Sheba, and Ophir, second on the basis of racial resemblances like skin colour, and third on the grounds of individual/personal relationships, such as the retrieval of a missing relative. It is crucial to note, however, that unlike Greek archaeology, which as the *most familiar strangeness* was adopted as an ideal the British could immediately identify with, in the case of the Zimbabwe ruins, as the initially *most unfamiliar strangeness*, a radical process of negation, reinterpretation, and assimilation was necessary before identification became possible.

7.6 Britain and Haggard's Zimboe

> [I]f the pigmy is shy he is anything but inoffensive, being the most vindictive little wretch in the world. If we have to hunt him out of his den, we may have some ticklish work before us, for he is human enough to defend himself with poisoned arrows and spears [...]. ›We need not worry our heads about poison‹, said Ryston; ›if we begin to count the danger, we shall end by giving up an expedition which may result in substantiating Darwin's theory of the descent of man.‹
> (*N'Shabé*: 66)

The fragility of this intentional history is reflected by the fact that the collective identification with the highly civilised former inhabitants of Great Zimbabwe may, on grounds of conceived racial, economic, and cultural similarities, culminate in an over-identification which, rather than offering a feeling of enduring comfort and stability, eventually results in a feeling of threat and uncertainty. Responsible for this feeling were fears of degeneration and doom evoked by the ruins of Great Zimbabwe as the remains of once highly civilised and powerful races, as implied by the *Wesleyan-Methodist Magazine*:

»We now pass to the extensive remains of a period centuries ago, when an unknown but powerful race built fortresses and temples upon the granite hills, and, burrowing into the quartz reefs, extracted gold, which, passing from their hands, doubtless represented the wealth of people far away: a race so obscure or living so far back in history that the present occupants of the region have not traditions whatever concerning it.« (*Wesleyan-Methodist Magazine*, Mar 1893: 178)

The question as to how it was possible that these advanced civilisations of the past could have disappeared and died out in spite of their cultural and political power is reflected in various fictional and factual texts dealing with the Zimbabwe ruins and underscores a general feeling of collective decline (cf. Arata 1990: 622). This latent fear, which persisted into the Edwardian period, is particularly prominent in people's reactions to the Zimbabwe ruins, as made explicit by Major Peter Carew's reflections upon looking at Great Zimbabwe in Page's *The Rhodesian*:

»Was it all, then, vanity, this building and striving? … This making of walls and fortification for another race, centuries afterwards, to look upon with cold wonder and curiosity? Three thousand years ago perhaps another man had stood even there and mourned his king that was dead. And so soon … so soon … he also died, and the massive walls became ruins, and the dynasty, or empire, or era, passed away into, oblivion. How soon might a similar fate over take his own great Empire! … and the beloved King, Edward the Peacemaker, be perhaps a legend to some strange new race.« (*Rhodesian*: 12-13)[1]

The identification with the Zimbabwe ruins for the beholder and the resulting fears and doubts underline the relevance Great Zimbabwe had for the Western explorer as the epitome of a fallen city and a doomed culture.[2] Besides the identification with these once powerful civilisations and their

1 Similar thougths are presented in *A Romance of N'Shabé* (1891), reflecting fears and uncertainties in regard to the aging Queen Victoria: »A great fear and terror seemed to have fallen upon N'Shabéland. The death of the female descendant of their ancient dynasty, following immediately capture of their hitherto impregnable strongholds by a few white magicians, created a sad and growing impression that the nation was doomed to extinction.« (*N'Shabé*: 282)

2 Cf. also the following passages from Walmsely's *The Ruined Cities of Zulu Land*: »The stars were very brilliant, and the soldier sat thinking of the past, and peopling in his imagination those fantastic masses of fallen ruins which had once at that hour rang with bustle and merriment.« (*Ruined Cities* Vol. I: 177); »The ruins of a large stone fort were crumbling away before them, the masses of fallen masonry gradually disappearing before the slow but steady action of time, besides being partially buried in the sand drifted up before the winter gales.« (Ibid.: 33-134); »›There lie the gold fields of Solomon somewhere in that neighbourhood; the ruined cities of the mighty old Egyptians, the ancient gold digger, crumbled into dust.‹« (Ibid: 135, cf. also Ibid.: 170)

downfall, attempts to provide reasons for the latter are prominent in most of the texts dealt with. One of the common ideas was that these ancient cultures were weakened by primitive forces, undermining civilisation and eventually bringing about their doom. Generally, these primitive forces are projected on indigenous African people and their ancestors, who are eventually identified as responsible for the degeneration of the once highly advanced inhabitants of the ruins. Haggard's *Elissa* is a case in point in this context. As already suggested by the full title of the novel, *Elissa: Or, the Doom of Zimbabwe*,[3] the novel deals with the city's downfall. In his introductory note to the novel, Haggard informs the reader of the following:

»However these things may be, there remains ample room for speculation both as to the dim beginnings of the ancient city and its still dimmer end, whereof we can guess only, when it became weakened by luxury and the mixture of races, that hordes of invading savages stamped out of existence beneath their blood-stained feet, as, in after ages, they stamped out the Empire of Monomotapa. In the following romantic sketch the writer has ventured – no easy task – to suggest incidents such as might have accompanied this first extinction of the Phœnician Zimbabwe.« (*Elissa*: viii)

The reason Haggard gives for the city's doom is the fact that it »became weakened by luxury and the mixture of races« (ibid.), which then made it possible for »hordes of invading savages [to] stamp[] [the city] out of existence beneath their blood-stained feet« (ibid.). That Haggard makes these aspects responsible for the doom of Zimboe is crucial, since they reflect contemporary fears of degeneration and doom in regard to Britain and its Empire, referred to by Spencer:

»Nor was [the] boundary [between human and ape] a matter of abstract speculation for civilized Europeans; for if humans could *evolve*, it was thought they could also *devolve or degenerate*, both as *nations and as individuals*. At what point in a down-

3 Note again here the connection between past and present. In the book the city is referred to as Zimboe, a name consequently connected with the past; Zimbabwe, however, very much refers to the modern name of the city and thus is associated with the present. Consequently, giving the book the title »the Doom of Zimbabwe« connects the past and present.

ward slide did a human being cross over the line into animality? [...] Such concerns underlay the tremendous public anxieties at the end of the century about the condition of the British Empire and the warning that, like the Roman predecessor, it could fall, and for what were popularly perceived as the same reasons – moral decadence leading to racial degeneration.« (Spencer 1992: 204, emphasis mine)

These fears were connected to the recession of political and economic growth of the British Empire towards the end of the nineteenth century after decades of prosperity and power. How these anxieties were stirred by the Zimbabwe ruins through Britain's identification with their alleged inhabitants is underlined in Wilmot's preface to *Monomotapa*, where Wilmot equates the power of the English to that of the ancient Phoenicians: »Gain and slaves were the objects of the voyages of this crafty, heartless, and adventurous race, who were the English of the ancient world without the English honour.« (Wilmot 2005: xvii) In spite of this feeling of omnipotence on the part of Britain, an underlying collective feeling of uncertainty prevailed, which is reflected in Haggard's *Elissa*: »What Haggard finds in the incipient mining industries of South Africa, and in the archaeology/mythology of the Great Zimbabwe, are potential analogues of the British process of modernisation and capital accumulation; what he turns them into is a corrective.« (Chrismann 1997: 301) Although this threat was diffuse, it was generally characterised by the fear of degeneration that many perceived as coming from beyond the British boundaries. This is made explicit by the famous passage in Haggard's *She* in which Ayesha informs Holly of her invasion plans in regard to Britain:

»›And now tell me of thy country – *'tis a great people, is it not? with an empire like that of Rome*! Surely thou wouldst return thither, and it is well, for I mean not that thou shouldst dwell in these caves of Kôr. Nay, when once thou art even as I am, we will go hence – fear not but that I shall find a path – and then shall we cross to this England of thine, and live as it becometh us to live. Two thousand years have I waited for the day when I should see the last of these hateful caves and this gloomy-visaged folk, and now it is at hand, and my heart bounds up to meet like a child's towards its holiday. For though shalt rule this England – ›But we have a queen already‹, broke in Leo, hastily. ›*It is naught, it is naught*‹, said Ayesha; ›*she can be overthrown.*‹« (*She*: 254, emphasis mine)

While on the one hand the greatness of the British Empire is emphasised by Ayesha in the first sentence, on the other hand it is weakened by the fact that Ayesha thinks it possible that Queen Victoria can be overthrown.[4] This points towards the ambivalent collective feeling of confidence and insecurity characteristic of late nineteenth-century Britain in general.

At the same time, it is striking that Ayesha considers her planned immigration to Britain as an escape from the »hateful caves and this gloomy-visaged folk« (ibid.), which is repeatedly referred to as the degenerated offspring of the once mighty people of ancient Kôr. Significantly, as Holly and the other men are informed by Ayesha, the initial reason given for Kôr's downfall is a pestilence that exterminated nearly all its inhabitants:

»Kôr is fallen! [...] Twenty and five moons ago did a cloud settle upon Kôr, and the hundred cities of Kôr, and out of the cloud came a pestilence that slew her people, old and young, one with another, and spared not. The pestilence slew and slew, and ceased not by day or by night, and those who escaped from the pestilence were slain of the famine.« (Ibid.: 182)

In connection with Egyptian archaeology, the pestilence referred to by Ayesha can be read in regard to contemporary fears of invisible germs recently discovered and made visible by scientific and technical insights and innovations (cf. Chapter 2.2). The pestilence, as an invisible and lethal danger, thus becomes the ultimate threat for even the most powerful and highly civilised culture. The remains of the once prosperous Kôr function as a reminder of this. What happens once the powerful structures of civilisations break down is illustrated by the further development of the city of Kôr, which is summed up by Ayesha:

»Time after time have nations, ay, and rich and strong nations, learned in the arts, been and passed away and been forgotten, so that no memory of them remains. This is but one of several; for Time eats up the works of man, unless, indeed, he digs in caves like the people of Kôr, and then mayhap the sea swallows them, or the earthquake shakes them in. Who knows what hath been on earth or what shall be? [...] Yet were not these people utterly destroyed, as I think. Some few remained in the

4 Cf. also above, where I discuss fears of reverse invasion in regard to the reception of Egyptian archaeology.

other cities, for their cities were many. But the barbarians from the south, or perchance my people the Arabs, came down upon them, and took their women to wife and the race of the Amahagger that is now is a bastard brood of the mighty sons of Kôr, and behold it dwelleth in the tombs with its fathers' bones.« (Ibid.: 183-184)

First of all this passage once more stresses the fact that even powerful and highly civilised nations such as Kôr have been extinguished and have disappeared from the earth completely, so that »no memory of them remains« (ibid.). Again, the idea of a culture being not only extinguished but also forgotten can be seen as one of the basic anxieties of humankind as such. However, what is even worse is the fact that the people of Kôr were not only extinguished by the plague, but that those few who survived succumbed to the »barbarians« (ibid.), who ultimately destroyed the noble race of Kôr by »[taking] their women to wife« (ibid.) and thus brought forth the primitive and uncivilised »bastard brood« (ibid.), the Amahagger now dwelling in the ruins of the once mighty city of Kôr (cf. ibid.). The degree of their degeneration is emphasised by the characterisation of the Amahagger and their customs (cf. Murphy 1999: 765) throughout the novel. Furthermore, their degeneration is supported in the above passage by the very fact that »the bastard brood of the mighty sons of Kôr« (*She*: 84) lives in the tombs of their ancestors. The barbarity of the »bastard brood« (ibid.) is most strikingly displayed in the novel when Holly and the other men witness a local instance of mummy burning:

»Scarcely were the words out of [Ayesha's] mouth when from every point we saw dark forms rushing up, each bearing with him what we at first took to be an enormous flaming torch. Whatever they were they were burning furiously, for the flames stood out a yard or more behind each bearer. On they came, fifty or more of them, carrying their flaming burdens and looking like so many devils from hell. Leo was the first to discover what these burdens were. ›Great heavens!‹ he said, ›they are corpses on fire!‹ I stared and stared again – he was perfectly right – the torches that were to light our entertainment were human mummies from the caves! [...] Heavens! how they roared and flared! No tar barrel could have burnt as those mummies did.« (Ibid.: 218)

Although the burning of mummies had been a common practice for centuries both in Africa and Europe due to their high flammability,[5] the fact that the Amahagger do not burn ›simply‹ any mummies but those of *their* forefathers seems to render the spectacle extremely terrifying to the onlookers. Furthermore, the idea of burning the remains of your ancestors as well as the fact that the Amahagger live in the tombs amongst the bones of their forefathers is disturbing since it layers further human anxieties, including fears of cannibalism. Tellingly, in this particular passage Leo is the first to recognise the ›torches‹ as »corpses on fire« (ibid.) and thereby identifies them as human bodies before the reader learns that they are mummies.[6] This is interesting in that the recognition of the ›torches‹ as human makes possible a more immediate identification with the ›torches‹ than is the case with a mummy. In other words, while a human corpse provides a relatively high degree of identification, the mummy is *per se* much more the ›other‹, as noted by Wieczorkiewicz in regard to the mummy as a public spectacle: »›My body‹ and ›my beliefs‹ are situated on one side of the discursive space, ›your curiosity‹ and ›your enjoyment‹ are located on the other.« (Wieczorkiewicz 2005: 54) In *She*, the men's horror upon recognising the ›torches‹ as human forms can also be read in this context as an implicit self-criticism of the British colonial practices.[7]

---

5 Since the early times, mummies (chopped up) had been exported to Europe both as combustibles and medicine (cf. also Chapter 3.3).

6 In this context it is interesting that Leo is the first to recognise the ›torches‹ as »corpses on fire« (*She*: 218), since he might actually be related to these ancient inhabitants of Kôr, as suggested by the lineage described on the Sherd of Amenartas.

7 Cf. the meaning of guilt in regard to the exploration and exploitation of ancient Egyptian tombs discussed in Chapter 7.4. Similarly, the following comment of Ithobal in *Elissa*, addressed at the rulers of Zimboe, can be understood as an implicit self-criticism on the part of the British colonisers: »›For many years I and the countless tribes which I rule have suffered much at the hands of your Phœnicians, who centuries ago settled here in my country as traders. That you should trade we are content, but not that you should establish yourselves as a sovereign power, pretending to be my equals who are my servants. Therefore, in the name of my nation, I demand that the tribute which you pay to me for the use of the mines of gold shall henceforth be doubled; that the defence of this city be

However, the main terror of the mummy burning witnessed by the men seems to stem from the fact that the Amahagger *consume* their ancestors, which turns them into cannibals. The identification of the Amahagger as potential cannibals is crucial since cannibalism is a prominent discourse in both colonial fiction in general and fiction dealing with the Zimbabwe ruins in particular.[8] One crucial reason for this immense terror and threat perceived as coming from cannibalism is the basic idea of being attacked from *within*, that is, by an individual who belongs to one's own group, associated with the *self* rather than the other: »[T]he difficulty of maintaining an opposition between self and other becomes highly problematic once the idea of cannibalism is evoked, for even as it functions as a marker of difference cannibalism also carries the reminder of how similar we are.« (Guest 2001: 112) In regard to the above-quoted passage this is significant, since here the identification with the burning mummies as human corpses functions as the first moment of recognition and identification. The second moment marks the subconscious realisation that the burning of the mummies is a symbolic act of cannibalism.

Significantly, it is the fear of cannibalism that is most often associated with degeneration and doom of powerful and civilised cultures, as emphasised by Stott, who points out that »[c]annibalism comes to represent the nadir of barbarism, a marker between civilized and uncivilized man« (Stott 2001: 156). As an inhuman practice it is most frequently associated with the ›uncivilised‹ tribes the white explorers were sure to discover in the African ›wilderness‹. This is also true of *Elissa*, where the highly advanced Phoenician town is constantly threatened by the »black-hearted savage« (*Elissa*: 75) Ithobal and his people:

»›Have you heard, Phœnician, that the chiefs of certain of my tribes love to decorate their spear-shafts with the hide of white men, and to bray their flesh into a medicine which gives courage to its eater?‹ With this pleasing and suggestive query Ithobal paused, and looked towards the door of the tent as though he were about to call his guard.« (Ibid.: 142)

---

thrown down; and that you cease to enslave the natives of the land to labour to your service.‹« (*Elissa*: 81)

8 Other Zimbabwe texts referring to cannibalism or practices involving human sacrifices are *A Romance of N'Shabé*, *King Solomon's Mines*, and *Elissa*.

While in *Elissa* the threat of cannibalism is conventionally associated with black – or at least not purely white – people, in *A Romance of N'Shabé* cannibalism is also dealt with in connection to a »white race« (*N'Shabé*: 9):

»›What a pity it is that tradition reveals so little of this people‹, I exclaimed. ›It would be most interesting to identify the very habitat of Solomon's Queen of Sheba!‹ ›Ah, I have something more to add!‹ said Clavering. ›It is that when the white race fled from its home, it travelled up the Zambesi to somewhere in the interior, where it exists to this day, though they are not so nice a set of people as formerly, for they married amongst the cannibals by whom they were surrounded, and adopted their fearful customs. The Mashonas tell me that once a year a number of fair-skinned natives come from some distant land to worship, and perform magical rites amongst the huge stone ruins. There, I think I have now told you everything I could gather concerning this vanished nation.‹« (Ibid.)

Here the legendary stone ruins are associated not only with the potential »habitat of Solomon's Queen Sheba« (ibid.) but also with a white race. Unlike other texts dealing with the Zimbabwe ruins, however, this white race had already undergone the process of degeneration by »marr[ying] amongst the cannibals by whom they were surrounded« (ibid.) and, even worse, by »adopt[ing] their fearful customs« (ibid.). The fact that the white people adopt the ›monstrous‹ customs attributed to local black peoples thus points towards a fundamental threat perceived by a highly civilised nation, such as Victorian and Edwardian Britain, of being invaded, dominated, and infected by the barbarous, the uncivilised, the ›other‹. As discussed in regard to mummy fiction, there is also always the latent fear that this barbarism actually exists as an invisible entity beyond the surface of civilised societies: »Thus the fear expressed in such texts is of barbarism beneath the façade of civilization which could be released at any point, a barbarism constituting primitive impulses and primitive sexuality, a barbarism identified in evolutionary theory with blackness and specifically with the black female as sexual being.« (Stott 1989: 75) This constant threat of the uncivilised as part of one's own civilised self is underlined by the following passage, in which the Western explorers encounter »a huge tribe of baboons« (*N'Shabé*: 32) which the British explorer Montfort comments on: »Our unexpected appearance sent them scampering away up the hill, and it was a sight to see the baby baboons on their mothers' backs with their little arms and legs clinging round them – a sight that brought Darwin's theory forcibly to my mind, for they looked more like human beings than monkeys!« (Ibid.: 33) Montfort's explicit reference to Darwin and the humanness he perceives in the monkeys is significant since it points towards the fact that at that moment the stabilising boundary between human and animal is blurred: »Scientific discourses, shaped by Darwin's distinction between ›civilized‹ and ›savage‹ races in the 1871 Descent of Man, reinforced comforting presumptions of European ascendancy on the evolutionary ladder.« (Murphy 2000: 23) For many people at the time the blurring of the line between human and animal would have been threatening, since it addressed the fear of degeneration by pointing out that man was not much different from animals and peoples traditionally conceived as primitive by Western society: »Africa itself is presented to the reader as a landscape inhabited by ›beastly‹ natives and wild animals galore. [...] The animal imagery helped to produce durable models of African identity and otherness which were compatible with current ideas of geography, race and human evolution.« (Sinha 2008: 29) In the novel, Clavering emphasises this threat by commenting on Montfort's statement: »The same idea struck us all, and Clavering told us of an event which happened in the old Colony many years ago, where a man baboon seized a young girl one evening, as she was walking home, and carried her off to the mountains.« (*N'Shabé*: 33) Here the threat of degeneration becomes externalised through the *man*[9] baboon abducting the woman (as a mate); at the same time the scene also implicitly refers to the fact that the monkey would not be interested in the woman if there were not a close kinship between man and monkey.[10]

Later in *A Romance of N'Shabé*, the white European Jules Decluse, who lives among the people of N'Shabé, utters the following:

»›*Mon Dieu, mon Dieu!* [...] What new catastrophe has Heaven in store for me? With the destruction of the arsenal has gone this ancient people's last chance of

9 The use of the word ›man‹ in this context once more emphasises the close kinship between the monkeys and humans. Cf. also the devolution of Ayesha at the end of *She* to the form of a monkey (cf. *She*: 292-293) and the fact that Holly is referred to as converting others to Darwin's »monkey theory« (cf. ibid.: 18).

10 The fact that the baboon abducts the woman is also reminiscent of the threat perceived as coming from intermarriage between whites and blacks (cf. below).

successfully withstanding the cannibal tribes of the north. I have laboured to preserve this race, for its history is replete with antiquarian interest; but Heaven has apparently willed it otherwise. Nevertheless, I must hasten to save what I can, and I will see about this change of guard on my return.‹« (Ibid.: 251-252)

Once more, the »cannibal tribes« (ibid.) pose the ultimate threat to the noble race Decluse is so anxious to preserve. However, it is crucial to consider the grounds on which Decluse points towards the people's »history [as being] replete with antiquarian interest« (ibid.). Considering that we are dealing with the descendants of an ancient *white* people, Decluse's motivation for preserving them seems obvious, since he would certainly not be interested in preserving an ancient *black* people. Thus, once more it is the identification with the people that makes the threat of cannibalism, connoting barbarism and degeneration, so immanent.

In this context, the above-quoted passages can again be read as the essential fear of reverse colonialisation leading to degeneration and doom of a highly advanced Empire – a fear very prominent in the late nineteenth and early twentieth centuries, as the popularity of invasion literature and texts dealing with reverse invasion illustrates. Furthermore, it is important to note how the civilised white races become infected with the barbarous customs of the black tribes through intermarriage/intercourse, as emphasised by Chennells:

»›Elissa‹ makes Great Zimbabwe's ruins into a cautionary tale. The ancient colony became decadent and the destruction of the city was the price the colonizers had to pay. All of the novels written after 1889 use the city's destruction to admonish, implicitly or explicitly, the new colonizers, and leisure competes *with sex between the races as a signifier of decadence.*« (Chennells 2007: 13, emphasis mine)

This is crucial insofar as intermarriage is one of the central aspects mentioned in texts dealing with the Zimbabwe ruins and their history. It is the deep-rooted fear of degeneration by becoming the ›other‹ both in terms of practices and appearance which eventually represents the ultimate threat to the survival of the white Western race. To prevent the doom of the white civilised Western cultures, it is therefore necessary to be not only defensive but also offensive, as underlined by Vincent's exclamation in the *City of Gold*: »›I do venture to assert my belief that the civilisation which hath no power to expand and subject inferior races to its sway is foredoomed to extinction.‹« (*City of Gold*: 219)[11] Clearly, these assumptions were meant to justify the aggressive British Imperialism in Africa and elsewhere.

Significantly, in *Elissa* it is the black Ithobal who threatens the Phoenician community by trying to force a marriage between him and the white Elissa, the daughter of the Governor of Zimboe. Ithobal, who himself is repeatedly referred to as being the offspring of a mixed relationship,[12] embodies the ultimate danger for the white Phoenician city of Zimboe. He functions as living proof of the evil that comes out of intermarriages: »But this Ithobal is the terror of our city, for if he chooses he can bring a hundred thousand savages upon us, shutting us within our walls to starve, and cutting us off from the working of the mines whence we win gold.« (*Elissa*: 33) As this passage shows, the savageness of Ithobal threatens not only the Zimboean civilisation but also the city's prosperity, secured by the nearby gold mines. This underlines how alleged barbarism was also implicitly conceived as posing a material threat to Western civilisation. In this respect it becomes even clearer why the uncivilised was generally considered as something that had to be shut out of prosperous and highly advanced societies, as it was seen not only as a moral threat to Western civilisation but eventually also as a material one. This also helps explain why nineteenth-century British society conceived intermarriage as a major threat, bringing about the degeneration and doom of highly advanced civilisations.

In *Elissa*, the eponymous heroine and her people fight with all their means against a marriage between the white Elissa and the black Ithobal, since this intermarriage would mean the victory of the uncivilised over the civilised and thus the doom of Zimboe. The consequences of an intermarriage with Ithobal are hinted at in the following description of Ithobal and his warriors besieging the city: »It was a wondrous and fearful sight, that of these hordes of plumed warriors, their broad spears flashing in the sunrise, and *their fierce faces alight with hereditary hate and the lust of slaughter*.«

11 Cf. also: »This was a state of affairs we could not allow to continue; for a nation that has once lost its self-respect would soon become a mere collection of slaves, and an easy prey to its more warlike neighbours.« (*N'Shabé:* 282)

12 Cf. »His tawny complexion marked him of mixed race« (*Elissa*: 28), »half-savage Ithobal« (ibid.: 39), »crossbred hound« (ibid.: 228), »half-bred barbarian« (ibid.: 74), »cross-bred savage« (ibid.: 145).

(Ibid.: 216, emphasis mine) The reference to »hate and lust of slaughter« (ibid.) as »hereditary« (ibid.) implies that their uncivilised behaviour is something that is inherited. In this context a marriage with Ithobal would mean the spread of his degeneration through heredity and thus can be read as the slow invasion of the civilised white society by the uncivilised blacks.

Even though Elissa eventually averts a marriage with Ithobal, the city's doom is finally brought about by her love for another man, Prince Aziel:

»And thus, because of the fateful and predestined love of Aziel the Prince, and Elissa the priestess and daughter of Sakon, three thousand years and more ago, the ancient city of Zimboe fell at the hand of King Ithobal and his Tribes, so that to-day there remain of it nothing but a desolate tower of stone, and, beneath, the crumbling bones of men.« (Ibid.: 244)[13]

Significantly, however, the reason for the disastrous consequences of Elissa and Aziel's love, leading to both their and the city's downfall, is not only the threat from Ithobal but also the religious conflict between Elissa and Aziel's disciples:

»This lady could marry, indeed she was expected to do so, but her husband must take the title of Shadid, and for her lifetime act as high priest of El. Therefore, thought Metem, if it could be brought about that Elissa should be chosen as the new Baaltis, it was obvious that there would be an end of the possibility of her marriage to Aziel. Then, in order to wed her, he must renounce his own religion – a thing which no Jew would do – and pose as the earthly incarnation of one whom he considered a false divinity or a devil.« (Ibid.: 122)

Thus, in the text both the threat through barbarism, in the person of Ithobal, and the conflict between Jewish and Pagan religions, embodied by Elissa and Aziel, eventually lead to Zimboe's doom. This is significant in regard to nineteenth and early twentieth-century Britain, since it reflects the subtle fear of the uncivilised and primitive as well as a certain anti-Semitism

13 Cf. also: »›You could not wed this woman, who is not of your race, or rank, or religion, and if you could, it would bring about a struggle that must cost thousands their lives, and this city its wealth.‹« (*Elissa*: 134)

caused by the Jewish immigration wave from Eastern Europe,[14] respectively. Both Ithobal with his savageness and Aziel as a Jew can thus be read as the materialised threat associated with the other.

It is worthwhile to note in this context that both of these threats, attributed to Ithobal and Aziel respectively, at first seem to come from the *outside* of the city. The savage Ithobal and his people, who »haunt the outskirts of the city, seeking to steal white women to be their wives« (ibid.: 22), on the one hand and the Jewish Prince Aziel on the other hand both enter the city and threaten to bring their customs and beliefs with them. However, very similar to Stoker's *Dracula*, this reverse invasion is only possible because the city of Zimboe as such is weak, because its inhabitants are not strong enough to fight off the enemy from outside. Consequently, the reason for Zimboe's doom is actually to be found *within* the city, as the dialogue between the Prince and his follower Metem underlines: »›These walls are strong, and we shall beat them back‹, said Aziel. ›Nay, Prince, for strong walls do not avail without *strong hearts* to guard them, and those of the *womanish citizens of Zimboe* and their hired soldiers are white with fear.‹« (Ibid.: 218, emphasis mine)[15] Here it becomes clear that even the strongest walls will fail to protect the community when its individuals as such are weak. This points towards the essential fear of the enemy within dealt with in *Dracula*. Ithobal and Aziel pose a threat to the Empire because they address weak points of British society as such. In this context, the comment of the physician in the *City of Gold* sounds like a warning addressed to any civilised society: »›Civilisation is ever prone to underrate the forces of barbarism. Think of the interests committed to thy charge, and do not peril their safety by too implicit a trust in thy outer defences.‹« (*City of Gold*: 173) Barbarism becomes a threat to civilisation because it is not

14 »Between 1881 and 1914, 120,000 to 150,000 East European Jews settled permanently in Great Britain, effecting a radical transformation in the character of Anglo-Jewry. Their poverty, occupation and foreignness drew unwanted attention to them and native-born Jews alike, fueling the fires of xenophobia and antisemtism.« (Edelmann 2002: 127)

15 Cf. also the following utterance of one of the City of Gold's inhabitants: »›[T]he social conditions which it hath been our boast and pride to maintain are threatened from the point whence alone destruction is possible – from within!‹« (*City of Gold*: 218)

taken seriously enough – as much as many threats were conceived as not being taken seriously enough in Britain at the time.

In addition to that, in both Ithobal and Aziel's case it is their (sexual) desire, stirred by Elissa, that turns out to be a threat to society as whole. Similar to *Dracula*, female sexuality and »the womanish citizens of Zimboe« (*Elissa*: 218) are thus the weak points of Zimboeian society in Haggard's novel.[16] Consequently, it is the enemy within that makes figures like Ithobal and Aziel so dangerous for the community and not these figures *per se* (cf. *Dracula Discourse*). This also implies that Zimboe is responsible for its own downfall, which could actually have been prevented if only its inhabitants had been strong enough. Tellingly, the emphasis is here on moral rather than on political or physical strength. The city eventually falls due to Elissa's inability to resist Aziel and due to its inhabitants' lack of courage. This is alluded to by Ithobal's threat to Aziel prior to his invasion of the city: »*Once let them march beneath yonder walls*, and before they leave it, Zimboe, city of gold, shall be nothing but a heap of ruins, and a habitation of the dead.« (Ibid.: 117, emphasis mine) The city's ruin is only possible once the enemy is *let* inside. In regard to contemporary British society this is significant, since it suggests that society as such is responsible for the invasion and creation of destructive evil forces.[17] Degeneration and doom become possible only if society and its individuals are weakened and *allow* themselves to be weakened. At the same time, there is also always the fear of discovering the self in the ›other‹, as is suggested in the *A Romance of N'Shabé*, where one of the bushmen suddenly turns out to be of European origin:

16 This is also true of the discourse of masculinity in *She*, which emphasises male strength and power while at the same time repeatedly questioning it through the powerful Ayesha and the homoerotic tendencies in the relationship of Holly and Leo: »The dominant code of manliness in the 1890s, so hostile to emotional expression and so intolerant of both androgyny and homosexuality, can be interpreted as a by-product of a raised imperial consciousness – especially with regard to the imperial frontier and the manly qualities required there.« (Tosh 1994: 196)

17 This also points to the assumption that evil is an inherent part of Victorian society (cf. *Jekyll and Hyde Discourse*).

»A man, taller than the rest of the Bushmen by quite a hand, separated himself from the conference of chiefs, and advanced towards us waving a white rag. [...] To our utter astonishment, the foe replied in such unmistakably good English that we both were convinced we were being addressed by a countryman who, judging by his dialect, hailed from Somersetshire.« (*N'Shabé*: 94)

To recognise the self among the enemy is as disturbing as to become aware of the enemy behind one's own lines. The Western explorers are implicitly confronted with both of these aspects when discovering the remains of a once highly advanced people: On the one hand it seems to fulfil the longing to discover the self in the realm of the ›other‹, while on the other it also underlines the fear of this very discovery.

In addition, these examples show how the alleged barbarism of indigenous African people is used as an explanation for the downfall of a once powerful and flourishing culture of which only ruins have survived. In the case of the Zimbabwe ruins, the imagined threat from the uncivilised is conceived as very concrete through the experiences the Western explorers have in the African ›wilderness‹. As a consequence, this feeling of immanent danger in the present is projected onto the ruins, that is, the past, and is thus used as an explanation for the doom of the ancient Zimbabweans. The experience of the ›wilderness‹, in combination with the confrontation with the remains of remote civilisations, renders the Zimbabwe ruins unique in that they represent both the highest achievements of civilisation *and* the immanent threat of degeneration and doom. This is underlined by the many fictional texts dealing with the Zimbabwe ruins in which these fears of degeneration and doom are central.

Interim Findings: The Mummy and Great Zimbabwe as the *most unfamiliar strangeness*

In the present chapter, I have shown how the archaeological space was increasingly perceived as destabilising and threatening. While in the examples discussed in the previous chapter order was only momentarly disturbed through the inversion of the object-subject relation and binary gender roles, the texts I have discussed in this chapter elucidate that stability in the present is no longer possible. The threat associated with the past is particulary strong in *Pharos* and *Iras*, where it culminates in fantasies of reverse invasion, thus reflecting the *Dracula Discourse.* While in *Pharos* the evil forces of the past materialising in Pharos's character at first seem to appear unmotivated and thus particularly disturbing, I have demonstrated in a further reading of *Pharos* how in an act of identification of the British with the past underlying feelings of guilt emerge in regard to the treatment of this past. A necessary prerequisite for this identification, however, is a certain affiliation with that past, which was given in the case of Egyptian archaeology. As what I have termed the *less familiar strangeness*, the Egyptian past represented both the self and the ›other‹. I have further demonstrated how in regard to the Zimbabwean past identification was only possible on the basis of a radical negation and assimilation of the Zimbabwe ruins, which in relation to Greek and Egyptian archaeology were *per se* perceived as the *most unfamiliar strangeness*. In an act of intentional history, the ruins were appropriated and reinterpretated by the British colonists as part of their own past. Unlike the Greek and Egyptian past, which was predominantly used as a valuable model, the Zimbabwean past was virtually invaded and usurped by the British. As a result, the colonists no longer conceived the Zimbabwe ruins as part of the ›other‹ but as part of their own culture, as I have shown in regard to their reception in fiction. This assimilation on the basis of negation, however, eventually culminated in an over-identification with the ruins, causing anxieties of doom and degeneration. On a symbolical level, the confrontation with the archaeological remains of these civilisations can also be read as functioning as a reminder of the Western explorer and archaeologist's incomplete present self as opposed to its former imagined and lost wholeness, in the Lacanian sense never to be regained.

Conclusion

On July 12, 1856, the *Illustrated London News* published the following account of the Crystal Palace exhibition:

»In the restoration of the architecture and decorative art of the Assyrian, Egyptian, Greek, Romans, and Moors of Spain, as well as of the period of the Decline and Fall, the beautiful painted transformations and the picturesque Renaissance, which, crab fashion, began with Gothic, and ended in Roman, we had, as it were, a practical course of architectural education for the *masses*. Only those at their case in pecuniary circumstances could travel to where the citrons bloom, and judge for themselves of the transitions from cycle to cycle of constructive art. A couple of visits to the Crystal Palace will teach a man more of *archæology* than he will find in all the ponderous tomes of Agincourt and Gally Knight. In the same spirit of *popular* instruction, comprising chronological order and choice selection, the directors of the Crystal Palace, are now about to extend their sphere of operation to the illustration of art as allied to manufacture, and with decorative fictile works they have made a most vigorous and successful initiative.« (*Illustrated London News*, Jul 12, 1856: 34, emphasis mine)

Focusing on »popular« (ibid.) »archæology« (ibid.) as received by the »masses« (ibid.), this book has analysed a text corpus comprised of Victorian and Edwardian novels, short stories, magazine and newspaper articles, and excavation accounts. On the basis of the study of the popular reception of Greek, Egyptian, and Zimbabwean archaeology in these texts, it has been argued that the archaeological discourse functioned as a means of identity formation.

Nineteenth-century Britain was characterised by profound and diverse changes which affected its contemporaries as never before in the country's history. As a reaction to this,

»[h]ere and there people were found carrying on past traditions when marked changes were taking place around them; art and literature continued, in various direction, to reflect the ideas and aspirations of an earlier day. But taken as a whole, the change that occurred in habits and thought, in style of dress in outward manners and customs, with the accession of Queen Victoria, was so obvious that even contemporaries must have realised that they were passing through a complete and drastic change.« (Chancellor 2007: 12)

Although these drastic and complex changes influenced the social classes in different ways and to different extents, they all had in common that they deeply affected people's conception of time and space. In the wake of industrialisation and urbanisation, social structures underwent radical changes, bringing forth the urban working classes as a new stratum of society. As a mixture of regulated governmental attempts to cultivate the emergent working classes and the concurrent development of an independent working-class culture, an unprecedented popular culture evolved in the cities as a Gramscian compromise equilibrium. The collective consumption of popular newspapers, magazines, books, and public spectacles determined people's leisure time and gave rise to a new mass culture.

Accompanying this, technological innovations, such as the expansion of the railway system, revolutionised public transport and deeply affected contemporaries' conception of time and space. Geological and biological insights gained in the further course of the century not only opened up prehistory as a hitherto unknown time and space, but also questioned man's singular and chosen position in the world. This was both reflected and enforced by the crisis of the Anglican Church, revealing the instability of traditional Christian belief systems in general. Besides these radical developments at home, the rapid growth of the British Empire, with its massive land reclamations, had a lasting influence on people's collective perception of space. Additionally, the encounter with indigenous peoples in the colonies, which, according to the common imperial ideology, were perceived as less advanced and as associated with earlier stages of humankind, emphasised the idea of time as a determining factor in regard to the evolution of modern men. At the same time, it fuelled anxieties of devolution and degeneration of modern society. Towards the turn of the century, advances in microbiology gave rise to fantasies of reverse invasion by invisible forces. This was further enforced by an underlying general fear of destabilisation evoked by a decrease in economic and political power as well as Britain's aging monarch and her death in 1901. Moreover, changes in regard to gender roles materialising in the suffragette movement and the appearance of the Decadent dandy and the New Woman challenged traditional gender concepts.

These radical changes and developments affecting all social classes had their lasting effect on Victoria's subjects, causing a collective feeling of alienation. People now found themselves confronted with a time characterised by instability and rapid pace. This resulted in the paradoxical situation that contemporaries on the one hand strongly favoured progress, while on the other they were subconsciously longing for an allegedly more stable and secure past. It was in the context of this popular interest in the past, spurred by recent insights and advances in the fields of geology, biology, and historical studies, that, in combination with a new urban mass culture, a popular archaeological discourse emerged. For one thing, the emergence of a popular archaeology was determined by the permeability of the antiquarian discourse as such, which can be seen as rooted in its close proximity to amateurism since its beginning in the sixteenth century. As a consequence of this permeability, a regular exchange between more professional antiquaries and amateurs had been quite common. Theoretically, the study of antiquity and early archaeology had been open to all who had the time and finances to dedicate themselves to the subject, which led not only wealthy connoisseurs to engage themselves in the field, but also enthusiastic adventurers and explorers, as prominently illustrated by figures such as Giovanni Belzoni. Through his double status as a showman and an amateur archaeologist, Belzoni assumed a crucial role as a populariser of archaeology and Egyptology. The popular presentation of knowledge that had hitherto been reserved to a small elite – the valuable *Descriptions of Egypt* being a case in point – meant that this knowledge was now accessible to the broader masses. The example of Belzoni is thus representative of the formation of a popular archaeological discourse out of what had been separated into ›high‹ and ›low‹ culture for a long time. Other popular archaeologists, such as Austen Henry Layard, Heinrich Schliemann, and Amelia B.

Edwards, followed Belzoni's example and helped through their specific status as mediators and transformers of archaeological knowledge to popularise the subject further in their own way. At the same time, these cases underscore how a popular archaeology emerged at the time as the result of an interplay between different interests and forces comprising the sensationalistic concern of the masses, a governmental endeavour to cultivate and educate them, and the personal issues of the popularisers, which could be both pertinent and self-serving. How the process of popularising archaeology was also prone to ideological abuse has become particularly clear in regard to the presentation of Zimbabwean archaeology by men like Cecil Rhodes and Theodore Bent.

On the basis of the assumption that the emergence of a popular archaeology in the nineteenth century resulted from an interplay of various versatile and tightly interwoven changes and developments, I have argued that archaeology functioned as a counter-discourse to Victorian and Edwardian society in which the collective identity of a traumatised and alienated self was (re)created and renegotiated. This also entailed the archaeological space serving as a projection screen for contemporary sociocultural concerns that were imported, transformed, and at times abandoned within that parallel discourse. In regard to the popular reception of archaeology, four dominant discursive threads have been identified, which in analogy to major archetypical Victorian myths have been referred to as (1) the *Frankenstein Discourse*, (2) the *Dracula Discourse*, (3) the *Jekyll and Hyde Discourse*, and (4) the *Sherlock Holmes Discourse*. In addition to their representative functions regarding contemporary discursive formations, the four discourses also reflect the negotiation of different notions of identity and alterity.

Drawing on a model of the formation of cultural identity proposed by the cultural theorist John Storey, I have further argued that the popular consumption of archaeology in Victorian and Edwardian times functioned as a displacement strategy that attempted to overcome a collective feeling of lack brought forth by the alienation of the individual through the substantial changes of the time. Following Storey's concept, which conceives of consumption as a way of producing identity, I have illustrated how the popular consumption of archaeology became a means of production in regard to a collective identity. On the basis of Aleida and Jan Assmann's notion of the collective memory, the popular archaeological texts and practices looked at confirmed the social-constructivist character of individual and collective memory. This has been underlined by the workings of Gehrke's intentional history, which prominently characterised the popular reception of archaeology. In addition, the texts and practices looked at in this study have confirmed the performative character of identity postulated by Butler in regard to gender identities, which was adopted by Storey in regard to identity in general. In this context, the study has shown how the archaeological space, represented by archaeological sites and artefacts as sites of memory in the broadest sense,[1] evidently satisfied both the search for what Storey terms the ›roots‹, that is, memory, and the ›routes‹, that is, the desire to attain completion, of cultural identity. On the one hand, the archaeological space, through its association with the past, thus functioned as a prosthetic memory in that it provided people with ›borrowed‹ memories in the form of archaeological texts and practices, while on the other hand, through its association with a former state of completion, it was seemingly capable of satisfying the collective desire to attain irretrievable wholeness. In this sense, the return to the past represented by the archaeological space very much reflects Victorian times in that it was directed both towards the past *and* the future (never to be attained).

Against this background, I have shown how popular archaeology, as an ambivalent and indefinite space essentially lacking structure and thus meaning, was initially perceived as the ›other‹, as opposed to a British collective identity. As a consequence of its ambivalent ›otherness‹, contemporaries experienced archaeology as both threatening and fascinating. In this context, the exploration of the spatial and temporal ›other‹ also functioned as the symbolic exploration of the Western explorer's own »underground self« (Said 1978: 3). Accordingly, the penetration of the archaeological space can be seen as a threefold confrontation with an unknown and unexplored space involving the spatial, the temporal, and the psychological ›other‹. Due to the lack of definite meaning of the archaeological space, in combination with its function as a site of memory offering stability and orientation in the past and the future, the archaeological space became a site for the (re)creation and stabilisation of an alienated and traumatised identity. On the basis of both the *Frankenstein Discourse* and the

1 This also includes the *memory industry*, which according to Storey also functions in terms of Landsberg's prosthetic memory as a site of memory.

Sherlock Holmes Discourse, the study has shown how the process of searching for and excavating archaeological sites and relics can be read as an attempt to excavate and recreate an individual and collective identity irretrievably lost in the process of modernisation. In this context, the study succeeded in verifying the relevance of performativity for identity formation. Moreover, focusing on the inversion and subversion of identities based on what was introduced as the *Jekyll and Hyde Discourse*, the study confirmed the constructed character of identity. Subsequently, I presented and discussed different ways of dealing with archaeology as an alterity, oscillating between rejecting and adopting it, against the background of what was identified as the *Dracula Discourse*.

In addition to these commonalities in regard to the popular reception of archaeology, I was also able to confirm the existence of specific tendencies as far as the reception of the different archaeologies is concerned. Ancient Greece had a long-standing tradition in Western culture, reaching its peak in Britain with eighteenth- and nineteenth-century Hellenism, initiated and strongly shaped by Johann Joachim Winckelmann and represented by the ›second generation‹ of the Romantic poets. In the Victorian Age the Hellenistic tradition was further cultivated by authors such as Matthew Arnold, Walter Pater, George Eliot, Algernon Charles Swinburne, and Oscar Wilde. Accordingly, the reception of Greek archaeology, as the *most familiar strangeness* (Hölscher 1994) representing the ›defining culture‹, was mainly characterised by identification on the part of the British, as aptly brought to the point by the *Edinburgh Review* in regard to Schliemann's excavations:

»Such a discovery was indeed calculated to arouse the attention, not only of archæologists and scholars, but of every cultivated person in the three kingdoms; for who is there that can pretend to that title, to whom the names Priam and Hecuba, of Hector and Andromache, are not as familiar as household words? Great as was the interest attached to such marvellous discoveries as those at Nineveh, which may be said to have brought to light again the existence of a buried empire, they were deficient in that highest source of interest which is derived from the association and connexion with persons well known in history, or in that poetical and legendary story, which is apt to impress itself more strongly on the mind than any true history.« (*Edinburgh Review*, Apr 1874: 506-507)

In accordance with the perception of ancient Greece as the *most familiar strangeness*, Greek archaeology was commonly received and used as a means of (re)creating a collective identity by recruiting on the assumption of an alleged cultural origin of Western culture in ancient Greece.

Although ancient Egypt had fascinated people before that time, the emerging *Egyptomania* in the wake of Napoleon's Egyptian Campaign was without comparison, as reflected by the prominent ancient Egyptian style influencing Victorian art, architecture, and fashion. In accordance with Said's *Orientalism*, the reception of ancient Egyptian archaeology in Britain was characterised by two different strands right from the beginning. On the one hand it was very much determined by stereotypical associations of the Orient with the beautiful, luring, sensual, and mysterious, while on the other hand it was characterised by an underlying uncanniness, evoked particularly by the Egyptian death cult prominently represented by the figure of the mummy. In addition, unlike modern Egypt, as a highly advanced culture ancient Egyptian culture was considered prestigious in Britain. It is this ambivalence, caused by the different perceptions of ancient Egypt, which essentially characterises the popular reception of Egyptian archaeology. Due to this oscillation between the familiar *and* the unfamiliar, Egyptian archaeology might, in analogy to Höscher's *most familiar strangeness*, be referred to as the *less familiar strangeness*. This is illustrated by the following account of a mummy unrolling given in the *Newcastle Weekly Courant*:

»On Wednesday afternoon, a large and distinguished company assembled in the botanical theatre of University College, London, to witness the unrolling of a mummy from Upper Egypt. […] Mr Budge unrolled the mummy, which was closely swathed in scores of yards of thick, yellowish linen of fine texture. […] At the beginning of the process of unrolling there was a very perceptible sickly smell of aromaties, which as the work went on gave place to a more pronounced and decidedly disagreeable odour. […] When at last the coverings had been removed the body was found to be of a very dark brown colour – so dark, indeed, as to be almost black. […] The features when disclosed stood out very clearly, and were those of a rather handsome person, but the sex could not be determined.« (*The Newcastle Weekly Courant*, Dec 21, 1889)

As suggested by the ambivalent description of the mummy's body – its »very dark and brown […] almost black« colour with its »sickly« and »disagreeable odour« on the one hand versus its »rather handsome« features on the other – in the figure of the mummy the familiar and the unfamiliar merge, which was often perceived as uncanny and threatening to British identity. Moreover, this ambivalence is enforced by the fact that the mummy's »sex could not be determined« (ibid.). This is underlined by the many mummy texts dealing with the inversion of identities and the invasion of Britain by forces from the Egyptian past. Characteristically, in many of these texts the protagonists fall in love with a beautiful woman of the ancient Egyptian past, representing the familiar, while at the same time they feel threatened by an evil force, standing for the unfamiliar, emerging from that very past. It is thus the familiar associated with Egyptian archaeology that is attractive, while that which is unfamiliar in it is perceived as dangerous. As far as the popular reception of Egyptian archaeology is concerned, this is further underlined by the increase in mummy texts featuring evil and destructive mummies that threaten modern Britain towards the end of the nineteenth century.

Since the Western exploration of former Rhodesia was only undertaken in the second half of the nineteenth century under the leading figure of Cecil Rhodes, ancient Zimbabwean history was considered for the most part as non-existent. Consequently, the ›discovery‹ of the ancient stone ruins in what was commonly seen as the African ›wilderness‹ was initially confusing for the colonisers. This was enforced by subsequent archaeological studies of Great Zimbabwe, which emphasised the old age of the ruins and the high level of cultural development of their builders. Considering the fact that within the imperial discourse Africa was conceived of as a ›dark continent‹ lacking both civilisation and history, the Zimbabwe ruins and their implication of a long history appeared as a confounder in the context of the imperial discourse. What made the discovery of the ruins even more disturbing was the fact that nothing whatsoever was known about their origin and their builders, which imbued the ruins with a deep sense of mystery. It was this mystery associated with the ruins which essentially made people perceive them as the ›other‹. This also meant that Zimbabwean archaeology, in differentiation to Greek and Egyptian archaeology, was conceived as what I have termed the *most unfamiliar strangeness*. In order to deal with the disturbing existence and the ›otherness‹ of the ruins in this otherwise undescribed space, hitherto conceived as a possession of the British Empire, the ruins had to be engrossed in the most radical way. This was accomplished by the aggressive and lasting colonial claim that the ruins had been built by white settlers in a remote past, as reflected in the following report by the *Glasgow Herald*:

»In summing up the results of his discoveries Mr Bent thought the first point that was obvious is that the ruins and the things in them are not in any way connected with any known African race; [...] The second obvious point is that the ruins formed a garrison for the protection of a gold-producing race in remote antiquity. So we must look around for such a race outside the limits of Africa, and it is in Arabia that we find the objects of our search.« (*Glasgow Herald*, Feb 23, 1892)

As emphasised by the respective reception of the Zimbabwe ruins in Victorian and Edwardian fiction, Britain was especially anxious to create an allegedly white past for Zimbabwe to thus legitimise and stabilise its own imperial identity. In other words, by negating Zimbabwe's black history in favour of a white past, contemporaries used archaeology as a means of constructing an intentional history. Consequently, the popular reception of the Zimbabwe ruins was characterised by the idea of the Zimbabwean past as a blank space in which the ruins only served as a point of contact to construct a feigned white history and thus identity for Zimbabwe.

As underlined by the different treatments of Greek, Egyptian, and Zimbabwean archaeology in popular cultural texts and practices in the Victorian and Edwardian Age, their reception was thus strongly connected to the degree of ›otherness‹ associated with the respective culture. In addition to the different tendencies in the reception of the individual archaeologies, there are also certain diachronic characteristics, as others have noted in regard to mummy fiction (cf. Daly 1999). While the popular mid-nineteenth-century reception of archaeology was still very much determined by the exploration of the past as associated with Western culture and identity, this gradually changed in the second half of the century, when the archaeological space was increasingly associated with the ›other‹. This eventually culminated in fantasies of reverse invasion, prominently dealt with in mummy fiction, towards the end of the nineteenth and at the beginning of the twentieth century, mirroring contemporary fears of doom and degeneration. At the beginning of the twentieth century, mummy stories of reverse invasion continued to appear alongside colonial texts focusing on the formation and definition of a new British (Empire) identity in the Edwardian era.

Bibliography

Academy: A Monthly Record of Literature, Learning, Science and Art, Feb 14, 1874. »Dr. Schliemann's Discoveries«. H. Schliemann. 174-76.

Academy: A Monthly Record of Literature, Learning, Science and Art, Mar 7, 1874. »Dr. Schliemann's Discoveries«. A.J. Sayce. 258.

Academy: A Monthly Record of Literature, Learning, Science and Art, Apr 11, 1874. »Dr. Schliemann's Excavations«. W.J. Stillman. 402.

Academy: A Monthly Record of Literature, Learning, Science and Art, May 2, 1874. »Society of Antiquaries«. 496-97.

Academy: A Monthly Record of Literature, Learning, Science and Art, July 3, 1875. »Society of Antiquaries«. 20.

Academy: A Monthly Record of Literature, Learning, Science and Art, Dec 15, 1877. »Fine Art«. A.S. Murray. 558-60.

Academy: A Monthly Record of Literature, Learning, Science and Art, Dec 22, 1877. »Dr. Schliemann's Trojan Antiquities«. A.S. Murray. 581.

Academy: A Monthly Record of Literature, Learning, Science and Art, May 4, 1878. »Mycenæ, Troy, and Ephesus«. 400.

Academy: A Monthly Record of Literature, Learning, Science and Art, Jan 11, 1879. »Magazines and Reviews«. 31.

Academy: A Monthly Record of Literature, Learning, Science and Art, Dec 11, 1880. »Fine Arts«. A.S. Murray. 429-30.

Academy: A Monthly Record of Literature, Learning, Science and Art, Apr 22, 1882. »Fine Arts«. D.B. Monro. 290.

Academy: A Monthly Record of Literature, Learning, Science and Art, Apr 29, 1882. »Book Review«. 300.

Academy: A Monthly Record of Literature, Learning, Science and Art, Nov 18, 1882. »Notes and News«. 361.

Academy: A Monthly Record of Literature, Learning, Science and Art, Nov 18, 1882. »Prof. Jebb on the Ruins of Troy«. A.J. Sayce. 368-69.

Academy: A Monthly Record of Literature, Learning, Science and Art, Nov 21, 1885. »Fine Arts«. J.P. Mahaffy. 345-46.

Academy: A Monthly Record of Literature, Learning, Science and Art, July 10, 1886. »Meetings of Societies«. 30-32.

Academy and Literature, Dec 1877. »*Mycenæ: A Narrative of Researches and Discoveries at Mycenæ and Tiryns*. By Dr. Henry Schliemann. The Preface by the Right Hon. W. E. Gladstone, M.P. (Murray)«. 819-21.

Adams, Amanda (2010): *Ladies of the Field: Early Women Archaeologists and Their Search for Adventure*, Vancouver/Berkeley: Greystone Books.

Addison, Joseph (1718 [1705]): *Remarks on Several Parts of Italy, &c. in the Years 1701, 1702, 1703*, London: Printed for J. Tonson.

After Work, Feb, 1879. »Notes Literary & C.«. 36.

Alan, Keith/Edward K. Brown (2009): *Concise Encyclopedia of Semantics*, Boston: Elsevier.

All the Year Round, June 13, 1874. »Dr. Schliemann's Early Career«. 199-205.

All the Year Round, June 2, 1877. »Mycenæ«. 328-32.

Allen, David Elliston (1994): *The Naturalists in Britain: A Social History*, Princeton: Princeton University Press.

Allen, Grant (1989 [1878]): »My New Year's Eve among the Mummies«. In: Peter Haining (ed.), *The Mummy: Stories of the Living Corpse*, New York: Severn, 32-46.

Allen, Grant (1884): »The Miraculous Explorer«. *English Illustrated Magazine* 12, 99-102.

Allen, Susan Heuck (1995a): »›Finding the Walls of Troy‹: Frank Calvert, Excavator«. *American Journal of Archaeology* 99, 379-407.

Allen, Susan Heuck (1995b): »In Schliemann's Shadow: Frank Calvert, the Unheralded Discoverer of Troy«. *Archaeology* May/June, 50-57.

Allen, Susan Heuck (1998): »A Personal Sacrifice in the Interest of Science: Calvert, Schliemann, and the Troy Treasures«. *The Classical World* 91.5, 345-54.

Allen, Susan Heuck (1999): *Finding the Walls of Troy: Frank Calvert and Heinrich Schliemann at Hisarlik*, Berkeley: University of California Press.

Allwood, John (1977): *The Great Exhibitions*, London: Studio Vista.

Althusser, Louis (2006 [1970]): »Ideology and Ideological State Apparatuses«. In: John Storey (ed.). *Cultural Theory and Popular Culture: A Reader*, Harlow et al.: Pearson, 336-47.

Altick, Richard (1978): *The Shows of London*, Cambridge: Belknap.

Altick, Richard (1998): *The English Common Reader: A Social History of the Mass Reading Public, 1800-1900*, Columbus: Ohio State University Press.

Anderson, A.A./A. Wall (1891): *A Romance of N'Shabé*, London: Chapman and Hall.

Arata, Stephen D. (1990): »The Occidental Tourist: ›Dracula‹ and the Anxiety of Reverse Colonization«. *Victorian Studies* 33.4, 621-45.

Arata, Stephen D. (1996): *Fictions of Loss in the Victorian* Fin de Siècle, Cambridge et al.: Cambridge University Press.

Arentzen, Wout (2001): »Frank Calvert, Henry Austen Layard and Heinrich Schliemann«. *Anatolian Studies* 51, 169-85.

Arnold, Matthew (1960): *Culture and Anarchy*, London: Cambridge University Press.

Assmann, Aleida (1999): *Erinnerungsräume: Formen und Wandlungen des kulturellen Gedächtnisses*, München: Beck.

Assmann, Aleida (2006): *Der lange Schatten der Vergangenheit: Erinnerungskultur und Geschichtspolitik*, München: Beck.

Assmann, Jan (2005 [1992]): *Das kulturelle Gedächtnis: Schrift, Erinnerung und politische Identität in frühen Hochkulturen*, München: Beck.

Assmann, Jan (2011): *Cultural Memory and Early Civilization: Writing, Remembrance, and Political Imagination*, Cambridge et al.: Cambridge University Press.

Athenæum, London Literary and Critical Journal, Nov 7, 1874. »Trojan Antiquities«. F. Calvert. 610-11.

Athenæum, London Literary and Critical Journal, Mar 27, 1875. »Schliemann's ›Troy‹«. 431.

Athenæum, London Literary and Critical Journal, Mar 10, 1877. »The Excavations at Mycenæ«. 327.

Athenæum, London Literary and Critical Journal, June 23, 1877. »Dr. Schliemann's Excavations at Mycenæ«. J. Gennadius. 807.

Athenæum, London Literary and Critical Journal, Dec 22, 1877. »Mycenae: A Narrative of Researches and Discoveries at Mycenæ and Tiryns«. 819-21.

Athenæum, London Literary and Critical Journal, Feb 2, 1878. »The Excavations at Mycenæ«. F.W. Percival. 163.

Athenæum, London Literary and Critical Journal, Aug 17, 1878. »Mycenæ«. H. M. Westropp. 217-18.

Athenæum, London Literary and Critical Journal, Dec 14, 1878. »Excavations at Ilium«. H. Schliemann. 767-69.

Athenæum, London Literary and Critical Journal, Mar 22, 1879. »Cyprus and Mycenæ«. E. Burnouf. 385.

Athenæum, London Literary and Critical Journal, Aug 21, 1880. »Book Review«. 244-56.

Athenæum, London Literary and Critical Journal, Dec 4, 1880. »Book Review«. 748-49.

Athenæum, London Literary and Critical Journal, Dec 25, 1880. »Fine-Art Gossip«. 872-73.

Athenæum, London Literary and Critical Journal, Jan 3, 1891. »Dr. Henry Schliemann«. J. P. Mahaffy. 27-28.

Attridge, Steve (2003): *Nationalism, Imperialism and Identity in Late Victorian Culture: Civil and Military Worlds*, Basingstoke et al.: Palgrave.

Aubrey, John (1984 [1740]): *Stonehenge: A Temple Restor'd to the British Druids*, New York/London: Garland.

Aubrey, John (1984 [1743]): *Avebury, A Temple of the British Druids*, New York/London: Garland.

Auerbach, Jeffrey Aaron (1999): *The Great Exhibition of 1851: A Nation on Display*, New Haven et al.: Yale University Press.

Bagnall, Lilian (1910): »In the Tombs of the Kings«. *The London Magazine* 25, 439-55.

Barrell, John (1991): »Death on the Nile: Fantasy and the Literature of Tourism 1840-1860«. *Essays in Criticism* 41.2, 97-127.

Beauvoir, Simone de (1973): *The Second Sex*, New York: Vintage.

Beaven, Brad (2005): *Leisure, Citizenship and Working-Class Men in Britain, 1850-1945*, Manchester: Manchester University Press.

Bednarski, Andrew (2005): *Holding Egypt: Tracing the Reception of the ›Description de l'Égypte‹ in Nineteenth-Century Great Britain*, London: Golden House Publications.

Belgravia: An Illustrated London Magazine, July, 1877. »Henry Schliemann«. K. Field. 76-91.

Belzoni, Giovanni Battista (1820): *Narrative of the Operations and Recent Discoveries Within the Pyramids, Temples, Tombs, and Excavations, in Egypt and Nubia; and of a Journey to the Coast of the Red Sea, in Search of the Ancient Berenice; and Another to the Oasis of Jupiter Ammon*, London: Murray.

Benjamin, Walter (2003): »The Work of Art in the Age of its Technological Reproducibility: Third Version«. In: Michael William Jennings/Marcus Paul Bullock (eds.), *Selected Writings, 1938-1940*, Cambridge et al.: Belknap Press, 251-83.

Bennett, Tony (2006 [1986]): »Popular Culture and the ›Turn to Gramsci‹«. In: John Storey (ed.), *Cultural Theory and Popular Culture: A Reader*, Harlow et al.: Pearson, 92-99.

Bent, Theodore J. (1896): *The Ruined Cities of Mashonaland Being a Record of Excavation and Exploration in 1891*, London et al.: Longmans.

Berger, Peter L./Thomas Luckmann (1996 [1966]): *Die gesellschaftliche Konstruktion der Wirklichkeit: Eine Theorie der Wissenssoziologie*, Frankfurt: Fischer.

Bergstein, Mary (2009): »Freud's Uncanny Egypt: Prolegomena«. *American Imago* 66.2, 185-210.

Berlanstein, Lenard R. (1992): *The Industrial Revolution and Work in Nineteenth-Century Europe*, London et al.: Routledge.

Bishop, James (1977): *Social History of Edwardian Britain*, London et al.: Angus & Robertson.

Black, Alistair/Peter Hoare (eds.) (2006): *The Cambridge History of Libraries in Britain and Ireland*, Vol. 3: 1850-2000, Cambridge: Cambridge University Press.

Blackwood, Algernon (1974 [1908]): »Nemesis of Fire«. In: Algernon Blackwood, *Ancient Sorceries and Other Stories*, London: Penguin, 127-203.

Blaseio, Gereon/Hedwig Pompe/Jens Ruchatz (eds.) (2005): *Popularisierung und Popularität*, Köln: DuMont.

Blount, Henry (1977 [1636]): *A Voyage into the Levant*, Norwood: Johnson.

Bogdan, Robert (1990): *Freak Show: Presenting Human Oddities for Amusement and Profit*, Chicago: University of Chicago Press.

Boorstin, Daniel J. (1983): *The Discoverer*, New York: Random.

Boothby, Guy (1899): *Pharos the Egyptian*, London: Ward, Lock & Co.

Boothby, Guy (1988 [1894]): »A Professor of Egyptology«. In: Peter Haining (ed.), *The Mummy: Stories of the Living Corpse*, New York: Severn. 95-113.

Bowden, Mark (1991): *Pitt Rivers: The Life and Archaeological Work of Lieutenant-General Augustus Henry Lane Fox Pitt Rivers, DCL, FRS, FSA*, Cambridge et al.: Cambridge University Press.

Bowdler, Sandra (1996): »Freud and Archaeology«. *Anthropological Forum* 7, 419-38.

Brantlinger, Patrick (1985): »Victorians and Africans: The Genealogy of the Myth of the Dark Continent«. *Critical Inquiry* 12.1, 166-203.

Brantlinger, Patrick (1988): »Imperial Gothic: Atavism and the Occult in the British Adventure Novel, 1880-1941«. In: Patrick Brantlinger (ed.), *Rule of Darkness: British Literature and Imperialism, 1830-1914*, Ithaca et al.: Cornell University Press, 227-54.

Brantlinger, Patrick (1990): *Rule of Darkness: British Literature and Imperialism, 1830-1914*, Ithaca et al.: Cornell University Press.

Bridges, Meilee D. (2008): »Tales from the Crypt: Bram Stoker and the Curse of the Egyptian Mummy«. *Victorians Institute Journal* 36, 137-65.

Briefel, Aviva (2008): »Hands of Beauty, Hands of Horror: Fear and Egyptian Art at the *Fin de Siècle*«. *Victorian Studies* 50.2, 263-71.

Briggs, Julia (1977): *Night Visitors*, London: Faber.

Bristow, Joseph (1995): *Effeminate England: Homoerotic Writings after 1885*, New York: Columbia University Press.

Browne, Ray B./Pat Browne (eds.) (1991): *Digging into Popular Culture: Theories and Methodologies in Archaeology, Anthropology, and Other Fields*, Bowling Green: Bowling Green State University Press.

Browne, Ray B./Ronald J. Ambrosetti (eds.) (1993): *Continuities in Popular Culture: The Present in the Past and the Past in the Present and Future*, Bowling Green: Bowling Green State University Press.

Brunton, Deborah (2008): *The Politics of Vaccination: Practice and Policy in England, Wales, Ireland and Scotland. 1800-1874*, Rochester: University of Rochester Press.

Buchan, John (1910): *Prester John*, London et al.: Thomas Nelson and Sons.

Buckley, Jerome Hamilton (1967): *The Triumph of Time: A Study of the Victorian Concepts of Time, History, Progress, and Decadence*, Cambridge: Belknap Press.

Bulkeley, Kelly (2005): *The Wondering Brain: Thinking about Religion with and Beyond Cognitive Neuroscience*, New York: Routledge.

Burrow, John W. (1981): *A Liberal Descent: Victorian Historians and the English Past*, Cambridge et al.: Cambridge University Press.

Butler, Judith (2007 [1990]): *Gender Trouble: Feminism and the Subversion of Identity*, New York et al.: Routledge.

Byron, Glennis (2007): »Bram Stoker's Gothic and the Resources of Science«. *Critical Survey* 19.2, 48-62.

Calder III, William M. (1972): »Schliemann on Schliemann: A Study in the Use of Sources«. *Greek, Roman, and Byzantine Studies* 13.3, 335-53.

Calder III, William M. (1986): »A New Picture of Schliemann«. In: William M. Calder III/David A. Traill (eds.), *Myth, Scandal, and History: The Heinrich Schliemann Controversy and a First Edition of the Mycenaean Diary*, Detroit: Wayne State University Press, 17-47.

Calder III, William M. (1990): »Apocolocyntosis: The Biographers and the Archaeologists«. In: William M. Calder III/Justus Cobet (eds.), *Heinrich Schliemann nach hundert Jahren*, Frankfurt: Klostermann, 15-21.

Calder III, William M. (1998): »›Pionier der ersten Stunde‹: Wolfgang Schindler and the Modern Study of Heinrich Schliemann«. In: John P. Harris/R. Scott Smith (eds.), *Men in Their Books: Studies in the Modern History of Classical Scholarship*, Hildesheim: Olms, 293-305.

Calder III, William M. (ed.) (2010): *Men in Their Books: Studies in the Modern History of Classical Scholarship*, Hildesheim: Olms.

Camden, William (1806 [1586]): *Britannia or a Chorographical Description of Flourishing Kingdoms of England, Scotland and Ireland and the Islands Adjacent; from the Earliest Antiquity*, Vol. 1-4, London: Stockdale.

Cannon, John/Anne Hargreaves (2009): »The Kings and Queens of Britain«. In: *Oxford Reference Online* (www.oxfordreference.com/views/ENTRY.html?subview=Main&entry=t44.e285). Accessed 29 May 2012.

Cantor, Geoffrey N./Sally Shuttleworth (eds.) (2004): *Science Serialized: Representation of the Sciences in Nineteenth-Century Periodicals*, Cambridge et al.: MIT Press.

Carroll, Scott T. (1988): »Solomonic Legend: The Muslims and the Great Zimbabwe«. *The International Journal of African Historical Studies* 21.2, 33-247.

Chambers, Mortimer (1990): »Schliemann and America«. In: William M. Calder III/Justus Cobet (eds.), *Heinrich Schliemann nach hundert Jahren*, Frankfurt: Klostermann, 397-414.

Champion, Sara (1998): »Women in British Archaeology«. In: Margarita Díaz-Andreu García/Marie Louise Stig Sørensen (eds.), *Excavating Women: A History of Women in European Archaeology*, London: Routledge, 175-97.

Champion, T.C. (2003): »Egypt and the Diffusion of Culture«. In: David G. Jeffreys (ed.), *Views of Ancient Egypt since Napoleon Bonaparte: Imperialism, Colonialism and Modern Appropriations*, London: UCL Press, 127-46.

Chennells, Anthony (2003): »Imagining and Living the Exotic: A Context for Early Rhodesian Novels«. *Journal of Literary Studies/Tydskrif vir Literatuurwetenskap* 19.2, 137-58.

Chennells, Anthony (2007): »Great Zimbabwe in Rhodesian Fiction«. In: Mbongeni Z. Malaba/Geoffrey V. Davis (eds.), *Zimbabwean Tradition: Essays on Zimbabwean Literature in English, Ndebele and Shona*, Amsterdam: Rodopi, 1-25.

Chew, Kenneth (1993): *Victorian Science and Engineering Portrayed in the* Illustrated London News, Stroud: Science Museum et al.

Chrisman, Laura (1990): »The Imperial Unconscious? Representation of Imperial Discourse«. *Critical Quarterly* 32.3, 38-58.

Clark, Ignatius Frederick (1997): *The Great War with Germany, 1890-1914: Fictions and Fantasies of the War-to-Come*, Liverpool: Liverpool University Press.

Clifford, James (1997): *Routes: Travels and Translation in the Late Twentieth Century*, Cambridge et al.: Harvard University Press.

Coates, John D. (2003): »The ›Spiritual Quest‹ in Rider Haggard's *She* and *Ayesha*«. *Cahiers Victoriens et Edouardiens* 57, 33-54.

Cobet, Justus (1990): »Troia vor Schliemann«. In: William M. Calder III/Justus Cobet (eds.), *Heinrich Schliemann nach hundert Jahren*, Frankfurt: Klostermann, 118-51.

Cobet, Justus (1991): *Schliemanns Troia: Ruhrlandmuseum, Essen, 21. März-20. Mai 1991, Museum Burg Linn, Krefeld, 16. Juni-25. August 1991*, Essen: Pomp.

Cobet, Justus (ed.) (1992a): *Archäologie und historische Erinnerung: Nach 100 Jahren Heinrich Schliemann*, Essen: Klartext.

Cobet, Justus (1992b): »Schliemanns Wirkungsgeschichte und die historische Erinnerung«. In: Joachim Herrmann (ed.), *Heinrich Schliemann: Grundlagen und Ergebnis moderner Archäologie 100 Jahre nach Schliemanns Tod*, Berlin: Akademie, 359-64.

Cobet, Justus (ed.) (2000): *Europa: Die Gegenwärtigkeit der antiken Überlieferung*, Aachen: Shaker.

Cobet, Justus (2006): »Schliemanns Troia«. In: Martin Zimmermann (ed.), *Der Traum von Troia: Geschichte und Mythos einer ewigen Stadt*, München: Beck, 149-64.

Cobet, Justus (2007): *Heinrich Schliemann: Archäologe und Abenteurer*, München: Beck.

Colby, Sasha (2009): *Stratified Modernism: The Poetics of Excavation from Gautier to Olson*, Oxford et al.: Lang.

Colla, Elliott (2007): *Conflicted Antiquities: Egyptology, Egyptomania, Egyptian Modernity*, Durham et al.: Duke University Press.

Contemporary Review, June 1874. »Homer's Place in History«. W.E. Gladstone. 1-22.

Cook, John M. (1973): *The Troad: An Archaeological and Topographical Study*, Oxford: Clarendon.

Cowie, Susan D./Tom Johnson (2002): *The Mummy in Fact, Fiction and Film*, Jefferson: McFarland.

Cox, Michael/Robert A. Gilbert (eds.) (2003): *The Oxford Book of Victorian Ghost Stories*, Oxford et al.: Oxford University Press.

Craft, Christopher (1984): »›Kiss Me with Those Red Lips‹: Gender and Inversion in Bram Stoker's *Dracula*«. *Representations* 8, 107-33.

Craig, J. Robert (2002): »From Karloff to Vosloo: The Mummy Remade«. *Popular Culture Review* 13.1, 101-10.

Craton, Lillian E. (2009): *The Victorian Freak Show: The Significance of Disability and Physical Differences in 19th-Century Fiction*, Amherst: Cambria.

Curl, James Stevens (1982): *The Egyptian Revival: An Introductory Study of a Recurring Theme in the History of Taste*, London: Allen.

Curl, James Stevens (1994): *Egyptomania: The Egyptian Revival*, Manchester: Manchester University Press.

D'Hancarville, Pierre François Hugues (1766-76): *The Collection of Etruscan, Greek, and Roman Antiquities from the Cabinet of the Honourable William Hamilton*, Naples: Morelli.

Daily Graphic, Jan 3, 1891. »Dr Schliemann«.

Daily News, June 25, 1875. »London, Friday, June 25«.

Daily News, Mar 23, 1877. »Dr. Schliemann and the Society of Antiquaries«.

Daily News, Dec 10, 1877. »Relics of Troy«.

Daily News, Dec 26, 1877. »Relics of Troy«.

Dalby, Richard (1999): »The Mummy«. *Book and Magazine Collector* 186, 52-65.

Daly, Nicholas (1994): »That Obscure Object of Desire: Victorian Commodity Culture and Fictions of the Mummy«. *Novel: A Forum on Fiction* 28.1; Fall, 24-51.

Daly, Nicholas (1999): *Modernism, Romance and the Fin de Siècle: Popular Fiction and British Culture, 1880-1914*, New York: Cambridge University Press.

Damrosch, David (2003): *What is World Literature?* Princeton: Princeton University Press.

Daniel, Glyn Edmund (1950): *A Hundred Years of Archaeology*, London: Gerald Duckworth.

Daniel, Glyn Edmund (1955): *Archaeology*, London: National Book League.

Daniel, Glyn Edmund/Colin Renfrew (1988): *The Idea of Prehistory*, Edinburgh: Edinburgh University Press.

Darwin, Charles (1839): *Journal and Remarks, 1832-1836*, London: H. Colburn.

Darwin, Charles (1959 [1859]): *The Origin of Species*, Morse Peckham (ed.), Philadelphia: University of Pennsylvania Press.

Darwin, Charles (1989 [1871]): *The Descent of Man, and Selection in Relation to Sex*, Paul H. Barrett and R.B. Freemann (eds.), Vol. 1, New York: New York University Press.

Daum, Andreas (1998): *Wissenschaftspopularisierung im 19. Jahrhundert: Bürgerliche Kultur, naturwissenschaftliche Bildung und die deutsche Öffentlichkeit, 1848-1914*, München: Oldenbourg.

Dawson, Warren R. (1934): »Pettigrew's Demonstrations upon Mummies: A Chapter in the History of Egyptology«. *The Journal of Egyptian Archaeology* 20.3/4, 170-82.

Day, David Howard (1997): *A Treasure Hard to Attain: Images of Archaeology in Popular Film with a Filmography*, Lanham: Scarecrow.

Dean, R. Dennis (1985): »›Through Science to Despair‹: Geology and the Victorians«. In: James Paradis/Thomas Postlewait (eds.), *Victorian Science and Victorian Values: Literary Perspectives*, New Brunswick: Rutgers University Press, 111-36.

Deane, Bradley (2008): »Mummy Fiction and the Occupation of Egypt: Imperial Striptease«. *English Literature in Transition, 1880-1920* 51.4, 381-410.

Demoor, Marysa (1995): »Male Monsters or Monstrous Males in Victorian Women's Fiction«. In: Valeria Tinkler-Villani/Peter Deavidson/Jane Stevenson (eds.), *Exhibited by Candlelight: Sources and Developments in the Gothic Tradition*, Amsterdam: Rodopi, 173-82.

Denon, Dominique Vivant (1998 [1802]): *Voyage dans la Basse et la Haute Egypte pendant les campagnes du général Bonaparte*, Paris: Promeneur.

Derby Mercury, Dec 27, 1899. »The Land of Ophir«.

Description de l'Egypte: publiée par ordre de Napoléon Bonaparte avec la collaboration de deux cents savants et artistes. 1809-1818, Paris: Imprimérie Impériale (Royale).

Di Piazza, Elio (2007): »Colonial Geographies in Two Novels by Rider Haggard«. In: Michele Bottalico/Maria Teresa Chialant/Eleonora Rao (eds.), *Literary Landscapes, Landscape in Literature*, Roma: Carocci, 88-98.

Díaz-Andreu García, Margarita/Marie Louise Stig Sørensen (eds.) (1998): *Excavating Women: A History of Women in European Archaeology*, London: Routledge.

Dictionary of Nineteenth-Century Journalism in Great Britain and Ireland (2009): Laurel Brake/Marysa Demoor et al. (eds.), London: British Library.

Döhl, Hartmut (1981): *Heinrich Schliemann: Mythos and Ärgernis*, München/Luzern: Bucher.

Döhl, Hartmut (1986): »Schliemann the Archaeologist«. In: William M. Calder III/David A. Traill (eds.), *Myth, Scandal and History: The Hein-*

rich Schliemann Controversy and a First Edition of the Mycenaean Diary, Detroit: Wayne State University Press, 95-121.

Dohrn-Van Rossum, Gerhard (1996): *History of the Hour: Clocks and Modern Temporal Orders*, Chicago et al.: University of Chicago Press.

Donajgrodzki, A.P. (ed.) (1977): *Social Control in Nineteenth-Century Britain*, London: Croom Helm.

Donaldson, Norman (1971): »R. Austin Freeman: The Invention of Inversion«. In: Francis M. Nevins Jr. (ed.), *The Mystery Writer's Art*, Bowling Green: Bowling Green State University Press, 79-87.

Douglas, Theo [pseud. Henrietta Dorothy Everett) (1896): *Iras, a Mystery*, New York: Harper & Brothers.

Dowling, Linda (1988): »Walter Pater and Archaeology: The Reconciliation with Earth«. *Victorian Studies* 31.2, 209-31.

Doyle, Arthur Conan (1892): »Lot No. 249«. In: Athur Conan Doyle, *The Great Keinplatz Experiment and Other Tales of Terror and Mystery*, Amsterdam: Fredonia, 179-224.

Doyle, Arthur Conan (2001 [1890]): »The Ring of Thoth«. In: Athur Conan Doyle, *The Great Keinplatz Experiment and Other Tales of Terror and Mystery*. Arthur Conan Doyle, Amsterdam: Fredonia, 139-62.

Doyle, Noreen. (2012): »Ancient Egypt in Fiction. A Bibliography in Progress«. (www.noreen-doyle.com/EgyptomaniaOrg//aef/Egyptfiction.html) Accessed 20 May 2012.

Dublin Review, Jan, 1894. »Mashunaland [sic] and its Neighbours«. 145-146.

Durbach, Nadja (2010): *Spectacle of Deformity: Freak Shows and Modern British Culture*, Berkeley: University of California Press.

Eaden, Jeanne van (2004): »The Colonial Gaze: Imperialism, Myths, and South African Popular Culture«. *Design Issues* 20.2, 18-33.

Eagleton, Terry (1991): *Ideology: An Introduction*, London et al.: Verso.

Easton, Donald F. (1984): »Schliemann's Mendacity – a False Trail?«. *Antiquity* 58, 197-204.

Easton, Donald F. (1991): »Troy before Schliemann«. *Studia Troica* 1, 111-29.

Easton, Donald F. (1992): »Was Schliemann a Liar?«. In: Joachim Herrmann (ed.), *Heinrich Schliemann: Grundlagen und Ergebnis moderner Archäologie 100 Jahre nach Schliemanns Tod*, Berlin: Akademie, 191-98.

Easton, Donald F. (1994): »Priam's Gold: The Full Story«. *Anatolian Studies* 44, 221-44.

Easton, Donald F. (1998): »Heinrich Schliemann: Hero or Fraud?« *Classical World* 91.5, 335-43.

Easton, Donald F. (2002): *Schliemann's Excavations at Troia 1870-1873*, Mainz: von Zabern.

Edinburgh Review, Apr, 1874. »Trojanische Alterthumer. Bericht uber [sic] die Ausgrabungen in Troja«. 506-44.

Edinburgh Review, Jan, 1878. »Mycenæ«. 220-56.

Edinburgh Review, Apr, 1881. »Ilios: The City and Country of the Trojans«. 514-47.

Edinburgh Review, Apr, 1892. »Schliemann's Excavations: An Archaeological and Historical Study«. 399-434.

Edwards, Amelia B. (1877): *A Thousand Miles up the Nile*, London: Longmans.

Edwards, Amelia B. (1891): *Pharaohs, Fellahs, and Explorers*, London: James R. Osgood.

Edwards, Robert (1998): »The Alien and the Familiar in *The Jewel of Seven Stars* and *Dracula*«. In: William Hughes/Andrew Smith (eds.), *Bram Stoker: History, Psychoanalysis and the Gothic*, London: Macmillan, 96-115.

Elegård, Alvar (1990): *Darwin and the General Reader: The Reception of Darwin's Theory of Evolution in the British Periodical Press 1859-1872*, Chicago: University of Chicago Press.

Endelmann, Todd M. (2002): *The Jews of Britain, 1656 to 2000*, Berkeley et al: University of California Press.

Engels, Eve-Marie and Thomas F. Glick (eds.) (2008): *The Reception of Charles Darwin in Europe*, London et al.: Continuum.

Erll, Astrid (2005): *Kollektives Gedächtnis und Erinnerungskulturen*, Stuttgart: Metzler.

Esposito, Elena (2002): *Soziales Vergessen: Formen und Medien des Gedächtnisses der Gesellschaft*, Frankfurt: Suhrkamp.

Eßbach, Wolfgang (ed.) (2000): *Wir, ihr, sie: Identität und Alterität in Theorie und Methode*, Würzburg: Ergon.

Etherington, Norman A. (1978): »Rider Haggard, Imperialism, and the Layered Personality«. *Victorian Studies* 22.1, 71-87.

Evans, Joan (1943): *Time and Chance: The Story of Arthur Evans and his Forebears*, London: Longman.

Evans, John (1860): *Flint Implements in the Drift: Being an Account of Their Discovery on the Continent and England, Communicated to the Society of Antiquaries by John Evans*, London: J. B. Nichols & Sons.

Fagan, Brian M. (1996): *The Oxford Companion to Archaeology*, Oxford: Oxford University Press.

Falconer, William (1824 [1762]): *The Shipwreck*, London: Jones and Company.

Feeney, Denis (2007): »On Not Forgetting the ›Literatur‹ in the ›Literatur und Religion‹: Representing the Mythic and the Divine in Roman Historiography«. In: Anton Bierl/Rebecca Lämmle/Katharina Wesselmann (eds.), *Literatur und Religion: Wege zu einer mythisch-rituellen Poetik bei den Griechen*, Berlin/New York: De Gruyter, 173-202.

Fink, Bruce (1997): *The Lacanian Subject: Between Language and Jouissance*, Princeton: Princeton University Press.

Fischer, Pascal (2007): »The Graphic *She*: Text and Image in Rider Haggard's Imperial Romance«. *Anglia* 125.2, 266-87.

Fludernik, Monika/Hans-Joachim Gehrke (eds.) (1999): *Grenzgänger zwischen Kulturen*, Würzburg: Ergon.

Fortnightly Review, Apr, 1883. »A Tour in the Troad«. R. C. Jebb. 514-29.

Fortnightly Review, Apr, 1884. »Homeric Troy«. R. C. Jebb. 433-52.

Franklin, Jeffrey J. (2001): »Memory as the Nexus of Identity, Empire, and Evolution in George Eliot's *Middlemarch* and H. Rider Haggard's *She*«. In: *Cahier victorine et edouardien: Une publication du Centre d'Etudes et de Rec* 53.13, 141-69.

Fraser's Magazine for Town and Country, July, 1877. »The Schliemannic Ilium«. W. Simpson. 1-16.

Fraser's Magazine for Town and Country, Dec, 1877. »Mycenæ«. W. Simpson. 675-93.

Frederick Ludvig (1795-98 [1755]): *Voyage d'Egypte et de Nubie*, Paris: De l'Imprimerie De Pierre Didot l'Ainé.

Freeman, Michael J. (2004): *Victorians and the Prehistoric: Tracks to a Lost World*, New Haven et al.: Yale University Press.

Freeman, Michael/Derek H. Aldcroft (1985): *The Atlas of British Railway History*, London et al.: Croom Helm.

Freeman, Richard Austin (1989 [1911]): *The Eye of Osiris*, Oxford: Oxford University Press.

Freeman's Journal and the Daily Commercial Advertiser, Aug 25, 1873. »The Ruins of Troy«.

Freud, Sigmund (2003 [1907]): *Delusion and Dream in Wilhelm Jensen's* Gradiva, Copenhagen/Los Angeles: Green Integer.

Fries, Jana Esther/Ulrike Rambuschek/Gisela Schulte-Dornberg (2007): *Science oder Fiction? Geschlechterrollen in archäologischen Lebensbildern*. Würzburg: Ergon.

Frost, Brian J. (2008): *The Essential Guide to Mummy Literature*. Lanham: Scarecrow.

Froude, James Anthony (1870 [1856–1870]): *History of England from the Fall of Wolsey to the Defeat of the Spanish Armada*, 12 Vols, London: Longmans, Green & Co.

Gallagher, Catherine/Stephen Greenblatt (2000): *Practicing New Historicism*. Chicago: University of Chicago Press.

Gardner, Philip (2006): »Literacy, Learning and Education«. In: Chris Williams (ed.), *A Companion to Nineteenth-Century Britain*, Malden et al.: Blackwell, 353-68.

Geertz, Clifford (1973): »Thick Description: Toward an Interpretive Theory of Culture«. In: Clifford Geertz (ed.), *The Interpretation of Cultures: Selected Essays*, New York: Basic Books, 3-32.

Geertz, Clifford (1983): *Local Knowledge: Further Essays in Interpretive Anthropology*, New York: Basic Books.

Gehrke, Hans-Joachim (1994): »Mythos, Geschichte, Politik – antik und modern«. *Saeculum* 45, 239-64.

Gehrke, Hans-Joachim (1999): »Einleitung: Grenzgänger im Spannungsfeld von Identität und Alterität«. In: Monika Fludernik/Hans-Joachim Gehrke (eds.), *Grenzgänger zwischen Kulturen*, Würzburg: Ergon, 15-24.

Gehrke, Hans-Joachim (ed.) (2001): *Geschichtsbilder und Gründungsmythen*, Würzburg: Ergon.

Gehrke, Hans-Joachim (2004): »Was heißt und zu welchem Ende studiert man intentionale Geschichte? Marathon und Troja als fundierende Mythen«. In: Gert Melville/Karl-Siegbert Rehberg (eds.), *Gründungsmythen, Genealogien, Memorialzeichen: Beiträge zur institutionellen Konstruktion von Kontinuität*, Köln: Böhlau, 21-36.

Gehrke, Hans-Joachim (2006): »Troia im kulturellen Gedächtnis«. In: Martin Zimmermann (ed.), *Der Traum von Troia: Geschichte und Mythos einer ewigen Stadt*, München: Beck, 211-25.

Gehrke, Hans-Joachim/Monika Fludernik (eds.) (2004): *Normen, Ausgrenzung, Hybridisierung und ›Acts of Identity‹*, Würzburg: Ergon.

Gehrke, Hans-Joachim/Miriam Sénécheau (2010): »Einleitung«. In: Hans-Joachim Gehrke/Miriam Sénécheau (eds.), *Geschichte, Archäologie, Öffentlichkeit: Für einen neuen Dialog zwischen Wissenschaft und Medien*, Bielefeld: transcript, 9-30.

Gentleman's Magazine, Jan, 1840. »Obituary William Smith, LL.D. F. G. S.«. 96-100.

Gentleman's Magazine, May, 1878. »Troja Fuit«. J. W. Males. 588.

Geyken, Frauke (2002): *Gentlemen auf Reisen: Das britische Deutschlandbild im 18. Jahrhundert*, Frankfurt: Campus.

Gilbert, Robert A. (ed.) (2001): *Rise of Victorian Spiritualism*, London: Routledge and Thoemmes P.

Gilbert, Sandra M. (1983): »Rider Haggard's Heart of Darkness«. In: George Edgar Slusser/Eric S. Rabkin/Robert Scholes (eds.) *Coordinates: Placing Science Fiction and Fantasy*, Carbondale: Southern Illinois University Press, 124-38.

Glanville, Ernest (1891): *The Fossicker: A Romance of Mashonaland*, London: Chatto & Windus.

Glasgow Herald, Dec 15, 1876. »The Discoveries at Mycenæ«.

Glasgow Herald, Feb 23, 1892. »The Great Zimbabwe Ruins«.

Gloag, John (1970 [1851]): *The Crystal Palace Exhibition, Illustrated Catalogue*, New York: Dover.

Glover, David (1992): »Bram Stoker and the Crisis of the Liberal Subject«. *New Literary History* 23, 983-1002.

Glover, David (1995): »The Lure of the Mummy: Science, Séance and Egyptian Tales in *Fin-de-Siècle* England«. *Australasian Victorian Studies* 1.1, 1-10.

Glover, David (1996): *Vampires, Mummies, and Liberals: Bram Stoker and the Politics of Popular Fiction*, Durham: Duke University Press.

Goetsch, Paul (2000): »Der Andere als Monster«. In: Wolfgang Eßbach (ed.), *Wir, ihr, sie: Identität und Alterität in Theorie und Methode*, Würzburg: Ergon, 279-95.

Goldmann, Stefan (1992a): »Die Homerische Welt als Symbol verschütterter Kindheit: Literaturpsychoanalytische Untersuchung von Heinrich Schliemanns Autobiographie (1869)«. In: William M. Calder III/Justus Cobet (eds.), *Heinrich Schliemann nach hundert Jahren*, Frankfurt: Klostermann, 191-205.

Goldmann, Stefan (1992b): »Der Mythos von Trojas Untergang in Schliemanns Autobiographie: Zur Archäologie eines Erinnerungsbildes«. In: Joachim Herrmann (ed.), *Heinrich Schliemann: Grundlagen und Ergebnis moderner Archäologie 100 Jahre nach Schliemanns Tod*, Berlin: Akademie, 37-47.

Goldsmith, Oliver (1982 [1766]): *The Vicar of Wakefield*, Hammondsworth: Penguin.

Gramsci, Antonio (2006 [1971]): »Hegemony, Intellectuals and the State«. In: John Storey (ed.), *Cultural Theory and Popular Culture: A Reader*, Harlow et al.: Pearson, 85-91.

Graphic, Apr 19, 1879. »Dr. Schliemann's Antiquities«.

Graphic, Apr 19, 1879. »Dr. Schliemann's Excavations at Troy – Some of the Discoveries«.

Graphic, Jan 20, 1877. »Dr. and Madame Schliemann«.

Graphic, Jan 20, 1877. »Dr. Schliemann's Discoveries at Mycenæ«.

Graphic, June 16, 1877. »Exhibition at Athens of the Mycenæ Treasures Discovered by Dr. Schliemann«.

Greaves, John (1646): *Pyramidographia: Or a Description of the Pyramids in Aegypt*, London: Badger.

Green, L. (1992): »Mummy Mania: The Victorian Fascination with Ancient Egypt's Mortal Remains«. *KMT: A Modern Journal of Ancient Egypt* 3, 34-37.

Greenblatt, Stephen (1988): *Shakespearean Negotiations: The Circulation of Social Energy in Renaissance England*, Oxford: Oxford University Press.

Griffith, George (1906): *The Mummy and Miss Nitocris: A Phantasy of the Fourth Dimension*, London: T. Werner Laurie.

Griffith, George (1974 [1903]): »The Lost Elixir«. In: George Griffith, *The Raid of ›Le Vengeur‹*, London: Ferret, 129-44.

Guest, Kristin (ed.) (2001): *Eating their Words: Cannibalism and the Boundaries of Cultural Identity*, Albany: State University of New York Press.

Gurney, Peter (2001): »An Approprated Space: The Great Exhibition, the Crystal Palace and the Working Class«. In: Louise Purbrick (ed.), *The Great Exhibition of 1851: New Interdisciplinary Essays*, Manchester et al.: Manchester University Press, 114-45.

Haggard, Henry Rider (1899): *Elissa: Or, the Doom of Zimbabwe*, London et al.: Hodder and Stoughton.

Haggard, Henry Rider (1920 [1906]): *Benita: An African Romance*, London et al.: Cassell.

Haggard, Henry Rider (1981 [1904]): »The Trade in the Dead«. In: Peter Haining (ed.), *The Best Short Stories of Rider Haggard*, London: Michael Joseph, 141-147.

Haggard, Henry Rider (2001 [1912-13]): »Smith and the Pharaohs«. In: John Richard Stephens (ed.), *Into the Mummy's Tomb*, New York: Berkley, 137-78.

Haggard, Henry Rider (2004 [1887]): *She: A History of Adventure*, London: Penguin.

Haggard, Henry Rider (2008 [1885]): *King Solomon's Mines*, Oxford: Oxford University Press.

Hahn, Hans-Werner (1990): »Wirtschaftliche Erfolge und wissenschaftlich-kulturelle Interessen: Entwicklungsprozesse im mitteleuropäischen Bürgertum vor dem Hintergrund der Biographie Heinrich Schliemanns«. In: William M. Calder III/Justus Cobet (eds.), *Heinrich Schliemann nach hundert Jahren*, Frankfurt: Klostermann, 309-25.

Halbwachs, Maurice (1941): *La topographie légendaire des évangiles en terre sainte. Étude de mémoire collective*, Paris: Presses universitaires de France.

Halbwachs, Maurice (1985 [1925]): *Das Gedächtnis und seine soziale Bedingungen*, Frankfurt: Suhrkamp.

Halbwachs, Maurice (1991 [1950]): *Das kollektive Gedächtnis*, Frankfurt: Fischer.

Hall, Martin (1995): »The Legend of the Lost City; Or, the Man with Golden Balls«. *Journal of South African Studies* 21.2, 179-99.

Hall, Martin (1996): »Heads and Tales«. *Representations* 54, 104-23.

Hamilton, William (1795): *Collection of Engravings from Ancient Vases of Greek Workmanship Discovered in Sepulchres in the Kingdom of the Two Sicilies but Chiefly in the Neighbourhood of Naples during the*

Course of the Years 1789 and 1790, Naples: Royal Acadamy of Painting.

Helmstadter, Richard J./Bernard Lightman (eds.) (1990): *Victorian Faith in Crisis: Essays on Continuity and Change in Nineteenth-Century Religious Belief*, Stanford: Stanford University Press.

Heron, E./H. [pseud. Hesketh V./Kate Pritchard] (1989 [1898]): »The Story of Baelbrow«. In: Peter Haining (ed.), *The Mummy: Stories of the Living Corpse*, New York: Severn, 81-94.

Herrmann, Joachim (ed.) (1992): *Heinrich Schliemann: Grundlagen und Ergebnis moderner Archäologie 100 Jahre nach Schliemanns Tod*, Berlin: Akademie.

Hibbert, Christoph (1975): *The* Illustrated London News' *Social History of Victorian Britain*, London: Angus & Robertson.

Hindman, Hugh D. (2009): *The World of Child Labor: An Historical and Regional Survey*, Armonk et al.: Sharpe.

Hines, John (2004): *Voices in the Past: English Literature and Archaeology*, Woodbridge: Brever.

Hinz, Evelyn J. (1972): »Rider Haggard's *She*: An Archetypal ›History of Adventure‹«. *Studies in the Novel* 4.3, 416-31.

Hölscher, Uvo (1994): *Das nächste Fremde: Von Texten der griechischen Frühzeit und ihrem Reflex in der Moderne*, München: Beck.

Homas, Peter (1989): *The Ability to Mourn: Disillusionment and the Social Origins of Psychoanalysis*, Chicago et al.: Univeryity of Chicago Press.

Hopkins, Lisa (1996): »›And I Shall Die, and this Unconquered?‹: Marlowe's Inverted Colonialism«. *Early Modern Literary Studies* 2.2, 1-23.

Hopkins, Lisa (1998): »Crowning the King, Mourning the Mother: *The Jewel of Seven Stars* and *The Lady of the Shroud*«. In: William Hughes/ Andrew Smith (eds.), *Bram Stoker: History, Psychoanalysis and the Gothic*, London: Macmillan, 134-49.

Hopkins, Lisa (2003): »Jane C. Loudon's *The Mummy!*: Mary Shelley Meets George Orwell, and They Go in a Balloon to Egypt«. In: *Cardiff Corvey: Reading the Romantic Text*. (http://www.cardiff.ac.uk/encap /journals/corvey/articles/cc10_n01.html). Accessed 29 May 2012.

Hornung, Erik (2001): *The Secret Lore of Egypt: Its Impact on the West*, Ithaca: Cornell University Press.

Hudson, Kenneth (1981): *A Social History of Archaeology: The British Experience*, Basingstoke: Macmillan.

Humbert, Jean-Marcel/Michael Pantazzi/Christiane Ziegler (eds.) (1994): *Egyptomania: Egypt in Western Art, 1730-1930*, Ottawa: Réunion des Musées Nationaux.

Hume, Fergus (1908): *The Green Mummy*, New York: G. W. Dillingham Company.

Hurley, Kelly (1993): »›The Inner Chambers of All Nameless Sin‹: The Beetle, Gothic Female Sexuality, and Oriental Barbarism«. In: Lloyd Davis (ed.), *Virginal Sexuality and Textuality in Victorian Literature*, Albany: State University of New York Press, 193-213.

Hurley, Kelly (1996): *The Gothic Body: Sexuality, Materialism, and Degeneration at the* Fin de Siècle, Cambridge: Cambridge University Press.

Hurley, Kelly (2008): »The Victorian Mummy-Fetish: H. Rider Haggard, Frank Aubrey, and the White Mummy«. In: Marlene Tromp (ed.), *Victorian Freaks: The Social Context of Freakery in Britain*, Columbus: Ohio State University Press, 180-99.

Hurley, Matthew/Daniel Clement Dennett/Reginald B. Adams (2011): *Inside Jokes: Using Humor to Reverse-Engineer the Mind*, Cambridge: MIT Press.

Hutton, James (1788): »Theory of the Earth; or an Investigation of the Laws Observable in the Composition, Dissolution, and Restoration of Land upon the Globe«. In: *Transactions of the Royal Society of Edinburgh*, Vol. 1, 209-304.

Hyde, Ralph (1988): *Panoramania! The Art and Entertainment of the ›All-Embracing‹ View*, London: Trefoil.

Hyne, Cutliffe C.J. (1904): »The Mummy of Thompson-Pratt«, In: C. J. Cutliffe Hyne, *Atoms of Empire*, London: Macmillan, 174-92.

Illustrated London News, July 20, 1850. »The Great Exhibiton of 1851«. 53-54.

Illustrated London News, July 12, 1856. »The Crystal Palace«. 34.

Illustrated London News, May 5, 1860. »Scientific News«. 434.

Illustrated London News, Jan 5, 1867. »Fire at the Crystal Palace«. 22-23.

Illustrated London News, Jan 12, 1867. »The Great Fire at the Crystal Palace«. 28 and 42-44.

Illustrated London News, Jan 11, 1873. »The Ophir of Scripture«. 46.

Illustrated London News, Mar 24, 1877. »Schliemann's Researches in Greece«. 281-84.

Illustrated London News, Mar 31, 1877. »Dr. Schliemann Giving an Account of His Discoveries at Mycenæ before the Society of Antiquaries at Burlington House«. 301.

Illustrated London News, Mar 31, 1877. »Dr. Schliemann's Researches at Mycenæ«. 306-8.

Illustrated London News, Apr 7, 1877. »The Ruins of Mycenæ«. 330-32.

Illustrated London News, Apr 14, 1877. »A Pilgrimage to Mycenæ«. 358.

Illustrated London News, Dec 29, 1877. »Dr. Schliemann's Collection«. 627-29.

Illustrated London News, Jan 5, 1878. »Dr. Schliemann's Excavations in the Troad«. 11-13.

Illustrated London News, Jan 12, 1878. »Views of the Troad«. 45-47.

Illustrated London News, Feb 23, 1878. »Grecian Cities of Antiquity«. 178

Illustrated London News, Apr 9, 1887. »Bank Holiday at the British Museum«. 411.

Illustrated London News, Jan 3, 1891. »The Late Dr. Schliemann«. 5.

Illustrated London News, May 4, 1907. »IZAL the Perfect Disinfectant: A Simple Lesson in Hygiene«. 705.

Illustrated London News, Special Number, May 13, 1967. »As Much of Life as the World Can Show«. 40-46.

Innes, Catherine L. (1994): »Virgin Territories and Motherlands: Colonial and Nationalist Representations of Africa and Ireland«. *Feminist Review* 47, 1-14.

Isaacson, Stephen (1845): *Barrow-Digging by a Barrow Knight*, London: John Ollivier.

Jackson's Oxford Journal, Mar 10, 1877. »The Recent Discoveries at Mycenæ«.

Jackson's Oxford Journal, Mar 31, 1877. »The Recent Discoveries at Mycenæ«.

Jeffreys, David (ed.) (2003): *Views of Ancient Egypt since Napoleon Bonaparte: Imperialism, Colonialism and Modern Appropriations*, London: UCL Press.

Jenkyns, Richard (1980): *The Victorians and Ancient Greece*, Oxford: Blackwell.

Jenkyns, Richard (1989): »Hellenism in Victorian Painting«. In: G. W. Clarke (ed.), *Rediscovering Hellenism: The Hellenic Inheritance and the English Imagination*, Cambridge: Cambridge University Press, 83-119.

Jensen, Wilhelm (1918 [1903]): *Gradiva: A Pompeian Fancy*, New York: Moffat, Yard and Company.

Jordan, John O. (ed.) (1995): *Literature in the Marketplace: Nineteenth-Century British Publishing and Reading Practices*, Cambridge: Cambridge University Press.

Journal of the Manchester Geographical Society, Apr 1, 1892. »The Ruins of Zimbabwe«. 188-189.

Joyce, Patrick (1994): *Democratic Subjects: The Self and the Social in Nineteenth-Century England*, Cambridge: Cambridge University Press.

Kinglake, William Alexander (1844): *Eothen: Traces of Travel Brought Home from the East*, London: John Ollivier.

Kirwan, Laurence P. (1938): »Recent Archaeology in British Africa«. *Journal of the Royal African Society* 37.149, 494-501.

Kitteringham, Guy Stuart (1981): *Studies in the Popularisation of Science in England, 1800-1830*, PhD Thesis. University of Kent at Canterbury.

Klancher, Jon P. (2009): *A Concise Companion to the Romantic Age*, Chichester/Malden: Wiley-Blackwell.

Korte, Barbara (2000a): »Agatha Was Here: An Exhibition Review of ›Agatha Christie und der Orient – Kriminalistik und Archäologie‹, Ruhrland Museum Essen«. *Public Archaeology* 1, 157-59.

Korte, Barbara (2000b): »›The Reassuring Science?‹ Archäologie als Sujet und Metapher in der Literatur Britanniens«. *Poetica* 32, 125-50.

Korte, Barbara (2002): »Archäologie in der viktorianischen Literatur: Faszination und Schrecken in der ›tiefen‹ Zeit«. In: Martin Middeke (ed.), *Zeit und Roman*, Würzburg: Königshausen & Neumann, 112-31.

Korte, Barbara/Catherine Matthias (2000): *English Travel Writing. From Pilgrimages to Postcolonial Explorations*, Basingstoke: Macmillan.

Korte, Barbara/Sylvia Paletschek (eds.) (2009): *History Goes Pop: Zur Repräsentation von Geschichte in populären Medien und Genres*, Bielefeld: transcript.

Krapp, Peter (2004): *Déjà Vu: Aberrations of Cultural Memory*, Minneapolis et al.: University of Minnesota Press.

Kretschmann, Carsten (2003a): *Wissenspopularisierung: Konzepte der Wissensverbreitung im Wandel*, Berlin: Akademie.

Kretschmann, Carsten (2003b): »Einleitung: Wissenspopularisierung – ein altes, neues Forschungsfeld«. In: *Wissenspopularisierung: Konzepte der Wissensverbreitung im Wandel*, Berlin: Akademie, 7-21.

Kretschmann, Carsten (2004): *Wissen in der Krise: Institutionen des Wissens im gesellschaftlichen Wandel*, Berlin: Akademie.

Kretschmann, Carsten (2009): »Wissenspopularisierung: Verfahren und Beschreibungsmodelle – ein Aufriss«. In: Petra Boden/Dorit Müller (eds.), *Populäres Wissen im medialen Wandel seit 1850*, Berlin: Kadmos, 17-34.

Kuspit, Donald (1989): »A Mighty Metaphor: The Analogy of Archaeology and Psychoanalysis«. In: Donald Kuspit (ed.), *Sigmund Freud and Art: His Personal Collection of Antiquities*, London: Thames and Hudson, 133-51.

Lambarde, William (1970 [1576]): *A Perambulation of Kent*, Bath: Adams & Dart.

Landsberg, Alison (1995): »Prosthetic Memory: *Total Recall* and *Blade Runner*«. *Body & Society* 1.3/4, 175-89.

Lane, Edward William (transl.) (1896 [1838-40]): *The Thousand and One Nights: Commonly Called in England the Arabian Night's Entertainment*, London: Bliss, Sands, and Foster.

Lane, Edward William (1904 [1836]): *An Account of the Manners and Customs of the Modern Egyptians, Written in Egypt during the Years 1833, -34 and -35, Partly from Notes Made During a former Visit to that Country in the Years 1825, -26, -27 and -28*, Vol. 1-2, London: Murray.

Lang, Andrew (1889): »She«. *The Grass of Parnassus Rhymes Old and New*, Ondon/New York: Longmans, Green and Co. 93.

Larsen, Ernest (2000): »The Mummy: Traffic in Mummies«. *Jump Cut* 43, 12-15.

Lascarides, A.C. (1977): *The Search for Troy, 1553-1874*, Bloomington: Lilly Library.

Layard, Austen Henry (1848-49): *Nineveh and Its Remains: With an Account of a Visit to the Chaldaean Christians of Kurdistan, and the Yezidis, or Devil-Worshipper; And an Enquiry into the Manners and Arts of the Ancient Assyrians*, London: Murray.

Leavis, Frank Raymond (2006 [1933]): »Mass Civilisation and Minority Culture«. In: John Storey (ed.), *Cultural Theory and Popular Culture: A Reader*, Harlow et al.: Pearson Longman, 12-19.

Ledger, Sally (1995): »New Woman and the Crisis of Victorianism«. In: Sally Ledger/Scott McCracken (eds.), *Cultural Politics at the Fin de Siècle*, Cambridge/New York: Cambridge University Press, 22-44.

Leeds Mercury, Mar 27, 1873. »Antiquarian Research at the Public Expense«.

Leeds Mercury, May 4, 1877. »Mr. Gladstone on Troy«.

Lehrer, Mark/David Turner (1989): »The Making of an Homeric Archaeologist: Schliemann's Diary of 1868«. *The Annual of the British School at Athens* 84, 221-68.

Leisure Hour: An Illustrated Magazine for Home Reading, July 11, 1874. »Trojan Antiquities«. W. F. Wilkinson. 438-45.

Leisure Hour: An Illustrated Magazine for Home Reading, July 7, 1877. »Dr. Schliemann's Greek Antiquities«. 423-26.

Leisure Hour: An Illustrated Magazine for Home Reading, July 14, 1877. »Dr. Schliemann's Greek Antiquities«. 440-44.

Leisure Hour: An Illustrated Magazine for Home Reading, July 28, 1877. »Varieties«. 480.

Leisure Hour: An Illustrated Magazine for Home Reading, Mar, 1891. »Henry Schliemann. Died at Naples, December 26, 1890«. Horace Groser. 320.

Leisure Hour: An Illustrated Magazine for Home Reading, Apr, 1892. »Varieties«. 426-30.

Leitgeb, Hanna (2001): »Schliemann der Eroberer: Wie ein Global Player zum antiken Troia fand und was er dort suchte«. *Literaturen* 10, 32-35.

Levine, Philippa (1986): *The Amateur and the Professional: Antiquarians, Historians and Archaeologists in Victorian England, 1838-1886*, Cambridge: Cambridge University Press.

Levine, Philippa (2007): *The British Empire: Sunrise to Sunset*, Harlow: Longman.

Lewis, Leslie Anne (2006): *Trading in Withered Flesh: Mummies, Movies and Modernity*. Diss. Northwestern University. *Dissertation Abstracts International, Section A: The Humanities and Social Sciences* 67.3, Sept, 757.

Lieven, Michael (1998): »Heroism, Heroics and the Making of Heroes: The Anglo-Zulu War of 1879«. *Albion* 30.3, 419-38.

Liverpool Mercury, June 8, 1877. »Dr. Schliemann«.

Livingstone, David N. (2008): *Adam's Ancestors: Race, Religion, and the Politics of Human Origins*, Baltimore: Johns Hopkins University Press.

London Society, 1867, Vol. 12, »Sketches from our Office Window. No.1. – Piccadily Circus«. 268-73.

Loudon, Jane C. Webb (1828 [1827]): *The Mummy! Or, A Tale of the Twenty-Second Century*, Vol. 1, London: Henry Colburn.

Loudon, Jane C. Webb (1828 [1827]): *The Mummy! Or, A Tale of the Twenty-Second Century*, Vol. 2, London: Henry Colburn.

Lübbe, Hermann (1992): *Im Zuge der Zeit*, Berlin: Springer.

Lubbock, John (1913 [1865]): *Prehistoric Times as Illustrated by Ancient Remains and the Manners and Customs of Savage*, New York/London: H. Holt and Williams & Norgate.

Luckhurst, Roger (2006): »The Mummy's Curse: A Genealogy«. In: Jarmila Mildorf/Hans Ulrich Seeber/Martin Windisch (eds.), *Magic, Science, Technology, and Literature*, Münster: LIT, 123-38.

Ludwig, Emil (1931): *Schliemann: The Story of a Gold-Seeker*, London: Little.

Lupton, Carter (2003): »›Mummymania‹ for the Masses – Is Egyptology Cursed by the Mummy's Curse?«. In: Sally MacDonald/Michael Rice (eds.), *Consuming Ancient Egypt*, London: UCL, 23-46.

Lyell, Charles (1997 [1830-33]): *Principles of Geology*, James A. Secord (ed.), London et al.: Penguin.

Lytton, Edward Robert Bulwer-Lytton (1863): *The Ring of Amasis: From the Papers of a German Physician*, London: Chapman and Hall.

Macaulay, Thomas Babington (1849-61): *The History of England from the Accession of James the Second*, 10 Vols, Leipzig: Tauchnitz.

MacDonald, Sally/Michael Rice (2003): *Consuming Ancient Egypt*, Portland: Cavendish.

MacFarlane, Karen E. (2010): »Mummy Knows Best: Knowledge and the Unknowable in Turn of the Century Mummy Fiction«. *Horror Studies* 1.1, 5-24.

Macherey, Pierre (1978): *A Theory of Literary Production*, London: Routledge & Kegan Paul.

Maclaren, Charles (1863): *The Plain of Troy Described and the Identity of Ilium of Homer with the New Ilium of Strabo Proved by Comparing the Poet's Narrative with the Present Topography*, Edinburgh: Black.

Maier, Franz Georg (1992): *Von Winckelmann zu Schliemann: Archäologie als Eroberungswissenschaft des 19. Jahrhunderts*, Opladen: Westdeutscher Verlag.

Malchow, Howard L. (1993): »Frankenstein's Monster and Images of Race in Nineteenth-Century Britain«. *Past and Present* 139, 90-130.

Malley, Shawn (1996a): *Nineteenth-Century Archaeology and the Retrieval of the Past: Carlyle, Scott, Bulwer-Lytton, Pater and Haggard.* Diss. University of British Columbia. *Dissertation Abstracts International, Section A: The Humanities and Social Sciences* 57.12, 5165.

Malley, Shawn (1996b): »Austen Henry Layard and the Periodical Press: Middle Eastern Archaeology and the Excavation of Cultural Identity in Mid-Nineteenth Century Britain«. *Victorian Review* 22.2, 152-70.

Malley, Shawn (1997): »›Time Hath No Power against Identity‹: Historical Continuity and Archaeological Adventure in H. Rider Haggard's *She*«. *English Literature in Transition, 1880-1920* 40.3, 275-97.

Malley, Shawn (2004): »Shipping the Bull: Staging Assyria in the British Museum«. *Nineteenth-Century Contexts* 26.1, 1-27.

Malley, Shawn (2012): *From Archaeology to Spectacle in Victorian Britain: The Case of Assyria, 1845-1854*, Farnham: Ashgate.

Manchester Times, Dec 29, 1877. »Dr. Schliemann's Collection«.

Mandeville, John (1983 [1356]): *The Travels of Sir John Mandeville*, Harmondsworth et al: Penguin.

Mannsperger, Brigitte (1992): »›Selbstinszenierung‹ Heinrich Schliemanns in der Darstellung der Mitarbeit seiner Frau Sophia«. In: Joachim Herrmann (ed.), *Heinrich Schliemann: Grundlagen und Ergebnis moderner Archäologie 100 Jahre nach Schliemanns Tod*, Berlin: Akademie, 65-72.

Markwick, Edward (1898): *The City of Gold: A Tale of Sport, Travel, and Adventure in the Heart of the Dark Continent*, London: Thacker.

Martineau, Harriet (1848): *Eastern Life, Present and Past,* Vol. 1-3, London: Edward Moxon.

Marx, John (1998): »Modernism and the Female Imperial Gaze«. *Novel* 32.1, 51-75.

Mauch, Karl Gottlieb (1969): *The Journals of Carl Mauch; His Travels in the Transvaal and Rhodesia, 1869-1872*, E. Bernhard/F.O. Bernhard/ E.E. Burke (eds.), Salisbury: National Archives of Rhodesia.

Maurer, Kathrin (2009): »Archaeology as Spectacle: Heinrich Schliemann's Media of Excavation«. *German Studies Review* 32.2, 303-17.

Mayhall, Laura E. Nym (2003): *The Militant Suffrage Movement: Citizenship and Resistance in Britain, 1860-1930*, Oxford: Oxford University Press.

McIver, Iver (1910): *An Imperial Adventure*, Edinburgh: William Blackwood & Son.

McKendry, Virginia (1994): »The *Illustrated London News* and the Invention of Tradition«. *Victorian Periodicals Review* 27, 1-24.

McLean, David (2006): *Public Health and Politics in the Age of Reform: Cholera, the State, and the Royal Navy in Victorian Britain*, London: Palgrave Macmillan.

Meier, Franz (2002): »Oscar Wilde and the Myth of the Femme Fatale in *Fin-de-Siècle* Culture«. In: Uwe Börker (ed.), *The Importance of Reinventing Oscar: Versions of Wilde During the Last 100 Years*, Amsterdam et al.: Rodopi, 117-134.

Mejer, Jørgen (1990): »Henrik Ibsens's *Peer Gynt* and Heinrich Schliemann: Fact and Fiction«. In: William M. Calder III/Justus Cobet (eds.), *Heinrich Schliemann nach hundert Jahren*, Frankfurt: Klostermann, 296-308.

Meyerhoff, Hans (1955): *Time in Literature*, Berkeley: University of California Press.

Micale, Mark S. (2008): *Hysterical Men: The Hidden History of Male Nervous Illness*, Cambridge: Harvard University Press.

Montrose, Louis Adrian (1989): »Professing the Renaissance: The Poetics and Politics of Culture«. In: Harold Aram Veeser (ed.), *The New Historicism*, New York: Routledge, 15-36.

Morgan, Kenneth (2004): *The Birth of Industrial Britain: Social Change, 1750-1850*, Harlow: Pearson.

Most, Glenn W. (2010): »Fragments«. In: Anthony Grafton/Glenn W. Most/Salvatore Settis (eds.), *The Classical Tradition*, Cambridge: Belknap Press, 371-77.

Murphy, Patricia (1999): »The Gendering of History in *She*«, *Studies in English Literature, 1500-1900* 39.4, 747-72.

Murphy, Patricia (2000): »The Fissure King: Parody, Ideology, and the Imperialist Narrative«. In: Jennifer Wagner-Lawlor (ed.), *The Victorian Comic Spirit: New Pespectives*, Aldershot: Ashgate, 23-42.

Murray, Tim (1993): »Archaeology and the Threat of the Past: Sir Henry Rider Haggard and the Acquisition of Time«. *World Archaeology* 15.2, 175-86.

Musson, Albert E./Eric Robinson (1969): *Science and Technology in the Industrial Revolution*, Manchester: Manchester University Press.

Neuhaus, Volker (1999): »Die Archäologie des Mordes«. In: Charlotte Trümpler (ed.), *Agatha Christie und der Orient*, Bern et al.: Scherz, 425-34.

Newcastle General Magazine, Aug, 1775. »An Account of the Late Discoveries of Antiquities at Herculaneum, Pompeii, &c. in a Letter from Camillo Paderni, Keeper of the Museum Herculanei, Dated Naples«.

Newcastle Weekly Courant, Dec 21, 1889. »The Unrolling of a Mummy«.

Noble, Louise (2011): *Medicinal Cannibalism in Early Modern English Literature and Culture*, New York: Palgrave Macmillan.

Nora, Pierre (1984-92): *Les lieux de mémoire*, Paris: Gallimard.

Nora, Pierre (1990): *Zwischen Geschichte und Gedächtnis*, Berlin: Wagenbach Norden.

Nordau, Max (1895): *Degeneration*, London: Heinemann.

Nowel, Ingrid (1982): *Entdeckungsreisen in Ägypten: 1815-1819 – In den Pyramiden, Tempeln und Gräbern am Nil. Mit einer Geschichte der Ägyptenreisen seit dem 16. Jahrhundert*, Köln: DuMont.

Official Descriptive and Illustrated Catalogue: Great Exhibition of the Works of Industry of all Nations. 1851. London: Spicer et al.

Oppenheim, Janet (1991): *›Shattered Nerves‹: Doctors, Patients and Depression in Victorian England*, New York, et al.: Oxford University Press.

Orrells, Daniel (2011): »Rocks, Ghosts, and Footprints: Freudian Archaeology«. In: Shelley Hales/Joanna Paul (eds.), *Pompeii in the Public Imagination from Its Rediscovery to Today*, Oxford et al.: Oxford University Press, 185-98.

Osterhammel, Jürgen (2009): *Die Verwandlung der Welt: Eine Geschichte des 19. Jahrhunderts*, München: Beck.

Owen, Alex (2004): *The Place of Enchantment: British Occultism and the Culture of the Modern*, Chicago: University of Chicago Press.

Page, Gertrude (1909): *The Rhodesian*, London: Hurst & Blackett.

Pall Mall Gazette, Dec 27, 1877. »The Tombs at Mycenæ«.

Pall Mall Gazette, May 16, 1891. »More Schliemann Reminiscences«.

Pall Mall Gazette, Oct 3, 1892. »Dr. Schliemann's Excavations«.

Parsons, Gerald (1988): »On Speaking Plainly: ›Honest Doubt and the Ethics of Belief«. In: Gerald Parsons (ed.), *Religion in Victorian Britain: Controversies*, Manchester: Manchester University Press, 191-219.

Pascoe, Judith (2006): *The Hummingbird Cabinet: A Rare and Curious History of Romantic Collections*, London: Cornell University Press.

Pater, Walter (1980 [1893]): *The Renaissance: Studies in Art and Poetry*, Donald L. Hill (ed.), Berkeley et al.: University of California Press.

Patterson, Bradley William (2011): *Redifining Reason: The Story of the Twentieth Century ›Primitive‹ Mentality Debate*, Bloomington: Xlibris.

Patton, Mark (2007): *Science, Politics and Business in the Work of Sir John Lubbock: A Man of Universal Mind*, Aldershot et al.: Ashgate.

Patzek, Barbara (1990): »Schliemann und die Geschichte der Archäologie im neunzehnten Jahrhundert: Von der Entstehung einer eigenen Wissenschaft zur archäologischen Sensation«. In: William M. Calder III/ Justus Cobet (eds.), *Heinrich Schliemann nach hundert Jahren*, Frankfurt: Klostermann, 31-55.

Patzek, Barbara (1993): »A. Christie: Krimi und Archäologie im Orient«. *Essener Universitätsberichte* 1.1993, 14-20.

Patzek, Barbara/Andreas Dudde/Regina Hauses (1999): »Der Detektiv und der Archäologe«. In: Charlotte Trümpler (ed.), *Agatha Christie und der Orient: Kriminalistik und Archäologie*, Bern et al.: Scherz, 391-409.

Pearce, Susan M. (2000): »Giovanni Battista Belzoni's Exhibition of the Reconstructed Tomb of Pharaoh Seti I in 1821«. *Journal of the History of Collection* 12.1, 109-25.

Pearce, Susan M. (2002): »Bodies in Exile: Egyptian Mummies in the Early Nineteenth Century and their Cultural Implication«. In: Sharon Ouditt (ed.), *Displaced Persons: Conditions of Exile in European Culture*, Aldershot: Ashgate, 54-71.

Pearl, Sharrona (2010): *About Faces: Physiognomy in Nineteenth-Century Britain*, Cambridge: Harvard University Press.

Pearson, Richard (2000): »Archaeology and Gothic Desire: Vitality beyond the Grave in H. Rider Haggard's Ancient Egypt«. In: Ruth Robbings/

Julian Wolfreys (eds.), *Victorian Gothic: Literary and Cultural Manifestations in the Nineteenth Century*, New York: Palgrave, 218-44.

Pemberton, Clive (2000 [1906]): »The Bulb«. In: John Pelan (ed.), *The Weird O'It*, Seattle: Midnight, 111-28.

Penny London Post or The Morning Advertiser, July 18, 1748. »News«.

Peter Parley's Annual: A Christmas and New Year's Present for Young People [date unknown]. »Dr. Henry Schliemann«.

Peterson, Linda H. (1986): *Victorian Autobiography: The Tradition of Self-Interpretation*, New Haven: Yale University Press.

Peterson, Linda H. (2001): *Traditions of Victorian Women's Autobiography: The Poetics and Politics of Life Writing*, Charlottesville et al.: University Press of Virginia.

Pettigrew, Thomas Joseph (1834): *A History of Egyptian Mummies*, London: Longman and Rees.

Pevsner, Nikolaus (1968): »The Egyptian Revival«. In: Nikolaus Pevsner, *Studies in Art, Architecture and Design*, London: Thames and Hudson, 213.

Pick, Daniel (1993): *Faces of Degeneration: A European Disorder, c. 1848 -c. 1918*, Cambridge: Cambridge University Press.

Pococke, Richard (1743/1745): *Description of the East and Some Other Countries*, Vol 1-2, London: W. Bowyer.

Pollock, Griselda (2006): »The Image in Psychoanalysis and the Archaeological Metaphor«. In: Griselda Pollock (ed.), *Psychoanalysis and the Image: Transdisciplinary Perspectives*, Oxford: Blackwell, 1-29.

Pollock, Griselda (2007a): »Freud's Egypt: Mummies and M/others«. *Parallax* 13.2, 56-79.

Pollock, Griselda (2007b): »Freud and the Object's Gaze«. In: Griselda Pollock (ed.), *Encounters in the Virtual Feminist Museum: Time, Space and the Archive*, London: Routledge, 67-86.

Porter, Andrew (2001): »Introduction: Britain and the Empire in the Nineteenth Century«. In: Andrew Porter (ed.), *The Oxford History of the British Empire: The Nineteenth Century*, Oxford et al.: Oxford University Press, 1-30.

Porter, Bernard (1982): »The Edwardians and Their Empire«. In: Donald Read (ed.), *Edwardian England*, New Brunswick: Rutgers University Press, 128-44.

Posnansky, Merrick (1982): »African Archaeology Comes of Age«. *World Archaeology* 13.3, 345-58.

Quarterly Review, Apr, 1874. »Trojanische Alterthumer: Bericht uber [sic] die Ausgrabungen in Troja«. 526-66.

Quarterly Review, Jan, 1878. »Mycenae; a Narrative of Researches and Discoveries at Mycenae and Tiryns«. 62-93.

Quarterly Review, July, 1881. »Ilios: The City and Country of the Trojans: The Results of Researches and Discoveries on the Site of Troy and throughout the Troad in the Years 1871-72-73-78-79. Including an Autobiography of the Author«. 205-39.

Quarterly Review, Apr, 1885. »Murray's Handbook for Travellers in Greece, Including the Ionian Islands, Continental Greece, the Peloponnese, the Islands of the Ægean, Crete, Albania, Thessaly, and Macedonia; and Detailed Description of Athens«. 298-322.

Quarterly Review, Jan, 1886. »Tiryns: The Prehistoric Palace of the Kings of Tiryns. The Results of the Latest Excavations«. 108-32.

Quarterly Review, July, 1889. »The Roman Poets of the Augustan Age. Virgil«. 98-131.

Quarterly Review, July, 1897. »Murray's Handbook for Travellers in Asia Minor & c.«. 64-87.

Quarterly Review, July, 1898. »The Mycenaean Age«. 90-112.

Rauch, Alan (1994): »Preface« and »Introduction«. In: Jane C. Webb Loudon, *The Mummy! A Tale of the Twenty-Second Century*, Ann Arbor: University of Michigan Press, v-xxxiv.

Read, Donald (1982): *Edwardian England*, New Brunswick: Rutgers University Press.

Reckwitz, Andreas (2000): *Die Transformation der Kulturtheorie*, Weilerswist: Velbrück.

Rehberg, Karl-Siegbert (2004): »Zur Konstruktion kollektiver ›Lebensläufe‹: Eigengeschichte als institutioneller Mechanismus«. In: Gert Melville/Karl-Siegbert Rehberg (eds.), *Gründungsmythen, Genealogien, Memorialzeichen: Beiträge zur institutionellen Konstruktion von Kontinuität*, Köln: Böhlau, 3-18.

Reliquary: Quarterly Archaeological Journal and Review, July, 1891. »Village Antiquities«. P. H. Ditchfield. 134-39.

Renfrew, Colin (1973): *Before Civilisation: The Radiocarbon Revolution and Pre-Historic Europe*, London: Cape.

Renfrew, Colin (1994): »Three Cambridge Prehistorians«. In: Richard Mason (ed.), *Cambridge Minds*, Cambridge: Cambridge University Press, 58-83.

Reynolds, Joshua (1842): *The Discourses of Sir Joshua Reynolds. Illustr. by Explanatory Notes & Plates by John Burnet*, London: Carpenter.

Rhind, Alexander Henry (1862): *Thebes, Its Tombs and Their Tenants, Ancient and Present*, London: Longman.

Rhodes, Cecil (2010 [1877]): »Confession of Faith, June 2, 1877«. In: William H. Worger/Nancy L. Clark/Edward A. Alpers (eds.), *Africa and the West: From the Slave Trade to Conquest, 1441-1905*, Oxford: Oxford University Press, 221-224.

Richards, Janet Radcliffe (2000): *Human Nature After Darwin: A Philosophical Introduction*, London et al.: Routledge.

Richards, Thomas (1991): *The Commodity Culture of Victorian England: Advertising and Spectacle, 1851-1914*, London et al.: Verso.

Richardson, Jonathan (1722): *An Account of the Statues, Bas-reliefs, Drawings, and Pictures in Italy, &c. with Remarks*, London: J. Knapton at the Crown.

Rickels, Laurence A. (1992): »Mummy's Curse«. *The American Journal of Semiotics* 9.4, 47-58.

Robinson, Jane (1990): *Wayward Women: A Guide to Women Travellers* Oxford et al.: Oxford University Press.

Robinson Schidkowski, Marcelle (1994): »Pioneer, Scholar, and Victim: An Appreciation of Frank Calvert (1828-1908)«. *Anatolian Studies* 44, 153-68.

Robinson Schidkowski, Marcelle (1995): »Frank Calvert and the Discovery of Troia«. *Studia Troica* 5, 323-41.

Robinson Schidkowski, Marcelle (2006): *Frank Calvert (1828-1908): Pioneer, Scholar, and Survivor*, Philadelphia: Xlibris Corporation.

Rohmer, Sax (1966 [1913]): »The Case of the Headless Mummies«. In: Sax Rohmer, *The Dream Detective: Tales of Perilous Adventure – Introducing Moris Klaw!*, New York: Pyramid, 137-54.

Rohmer, Sax (1989 [1903]): »The Mysterious Mummy«. In: Peter Haining (ed.), *The Mummy: Stories of the Living Corpse*, New York: Severn, 114-25.

Rohmer, Sax (2000 [1914]): »The Cat«. In: Jack Adrian (ed.), *The Ash-Tree Press Annual Macabre 2000*, Ashcroft: Ash-Tree, 79-98.

Royal Geographical Society (Great Britain) (1892): *Proceedings of the Royal Geographic Society*, Vol. 14, London: Edward Stanford.

Ruchatz, Jens (2009): »Vorträge sind Silber, Dias sind Gold: Medienkonkurrenz im Projektionsvortrag«. In: Petra Boden/Dorit Müller (eds.), *Populäres Wissen im medialen Wandel seit 1850*, Berlin: Kadmos, 101-18.

Rudwick, Martin J.S. (2005): *Lyell and Darwin, Geologists*, Aldershot: Ashgate Variorum.

Runnels, Curtis (2002): *The Archaeology of Heinrich Schliemann: An Annotated Bibliographic Handlist*, Boston: Archaeological Inst. of America.

Said, Edward W. (1978): *Orientalism*, New York: Pantheon.

Samida, Stefanie (2009): »Heinrich Schliemann, Troia und die deutsche Presse: Medialisierung, Popularisierung, Inszenierung«. In: Petra Boden/Dorit Müller (eds.), *Populäres Wissen im medialen Wandel seit 1850*, Berlin: Kadmos, 135-51.

Samida, Stefanie (2010a): »Schliemanns Erbe? Populäre Bilder von Archäologie in der Öffentlichkeit«. In: Hans-Joachim Gehrke/Miriam Sénécheau (eds.), *Geschichte, Archäologie und Öffentlichkeit: Für einen neuen Dialog zwischen Wissenschaft und Medien*, Bielefeld: transcript, 31-48.

Samida, Stefanie (2010b): »Objekte der Begierde: Archäologische Dinge zwischen Forschung und Kommerzialisierung«. In: Elisabeth Thiethmeyer/Claudia Hirschberger/Karoline Noack (eds.), *Die Sprache der Dinge. Kulturwissenschaftliche Perspektiven auf die materielle Kultur*, Münster: Waxmann, 89-98.

Samida, Stefanie (2010c): »Ausgräber und Entdecker, Abenteurer und Held: Populäre Geschichtsvermittlung in archäologischen Fernsehdokumentationen«. In: Klaus Arnold/Walter Hömberg/Susanne Kinnebrock (eds.), *Geschichtsjournalismus: Zwischen Information und Inszenierung*, Münster: LIT, 219-34.

Samida, Stefanie (2011a): »Vom Heros zum Lügner? Wissenschaftliche ›Medienstars‹ im 19. Jahrhundert«. In: Stefanie Samida (ed.), *Inszenierte Wissenschaft: Zur Popularisierung von Wissen im 19. Jahrhundert*, Bielefeld: transcript, 245-72.

Samida, Stefanie (2011b): »Literatur, Geschichte und Archäologie im 19. Jahrhundert: Der Burghügel von Hisarlik«. In: Stefan Burmeister (ed.),

Fluchtpunkt Geschichte: Archäologie und Geschichtswissenschaft im Dialog, Münster: Waxmann, 73-92.

Samida, Stefanie (2012): *Heinrich Schliemann*, Stuttgart: UTB.

Sandy, George (1973 [1610]): *A Relation of a Journey Begun an. Dom. 1610*, Amsterdam/New York: Theatrum Orbis Terrarum and Da Capo Press.

Sarasin, Philipp (ed.) (2007): *Bakteriologie und Moderne: Studien zur Biopolitik des Unsichtbaren 1870-1920*, Frankfurt: Suhrkamp.

Sayce, Archibald Henry (1923): *Reminiscences*, London: Macmillan.

Scandura, Joseph M. (1996): »Deadly Professions: *Dracula*, Undertakers, and the Embalmed Corpse«. *Victorian Studies* 40.1, 1-30.

Schaffer, Talia (1994): »›A Wilde Desire Took Me‹: The Homoerotic History of Dracula«. *ELH* 61.2, 381-425.

Scheick, William J. (1991): »Adolescent Pornography and Imperialism in Haggard's *King Solomon's Mines*«. *English Literature in Transition, 1880-1920* 34.1, 19-30.

Scheuer, Helmut (1990): »Heinrich Schliemanns ›Selbstbiographie‹: Zur Gattungstypologie der Autobiographik in der zweiten Hälfte des neunzehnten Jahrhunderts«. In: William M. Calder III/Justus Cobet (eds.), *Heinrich Schliemann nach hundert Jahren*, Frankfurt: Klostermann, 346-59.

Schindler, Wolfgang (1976): »Heinrich Schliemann: Leben und Werk im Spiegel der neuen biographischen Forschungen«. *Philologus* 120, 271-89.

Schindler, Wolfgang (1980): »Schliemanns Selbstporträt: Die Inszenierung eines Self-made-man«. *EAZ* 21, 655-58.

Schindler, Wolfgang (1990): »Dichtung und Wahrheit: Schliemanns Selbstbiographie im Kontext«. In: William M. Calder III/Justus Cobet (eds.), *Heinrich Schliemann nach hundert Jahren*, Frankfurt: Klostermann, 152-69.

Schivelbusch, Wolfgang (1993): *Geschichte der Eisenbahnreise: Zur Industrialisierung von Raum und Zeit im 19. Jahrhundert*, Frankfurt: Fischer.

Schliemann, Heinrich (1869): *Ithaka, der Peloponnes und Troja*, Leipzig: Brockhaus.

Schliemann, Heinrich (1886): *Tiryns – The Prehistoric Palace of the Kings of Tiryns*, London: Murray.

Schliemann, Heinrich (1953): *Briefwechsel von 1842-1875*, Ernst Meyer (ed.), Berlin: Mann.

Schliemann, Heinrich (1957): *Briefwechsel von 1876-1890*, Ernst Meyer (ed.), Berlin: Mann.

Schliemann, Heinrich (1958): *Abenteuer meines Lebens: Heinrich Schliemann erzählt*, Leipzig: F.A. Brockhaus.

Schliemann, Heinrich (1972 [1884]): *Troja: Results of the Latest Researches and Discoveries on the Site of Homer's Troy, and the Heroic Tumuli and Other Sites, Made in the Year 1882; and a Narrative of a Journey in the Troad in 1881*, Chicheley: Paul P. B. Minet.

Schliemann, Heinrich (1976a [1875]): *Troy and Its Remains: A Narrative of Researches and Discoveries Made on the Site of Illium, and in the Trojan Plain*, Philip Smith (ed.), New York: Arno.

Schliemann, Heinrich (1976b [1878]): *Mycenæ: A Narrative of Researches and Discoveries at Mycenæ and Tiryns*, New York: Arno.

Schliemann, Heinrich (1976c [1880]): *Ilios the City and the Country of the Trojans: The Result of Researches and Discoveries on the Site of Troy and throughout the Troad in the Years 1871, 72, 73, 78, 79. Including an Autobiography of the Author*, New York: Arno.

Schliemann, Heinrich (1990): *Heinrich Schliemann, Wegbereiter einer neuen Wissenschaft: Mit Auszügen aus Autobiographie und Briefwechsel sowie Testament und Lobreden*, Joachim Herrmann (ed.), Berlin: Akademie.

Schliemann, Heinrich/Sophie Kastromenos Schliemann (1949 [1892]): *Heinrich Schliemann: Selbstbiographie bis zu seinem Tode vervollständigt*, Leipzig: F.A. Brockhaus.

Schmied, Wieland (ed.) (1990): *Troja und Homer: Porträt eines Enthusiasten*, Frankfurt: Insel.

Scott, Joan Wallach (1999): *Gender and the Politics of History*, New York: Columbia University Press.

Scott, Walter (2000 [1820]): *Ivanhoe*, London: Penguin.

Senf, Carol A. (1979): »*Dracula*: The Unseen Face in the Mirror«. *Journal of Narrative Techniques* 9, 160-70.

Senf, Carol A. (1982): »*Dracula*: Stoker's Response to the New Woman«. *Victorian Studies* 26, 33-49.

Shanks, Michael (1996): *Classical Archaeology of Greece: Experiences of the Discipline*, London/New York: Routledge.

Shelley, Mary Wollstonecraft (2008 [1818]): *Frankenstein or The Modern Prometheus*, New York: Oxford University Press.

Shelley, Percy Bysshe (1821): *Hellas: A Lyrical Drama*, London: Charles and James Ollier.

Shelley, Percy Bysshe (1855 [1817]): »Ozymandias«. In: Mary Shelley (ed.), *The Poetical Works of Percy Bysshe Shelley*, Vol. 2, Boston: Houghton, 95.

Siebler, Michael (1990): *Troia, Homer, Schliemann: Mythos und Wahrheit*, Mainz: von Zabern.

Sinha, Madhudaya (2008): »Triangular Erotics: Politics of Masculinity, Imperialism and Big-Game Hunting in Rider Haggard's *She*«. *Critical Survey* 20.3, 29-43.

Sinnema, Peter W. (1998): *Dynamics of the Pictured Page: Representing the Nation in the* Illustrated London News, Aldershot: Ashgate.

Sklenár, Karel (1983): *Archaeology in Central Europe: The First 500 Years*, Leicester: Leicester University Press.

Smiles, Samuel [1852] (2002): *Self-Help, with Illustrations of Character, Conduct, and Perseverance*, Peter W. Sinnema (ed.), Oxford: Oxford University Press.

Smith, Andrew (2004): »Love, Freud, and the Female Gothic«. *Gothic Studies* 6.1; May, 80-89.

Speaker: The Liberal Review, Feb 5, 1892. »Schliemann's Excavations«. 175.

Spencer, Kathleen L. (1992): »Purity and Danger: *Dracula*, the Urban Gothic, and the Late Victorian Degeneracy Crisis«. *ELH* 59.1, 197-225.

Starkey, Paul/Janet Starkey (1998): *Travellers in Egypt*, London/New York: I.B. Tauris.

Stevenson, Robert Louis (1979 [1886]): »The Strange Case of Dr Jekyll and Mr Hyde«. In: Robert Louis Stevenson, *Dr Jekyll and Mr Hyde and Other Stories*, London: Penguin, 29-97.

Stewart, Aubrey (1886): *The Tale of Troy*, London: Macmillan.

Stiebel, Lindy (2001): »Creating a Landscape of Africa: Baines, Haggard and Great Zimbabwe«. *English in Africa* 28.2, 123-33.

Stoker, Bram (2008 [1897]): *Dracula*, New York: Oxford University Press.

Stoker, Bram (2008 [1903]): *The Jewel of Seven Stars*, London: Penguin.

Storey, John (2006): *Cultural Theory and Popular Culture: An Introduction*, Harlow et al.: Pearson Prentice Hall.

Storey, John (2008): *Inventing Popular Culture: From Folklore to Globalization*, Oxford et al.: Blackwell.

Stott, Rebecca (1989): »The Dark Continent: Africa as Female Body in Haggard's Adventure Fiction«. *Feminist Review* 32, 69-89.

Stott, Rebecca (2001): »›Scaping the Body‹: Of Cannibal Mothers and Colonial Landscape«. In: Angelique Richardson/Chris Willis (eds.), *The New Woman in Fiction and in Fact*, New York: Palgrave, 150-66.

Stuart, James/Nicholas Revett [1762] (2007): *The Antiquities of Athens*, New York: Princeton Architectural Press.

Stubbs, William (1874-78 [1867-76]): *The Constitutional History of England in Its Origin and Development, Edward Augustus Freeman History of the Norman Conquest*, 3 Vols, Oxford: Clarendon Press.

Suffling, Ernest R. (1896): »The Strange Discovery of Dr. Nosidy«. In: Ernest R. Suffling, *The Story Hunter or Tales of the Weird and Wild*, London: Jarrold & Sons, 15-35.

Sulcer, Robert (2007): »Ungentlemanly Scholars«. *Victorians Institute Journal* 35, 137-170.

Sullivan, Robert E. (2009): *Macaulay: The Tragedy of Power*, Cambridge: Belknap Press.

Tangri, Daniel (1990): »Popular Fiction and the Zimbabwe Controversy«. *History in Africa* 17, 293-304.

Thomas, Ronald R. (2001): »Detection in the Victorian Novel«. In: David Deirdre (ed.), *The Cambridge Companion to the Victorian Novel*, Cambridge et al.: Cambridge University Press, 169-91.

Thompson, Paul R. (1975): *The Edwardians*, Bloomington et al.: Indiana University Press.

Thomsen, Christian Jürgensen (1848): *A Guide to Northern Antiquities*, Francis Egerton Ellesmere (ed.), London: Bain.

Times, Mar 31, 1875. »Troy«. 7.

Times, June 26, 1875. »Homeric Troy«. 9.

Times, June 5, 1876. »The Troad«. 12.

Times, July 24, 1876. »Troy«. 6.

Times, Nov 6, 1876. »The Emperor of Brazil At Troy«. 4.

Times, Dec 7, 1876. »Treasure Trove at Mycenæ«. 8.

Times, Dec 22, 1876. »The Tomb of Agamemnon«. 4.

Times, Mar 17, 1877. »The Discoveries at Mycenæ«. 12.

Times, Mar 23, 1877. »Dr. Schliemann at Burlington House«. 10.

Times, Apr 12, 1877. »Troy and Mycenæ«. 8.

Times, June 4, 1877. »Mycenæ«. 5.

Times, June 9, 1877. »The Archæological Institute«. 9.

Times, June 16, 1877. »Dr. Schliemann and The Royal Historical«. 9.

Times, June 23, 1877. »The Emperor and Empress of Brazil«. 11.

Times, Aug 16, 1877. »Troy«. 3.

Times, Aug 20, 1877. »Troy«, Letter to the Editor J. Lubbock. 6.

Times, Dec 8, 1877. »The Birkbeck Institution«. 6.

Times, Apr 18, 1878. »Troy«. 7.

Times, Nov 27, 1878. »Excavations at Troy«. 4.

Times, Dec 24, 1881. »Buried Treasure«. 3.

Times, May 26, 1890. »Recent Excavations in Greece«. 10.

Times, Dec 29, 1890. »The Late Dr. Schliemann«. 6.

Times, July 23, 1891. »South Africa«. 5.

Tomes, Nancy (1990): »The Private Side of Public Health: Sanitary Science, Domestic Hygiene, and the Germ Theory, 1870-1900«. *Bulletin of the History of Medicine* 64, 509-39.

Topham, Jonathan R. (1998): »Beyond the ›Common Context‹: The Production and Reading of the Bridgewater Treatises«. *Isis* 89, 233-62.

Tosh, John (1994): »What Should Historians Do with Masculinity? Reflections on Nineteenth-Century Britain«. *History Workshop* 38, 179-202.

Trafton, Scott (2004): *Egypt Land: Race and Nineteenth-Century American Egyptomania*, Durham: Duke University Press.

Traill, David A. (1982): »Schliemann's American Citizenship and Divorce«. *Classical Journal* 77, 336-42.

Traill, David A. (1983): »Schliemann's Discovery of ›Priam's Treasure‹«. *Antiquity* 57, 181-86.

Traill, David A. (1984a): »Schliemann's Discovery of ›Priam's Treasure‹: A Re-Examination of the Evidence«. *JHS*, 104: 96-115.

Traill, David A. (1984b): »Further Evidence of Fraudulent Reporting in Schliemann's Archaeological Works«. *Boreas* 7, 295-316.

Triall, David A. (1985): »Schliemann's Dream of Troy«. *Classical Journal* 81, 13-24.

Traill, David A. (1986a): »Schliemann's Mendacity: A Question of Methodology«. *Anatolian Studies* 36, 91-98.

Traill, David A. (1986b): »Appendix B«. In: William M. Calder III/David A.Traill (eds.), *Myth, Scandal and History: The Heinrich Schliemann*

Controversy and a First Edition of the Mycenaean Diary, Detroit: Wayne State University Press, 240-56.

Traill, David A. (1990): »Schliemann and his Academic Employees«. In: William M. Calder III/Justus Cobet (eds.), *Heinrich Schliemann nach hundert Jahren*, Frankfurt: Klostermann, 237-55.

Traill, David A. (1991): »Schliemann on Dickens«. *The Classical World* 85.2, 105-96.

Traill, David A. (1993a): »The Archaeological Career of Sophia Schliemann«. In: David A. Traill (ed.), *Excavating Schliemann*, Atlanta: Scholars, 235-43.

Traill, David A. (1993b): »Schliemann's ›Dream of Troy‹: The Making of a Legend«. In: David A. Traill (ed.), *Excavating Schliemann*, Atlanta: Scholars, 29-40.

Traill, David A. (1993c): »Hartmut Döhl's *Heinrich Schliemann – Mythos und Ärgernis*: A Review«. In: David Traill (ed.), *Excavating Schliemann*, Atlanta: Scholars, 255-59.

Traill, David A. (1993d): *Excavating Schliemann*, Atlanta: Scholars

Traill, David A. (1995): *Schliemann of Troy: Treasure and Deceit*, London: Murray.

Traill, David A./Paul C. Appleton (2002): »Letters from Troy«. *Archaeology* 55.1, 54-58.

Trigger, Bruce G. (1989): *A History of Archaeological Thought*, Cambridge: Cambridge University Press.

Trigger, Bruce G. (1998): »Archaeology and Epistemology: Dialoguing across the Darwin Chasm«. *American Journal of Archaeology* 102.1, 1-34.

Tromp, Marlene (2008): *Victorian Freaks: The Social Context of Freakery in Britain*, Columbus: Ohio State University Press.

Trümpler, Charlotte (2008): *Das Große Spiel: Archäologie und Politik zur Zeit des Kolonialismus (1860-1940)*, Köln: DuMont.

Tucker, Jennifer (1997): »Photography as Witness, Detective, and Impostor: Visual Representation in Victorian Science«. In: Bernard V. Lightman (ed.), *Victorian Science in Context*, Chicago: University of Chicago Press, 378-408.

Turner, Frank Miller (1981): *The Greek Heritage in Victorian Britain*, New Haven: Yale University Press.

Turner, Frank Miller (1989): »Why the Greeks and not the Romans in Victorian Britain?«. In: C.W. Clarke (ed.), *Rediscovering Hellenism: The Hellenic Inheritance and the English Imagination*, Cambridge: Cambridge University Press, 61-81.

Turner, Frank Miller (1990): »The Victorian Crisis of Faith and the Faith that Was Lost«. In: Bernard V. Lightman/Richard J. Helmstadter (eds.), *Victorian Faith in Crisis: Essays on Continuity and Change in Nineteenth-Century Religious Belief*, Stanford: Stanford University Press, 9-38.

Tylor, Sir Edward Burnett (1871): *Primitive Culture: Researches into the Development of Mythology, Philosophy, Religion, Language, Art, and Custom*, Vol. 1, London: Murray.

Vaio, John (1992): »Schliemann and Gladstone: New Light from Unpublished Documents«. In: Joachim Herrmann (ed.), *Heinrich Schliemann: Grundlagen und Ergebnisse moderner Archäologie 100 Jahre nach Schliemanns Tod*, Berlin: Akademie, 73-76.

Van Wyhe, John (2004): *Phrenology and the Origins of Victorian Scientific Naturalism*, Aldershot et al.: Ashgate.

Vance, Norman (1997): *The Victorians and Ancient Rome*, Oxford et al.: Blackwell.

Vann, Jerry Don/Rosemary T. van Arsdel (1994): *Victorian Periodicals and Victorian Society*, Aldershot: Scholar.

Vansina, Jan (1985): *Oral Tradition as History*, Madison: University of Wisconsin Press.

Veeser, Harold Aram (1989): *The New Historicism*, New York et al.: Routledge.

Venuti, Marcello (1750 [1748]): *A Description of the First Discoveries of the Antient City of Heraclea Found Near Portici, a Country Palace Belonging to the King of the Two Sicilies*, London: Printed for R. Baldwin.

Venuti, Marcello (1773): *The Antiquities of Herculaneum*, London: Printed for S. Leacroft.

Vicari, Justin (2012): *Mad Muses and the Early Surrealists*, Jefferson et al.: McFarland.

Vries, Leonard de (ed.) (1976): *Panorama 1842-1865: The World of the Early Victorians as Seen through the Eyes of the* Illustrated London News, London: Murray.

Vries, Leonard de (ed.) (1979): *History as Hot News 1865-1897: The Late Nineteenth Century World as Seen through the Eyes of* The Illustrated London News *and* The Graphic, New York: St. Martin's.

Wallace, Jennifer (2004): *Digging the Dirt: The Archaeological Imagination*, London: Duckworth.

Walmsely, Mullenex Hugh (1869): *The Ruined Cities of Zulu Land*, Vol. 1, London: Chapman & Hall.

Walmsely, Mullenex Hugh (1869): *The Ruined Cities of Zulu Land*, Vol. 2, London: Chapman & Hall.

Webb, Timothy (1993): »Romantic Hellenism«. In: Stuart Curran (ed.), *The Cambridge Companion to British Romanticism*, Cambridge et al.: Cambridge University Press, 157-58.

Werner, Alex (2003): »Egypt in London – Public and Private Display in the 19th Century Metropolis«. In: J.M. Humbert/C.A. Price (eds.), *Imhotep Today: Egyptianizing Architecture*, London: UCL Press, 75-104.

Weselyan-Methodist Magazine, March, 1893. »Mashonaland, Past and Present«. John Aldred. 175-81.

Whitehead, Christopher (2005): *The Public Art Museum in Nineteenth Century Britain: The Development of the National Gallery*, Aldershot et al.: Ashgate.

Whitley, James (2001): *The Archaeology of Ancient Greece*, Cambridge/New York: Cambridge University Press.

Wieczorkiewicz, Ann (2005): »Unwrapping the Mummies and Telling Their Stories: Egyptian Mummies in Museum Rhetoric«. In: Mary Bouquet (ed.), *Science, Magic and Religion: The Ritual Process of Museum Magic*, Porto: Berghahn, 51-71.

Wilkinson, John Gardiner (1835): *Topography of Thebes, and General View of Egypt*, London: Murray.

Wilkinson, John Gardiner (1837): *Manners and Customs of the Ancient Egyptians*, London: Murray.

Wilkinson, John Gardiner (1847): *Hand-Book for Travellers in Egypt*, London: Murray.

Williams, Raymond (2002 [1958]): »Culture is Ordinary«. In: Ben Highmore (ed.), *The Everyday Life Reader*, London/New York: Routledge, 92-100.

Williams, Raymond (2006 [1961]): »The Analysis of Culture«. In: John Storey (ed.), *Cultural Theory and Popular Culture: A Reader*, Harlow et al.: Pearson, 32-49.

Williamson, Alice Muriel/C. N. Williamson (1914): *It Happened in Egypt*, New York: Grosset & Dunlap.

Wilmot, Alexander (1897-99): *The History of Our Own Times in South Africa*, Vol. 1-3, London: Juta.

Wilmot, Alexander (2005 [1896]): *Monomotapa Rhodesia: Its Monuments, and Its History from the Most Ancient Times to the Present*, London: Elibron Classic Series.

Wilson, Andrew N. (2002): *The Victorians*, London: Hutchinson.

Wilson, Lucy Sarah Atkins (1843 [1821]): *Fruits of Enterprise in the Adventures of Belzoni in Egypt and Nubia; With an Account of his Discoveries in the Pyramids, among the Ruins of Cities, and in the Ancient Tomb*, New York: Charles S. Francis.

Wilton, Andrew/Bignamini Ilaria (eds.) (1996): *Grand Tour: The Lure of Italy in the Eighteenth Century*, London: Tate Gallery.

Winckelmann, Johann Joachim (1764): *Geschichte der Kunst des Altertums*, Dresden: Waltherische Hofbuchhandlung.

Winckelmann, Johann Joachim (1808 [1755]): »Gedanken über die Nachahmung der griechischen Werke in der Malerei und Bildhaukunst«, In: C.L. Fernow (ed.), *Winckelmann's Werke. Erster Band: Welcher die Schriften über die Nachahmung der Griechen, die kleinen Aufsätze, und die Anmerkungen über die Baukunst der Alten enthält. Mit dem Portrait Winckelmanns und 16 Kupfern*, Dresden: Walther. 5-62.

Winckelmann, Johann Joachim (2011 [1762]): »Open Letter on the Discoveries Made at Herculaneum«. In: Carol C. Mattusch (ed.), *Letter and Report on the Discoveries at Herculaneum*, Los Angeles: J. Paul Getty Museum, 78-143.

Winslow, William Copley (1892): *The Queen of Egyptology (Amelia B. Edwards)*, New York: s.n.

Wohlleben, Joachim (1990): »Homer vor Schliemann«. In: William M. Calder III/Justus Cobet, *Heinrich Schliemann nach hundert Jahren*, Frankfurt: Klostermann, 27-30.

Woman's Herald, Oct 26, 1893. »Personal Notes«. 575.

Wood, Anthony (1960): *Nineteenth Century Britain, 1815-1914*, London: Longmans.

Woodham, Jonathan (1989): »Images of Africa and Design at the British Empire Exhibition between the Wars«. *Journal of Design History* 2.1, 15-33.

Woolf, Virginia (1966 [1923]): »Modern Fiction«. In: *Collected Essays II*, London: Hogarth Press, 103-10.

Woolf, Virginia (1966 [1925]): »Mr. Bennett and Mrs. Brown«. In: *Collected Essays II*, London: Hogarth Press, 319-37.

Woolls, Charles (1839): *The Barrow-Diggers: A Dialogue in Imitation of the Grave Diggers in* Hamlet, *with Numerous Explanatory Notes*, London: Whittaker.

Worboys, Michael (2000): *Spreading Germs: Diseases, Theories, and Medical Practice in Britain, 1865-1900*, Cambridge: Cambridge University Press.

Wordsworth, William (2002 [1802]): »Preface to *Lyrical Ballads*«. In: H. Meyer Abrams/Stephen Greenblatt (eds.), *The Norton Anthology of English Literature*, Vol. 2, New York et al.: Norton, 141-52.

Wortham, John David (1971): *The Genesis of British Egyptology, 1549-1906*, Norman: University of Oklahoma Press.

Young, Shannon (2005): »Myths of Castration: Freud's ›Eternal Feminine‹ and Rider Haggard's *She*«. *The Victorian Newsletter* 108, 21-30.

Zintzen, Christiane (1998): *Von Pompeij nach Troja: Archäologie, Literatur und Öffentlichkeit im 19. Jahrhundert*, Wien: WUV.

Zivie, Alain/Roger Lichtenberg (2005): »The Cats of the Goddess Bastet«. In: Salmia Ikram (ed.), *Divine Creatures: Animal Mummies in Ancient Egypt*, Cairo: American University in Cairo Press, 106-19.

Zulli, Tania (2006): »›Mapping the Unknown‹: Rider Haggard between Realism and Imagination«. In: Mariaconcetta Costantini/Renzo D'Agnillo/Francesco Marroni (eds.), *La letteratura vittorinia e i mezzi di trasporto*, Rome: Aracne, 24-32.

Index